Studies in Fuzziness and Soft Computing   283

Editor-in-Chief

Prof. Janusz Kacprzyk
Systems Research Institute
Polish Academy of Sciences
ul. Newelska 6
01-447 Warsaw
Poland
E-mail: kacprzyk@ibspan.waw.pl

For further volumes:
http://www.springer.com/series/2941

Krassimir T. Atanassov

On Intuitionistic Fuzzy Sets Theory

 Springer

Author
Prof. Krassimir T. Atanassov, DSc, DSc, PhD
Department of Bioinformatics and Mathematical Modelling
Institute of Biophysics and Biomedical Engineering
Bulgarian Academy of Sciences
Acad. G. Bonchev Str., Block 105
Sofia-1113, Bulgaria
E-mail: krat@bas.bg

ISSN 1434-9922   e-ISSN 1860-0808
ISBN 978-3-642-44259-9   ISBN 978-3-642-29127-2 (eBook)
DOI 10.1007/978-3-642-29127-2
Springer Heidelberg New York Dordrecht London

Printed on acid-free paper

Springer is part of Springer Science+Business Media (www.springer.com)

Preface

The bipolar model for evaluation of the events and objects around us has deeply underlined the mankind's conscience since ancient times. For instance, the Bulgarian thracologists Alexander Fol and Ivan Marazov point out in [231, 367] that ever since the Paleolithic Age, people have been using the opposition pairs good–evil, right–left, up–down, white–black, man–woman, day–night, Sun–Moon, etc. Each of these opposition pairs happens to be a base of a bipolar model.

A very good illustration of the idea of "bipolarity" is the following Bulgarian anecdote from the beginning of 1980s:

> In Bulgaria there is no unemployment, but no one works.
> No one works, but the plan is accomplished.
> The plan is accomplished, but there is nothing in the shops.
> There is nothing in the shops, but at home there is everything.
> At home there is everything, but all are dissatisfied.
> All are dissatisfied, but all vote "Yes".

Centuries had to pass until the saying *"The truth lies somewhere in the middle"* managed to gain the recognition it deserved. The three-valued propositional calculus, introduced by Jan Łukasiewicz who had been the precursor of many discoveries in science and practice of the 20-th century and 21-st century. This concept was successfully developed to multi-valued logic (by Łukasiewicz himself) and to fuzzy logic (by Lotfi A. Zadeh), both of which have at present a plethora of variants and extensions. In 1983, one of these extensions, the *intuitionistic fuzzy set*, was defined and analyzed by me. In 1999, I published my results collected in a course of 16 years in a book[39]. In the next 12 years, my research has been expanded into many directions in the intuitionistic fuzzy sets with many new ideas that are presented in this book. Hence, it may be regarded both as a continuation and extension of my previous book [39].

Throughout all those years, since the book mentioned has appeared, many colleagues of mine have sent me many comments and suggestion on how to clarify some matters, improve and correct some parts of the exposition given in [39]. I, myself, have also found many possibilities to improve the contents and exposition given in my previous book, as well as to introduce new definitions, properties, etc. They will be included in the present book. I hope that a new, extended and re-focused way of presentation and those new elements will make the book better, easier to read, more useful to the potential readers, opening them new vistas.

I have recorded more than 1500 papers on the intuitionistic fuzzy sets, written by researchers from more than 50 countries. Most of these papers discuss important issues related to the theory or applications of the intuitionistic fuzzy sets. For this reason, I realized that it would be virtually impossible to make a survey of the others' results and decided to limit the presentation to my own works and ideas only, with a proper reference to other contributions. I am convinced that this may work well and will trigger much research from many colleagues in the field. To the best of my knowledge, only one survey of the publications on the intuitionistic fuzzy sets exists by now [381], and another is being prepared in the last couple of years but has not been published so far.

I have used two words "I" and "the author" in this book. The former is applied when I wish to discuss my personal opinions and concepts introduced by myself, and the latter form is used in a more general context.

I am very thankful to my colleagues and students who motivated me to prepare the present book. The list of their names is long, but I must note (in the alphabetical order of their countries and in the alphabetical order of their first names) at least the names of my coauthors:

- from Australia: Anthony Shannon
- from Belgium: Chris Cornelis, Etienne E. Kerre, Glad Deschrijver
- from Bulgaria: Anton Antonov, Boyan Djakov, Boyan Kolev, Daniela Orozova, Desislava Peneva, Diana Boyadzhieva, Dimiter Dimitrov, Dinko Dimitrov, Evdokia Sotirova, George Gluhchev, George Mengov, Hristo Aladjov, Kalin Georgiev, Lilija Atanassova, Ludmila Todorova, Mariana Nikolova, Nikolai Nikolov, Olympia Roeva, Pavel Cheshmedjiev, Pencho Marinov, Peter Georgiev, Peter Vassilev, Sotir Sotirov, Stefan Hadjitodorov, Stefka Fidanova, Stefka Stoeva, Tania Pencheva, Trifon Trifonov, Valeri Gochev, Vassia Atanassova, Vihren Chakarov, and Violeta Tasseva
- from Germany: Dinko Dimitrov, Evelina Kojcheva,
- from India: Parvathi Rangasamy,
- from Italy: Gabriella Pasi,
- from Korea: Soon-Ki Kin and Taekyun Kim,
- from Morocco: Said Melliani,
- from Poland: Eulalia Szmidt, Janusz Kacprzyk and Maciej Krawczak,
- from Portugal: Pedro Melo-Pinto,
- from Romania: Adrian Ban,
- from Slovakia: Beloslav Riečan, Magdalena Renčova,

- from Spain: Humberto Bustince Sola,
- from the UK: Chris Hinde, Elia El-Darzi, Ilias Petrounias, Ludmila Kuncheva, Panagiotis Chountas, Stefan Danchev, and Vassilis Kodogianis,
- from USA: Cecilia Temponi, Dimiter Sasselov, Ronald R. Yager, and Vladik Kreinovich.

My post-graduate student Peter Vassilev, my Indian colleague Parvathi Rangasamy and my daughter Vassia Atanassova have read and corrected the text of the manuscript. They deserve my deep thanks and appreciation.

Special thanks are due to two persons. Firstly, the founder of the area of my research – Lotfi Zadeh, whose interest in my results motivated me during the last ten years. Special thanks are also due to one of the most active activists among the people in the world, interested in fuzzy sets, but also active researcher and my friend and coauthor – Janusz Kacprzyk, who urged me to finalize my several-year-long work on the theory and applications of intuitionistic fuzzy sets, and prepare my previous and present books.

I am grateful for the support provided by the projects DID-02-29 "Modelling processes with fixed development rules" and BIn-2/09 "Design and development of intuitionistic fuzzy logic tools in information technologies" funded by the National Science Fund, Bulgarian Ministry of Education.

Last but not least, I would like to thank the three most important ladies in my life: my mother, my wife, and my daughter, who have been encouraging, stimulating and supporting me in all my scientific research.

January 2012   Krassimir T. Atanassov
Sofia, Bulgaria.

Contents

1

On the Concept of Intuitionistic Fuzzy Sets

1.1 Definition of the Concept of Intuitionistic Fuzzy Set (IFS)

The origin of my idea of intuitionistic fuzziness was a happenstance – as a mathematical game. I read the Russian translation of A. Kaufmann's book [301][1] and decided to add to the definition, a second degree (degree of non-membership) and studied the properties of a set with both degrees. Of course, I observed that the new set is an extension of the ordinary fuzzy set, but I did not immediately notice that it has essentially different properties. So the first research works of mine in this area followed, step-by-step, the existing results in fuzzy sets theory. Of course, some concepts are not so difficult to extend formally. It is interesting to show that the respective extension has specific properties, absent in the basic concept.

Only when I convinced myself that the so-constructed sets do have worthy properties, I discussed them with my former lecturer from the time when I was a student at the Mathematical Faculty of Sofia University – George Gargov (7 April 1947 – 9 Nov. 1996) - one of the most vivid Bulgarian mathematicians, and a person with various interests in science - mathematics, physics, biology, philosophy, linguistics, psychology, sociology, etc., and arts - literature, music, theatre, cinema, pictural arts. He proposed the name *"Intuitionistic Fuzzy Set' (IFS)*, because the way of fuzzification holds the idea of intuitionism (see, e.g. [260]).

Let a (crisp) set E be fixed and let $A \subset E$ be a fixed set. An IFS A^* in E is an object of the following form

$$A^* = \{\langle x, \mu_A(x), \nu_A(x) \rangle | x \in E\}, \qquad (1.1)$$

---

[1] In early 1980's, only Russian translations of the books [217, 301, 592] were available in Bulgaria and these books influenced the development of the first steps of IFS theory.

K.T. Atanassov: On Intuitionistic Fuzzy Sets Theory, STUDFUZZ 283, pp. 1–16.
springerlink.com   © Springer-Verlag Berlin Heidelberg 2012

where functions $\mu_A : E \to [0,1]$ and $\nu_A : E \to [0,1]$ define the *degree of membership* and the *degree of non-membership* of the element $x \in E$ to the set A, respectively, and for every $x \in E$

$$0 \leq \mu_A(x) + \nu_A(x) \leq 1. \tag{1.2}$$

Obviously, every ordinary fuzzy set has the form

$$\{\langle x, \mu_A(x), 1 - \mu_A(x)\rangle | x \in E\}.$$

If

$$\pi_A(x) = 1 - \mu_A(x) - \nu_A(x),$$

then $\pi_A(x)$ is the *degree of non-determinacy* (uncertainty) of the membership of element $x \in E$ to set A. In the case of ordinary fuzzy sets, $\pi_A(x) = 0$ for every $x \in E$.

Let us note that every set in the sense of the set theory (see, e.g., [232, 256, 329]) is a fuzzy set and, hence, an IFS. What is more important is that the converse also holds: every IFS, and hence every fuzzy set, is a set in the sense of the classical set theory.

Let the universe E be fixed and let us define the IFS A^* for a fixed set $A \subseteq E$ by (1.1) with inequality (1.2).

Let $E^* = E \times [0,1] \times [0,1]$. Obviously, $A^* \subseteq E^*$ and we can define the characteristic function $\Omega_{A^*} : E^* \to \{0,1\}$ by:

$$\Omega_{A^*}(\langle x, a, b\rangle) = \begin{cases} 1, & \text{if } \mu_A(x) = a, \nu_A(x) = b \text{ and } \langle x, a, b\rangle \in A^*, \\ 0, & \text{otherwise} \end{cases}$$

i.e., A^* is a set in the sense of the set theory in the universe E^*. The backward transformation of this result to the ordinary fuzzy sets is trivial.

For simplicity, below we write A instead of A^*.

1.2   First Example

Let us start with a little funny example, a story with Johnny and Mary, characters in many Bulgarian anecdotes. They bought a box of chocolates with 10 pieces inside. Being more nimble, Johnny ate seven of them, while Mary – only two.

One of the candies fell into the floor.

In this moment, a girl friend of Mary came and Mary said, *"We can't treat you with chocolates, because Johnny ate them all"*.

Let us estimate the truth value of this statement at the moment of speaking, i.e. before we have any knowledge of subsequent events.

From classical logic point of view, which uses for estimations the members of the set $\{0,1\}$, the statement has truth value of 0, since Mary has also taken part in eating the candies and Johnny was not the only one who has eaten them. On the other hand, we are intuitively convinced that the statement is more true than false.

In 1926, Jan Lukasiewicz suggested the concept of ternary logics with its estimation values from the set $\{0, \frac{1}{2}, 1\}$ (see, e.g., [139]). If we estimate our statement in the terms of ternary logics, its truth value should be $\frac{1}{2}$. Thirty years later, Lukasiewicz generalized his idea to the concept of many-valued logics (see, e.g., [139]). For instance, if we use 11-valued logics, taking estimations from the set of elements $\{0, \frac{1}{10}, \frac{2}{10}, ..., \frac{9}{10}, 1\}$, our problem is again easy to solve: the statement's truth estimation is exactly $\frac{7}{10}$. But if we take 6-valued logics, with values of the estimating set being $\{0, \frac{1}{5}, \frac{2}{5}, \frac{3}{5}, \frac{4}{5}, 1\}$, then we would not be able to correctly evaluate the truth value of the above statement. We would hesitate between $\frac{3}{5}$ and $\frac{4}{5}$ but none of these would be correct, because

$$\frac{4}{5} - \frac{7}{10} = \frac{1}{10} = \frac{7}{10} - \frac{3}{5},$$

i.e. both values would be equally distant from the authentic truth value $\frac{7}{10}$.

Of course, there exists many other, practically infinite, evaluation sets, that will generate the same problem. Thus, we reach for Lotfi Zadeh's idea of fuzzy sets, using for evaluation the numbers from the interval $[0, 1]$ (see [593]).

Let us return to the problem as Mary greets her guest – statement's truth value is obviously 0.7. However, at the next moment Johnny can take the fallen candy and place it back into the box of treat the guest, preserving the truth value of Mary's statement at 0.7, and falsity 0.3. But he can always take advantage of the distraction and eat the last candy which would result in truth value of 0.8 and falsity of 0.2. In this sense the statement depends to a great extent on Johnny's actions. Therefore, the apparatus of IFSs gives us the most accurate answer to the question: $\langle 0.7, 0.2 \rangle$. The degree of uncertainty now is 0.1 and it corresponds to our ignorance of the boundaries of Johnny's gluttony.

1.3   Some Comments on the Concept of IFS

1.3.1. For many years I have been considering a change in the above definition (1.1), but I never found the strength to do it. Probably, 28 years are enough and I will here rewrite the above set to the form:

$$A = \{\langle x, \mu_A(x), \nu_A(x) \rangle \mid x \in E \ \& \ 0 \le \mu_A(x) + \nu_A(x) \le 1\},$$

where E is a fixed set and $A \subseteq E$. This form of the record of the IFS is more precise from set-theoretical point of view, but, of course, longer and more tedious to work with. So, I think, in future the shorter form will be preserved, but I wanted to be the one to clarify and correct the notation.

1.3.2. When I defended my Doctor of Science dissertation[2] on IFSs, two of my referees criticized me that I had not introduced the IFS as the couple $\langle A^+, A^- \rangle$. Of course, they were right that this form of the IFSs was possible, but this idea was formulated by Toader Buhaescu 10 years before my opponents, without criticism against the form used by me. Already 12 years I have been thinking about this story and I cannot find the answer whether had I used the shorter form I would have seen some mathematical facts that I did see. On the other hand, probably, I might have seen other facts... My criticasters asserted that from the shorter form it is clear that the IFS is a couple of ordinary fuzzy sets and by this reason it is not an interesting object. My answer was that the complex numbers are also ordered couples of real numbers and they should think twice before making the statement that the set of complex numbers was a trivial extension of the set of real numbers.

1.3.3. As it is well-known, in the beginning of the last century L. Brouwer introduced the concept of the *intuitionism* [141]. He appealed to the mathematicians to remove Aristoteles' law of excluded middle. In [142] he wrote:

"An immediate consequence was that for a mathematical assertion the two cases of truth and falsehood, formerly exclusively admitted, were replaced by the following three:
(1) has been proved to be true; (2) has been proved to be absurd; (3) has neither been proved to be true nor to be absurd, nor do we know a finite algorithm leading to the statement either that is true or that is absurd."

Therefore, if we have a proposition A, we can state that either A is true, or A is false, or that we do not know whether A is true or false. On the level of first order logic, the proposition $A \vee \neg A$ is always valid. In the framework of G. Boole's algebra this expression has truth value *"true"* (or 1). In the ordinary fuzzy logic of L. Zadeh, as well as in many-valued logics (starting with that of J. Lukasiewicz) the above expression *can* possess value smaller than 1. The same is true in the case of IFS, but here the situation occurs on semantical as well as on estimations' level. Practically, we fuzzify our estimation in Brouwer's sense, accounting for the three possibilities. This was Gargov's reason to offer the name "intuitionistic fuzzy set".

So far, I have not discussed, in detail, the connections between the IFS theory and Brouwer's intuitionism. But, I have seen that a serious research on this theme is much necessary, even delayed.

1.3.4. In March 1983, it turned out that the new sets allow the definition of operators which are, in a sense, analogous to the modal ones (in the case of ordinary fuzzy sets such operators are meaningless, since they reduce to

---

[2] In Bulgaria, similarly to other countries, there are two doctor degrees - PhD and Doctor of Science (DSc) or Second Doctor degree; I defended PhD thesis in the area of mathematics (on an extension of Petri nets, called Generalized Nets (GNs)); one Doctor of Science dissertation in the area of technical (computer) sciences again on GNs, and a second Doctor of Science dissertation in the area of mathematical sciences (on IFSs).

identity). It was then when I realized that I had found a promising new direction of research and published the results in [11].

1.3.5. In March 1991, I understood about the notion of an "IFS", as proposed by G. Takeuti and S. Titani [510]. However, they just put a very different meaning in the same term.

My first two communications appeared in June 1983 in Bulgarian [11] and English (with some extensions, written together with S. Stoeva) in August 1983 [102]. By the time when Takeuti and Titani's paper was printed, I had published four other papers on IFSs.

1.3.6. In 2005, a discussion started about the accuracy of the name of the IFS in the sense I had put into it (see [216]). In my answer [50] I noted that an analogue of Takeuti and Titani's research is constructed by Trifon Trifonov and me in [108], being an essential extension of their results.

About six years ago, the question about the name was launched for the first time. Now there are concepts *"vague set"*, *"neutrosophic set"* and others that practically denote one and the same object. In [159], H. Bustince and P. Burillo proved the coincidence of vague sets with IFSs, while in [236] K. Georgiev showed that on the one hand the neutrosophic sets have been incorrectly defined and on the other hand, even, if they had a correct definition, they would again coincide with the IFSs and they would not be IFS-extension, as their author F. Smarandache claimed, e.g., in [439].

1.3.7. Circa 1987, I saw for the first time the concept of *"Interval-Valued Fuzzy Set" (IVFS)* in [241], but my information concerning the original authors of the research was not precise. Later, in [16] and [93], together with G. Gargov, we discussed the equipolence (in the sense of [314]) of this concept with IFS. From our construction it is seen that each IFS can be represented by an IVFS and each IVFS can be represented by an IFS. I mention all of these years to emphasize that then I believed IFS were defined prior to IVFS. Now, I know (merely as a fact, without having seen the original texts) that IVFS are essentially older. I am preparing a detailed comparison between both concepts, that will be published soon.

1.3.8. As noted above, IFSs are an extension of the standard fuzzy sets. All results, which hold for fuzzy sets, have their interpretation here. Also, any research based on fuzzy sets, can be described in terms of IFS.

First, we discuss the relations between ordinary fuzzy sets and IFSs from two aspects: geometrical and probabilistical. Initially, I would like to note that some authors discuss the fact that in the case of ordinary fuzzy sets

$$\mu \vee \neg\mu \leq 1$$

as a manifest of the idea of intuitionism. Really, this inequality, in its algebraic interpretation of "$\vee$" by "max", does not satisfy the Law of Excluded Middle (LEM). But this is not the situation in a geometrical interpretation. Having in mind that in fuzzy set theory $\neg\mu = 1 - \mu$, we obtain that the geometrical interpretation as follows:

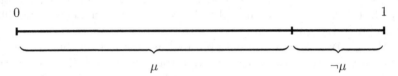

The situation in the IFS case is different and as follows:

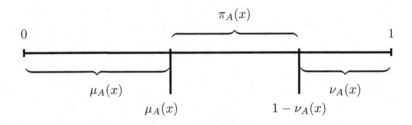

Now, the geometrical sums of both degrees can *really* be smaller than 1, i.e., LEM is not valid here. From probabilistic point of view, for case of the ordinary fuzzy sets, if $\mu \& \neg\mu = 0$, then the probability

$$p(\mu \vee \neg\mu) = p(\mu) + p(\neg\mu) = 1,$$

like in the geometrical case, while in IFS case we have the inequality

$$p(\mu \vee \neg\mu) \leq 1$$

that for proper IFS-elements will be strong.

It is important to note, that all of these constructions are only on the level of the definition of the set (fuzzy set or IFS), i.e., not related to the possible operations that can be defined over these sets.

1.4   First Example from Number Theory

Now, following [42], we construct an example from the area of the number theory.

It is related to the well known (see, e.g., [378]) Euler's totient function φ that determines the number of the natural numbers smaller than a fixed number n, which do not have common divisors with n. Therefore, if for a fixed natural number $n \geq 2$ we define the set

$$F(n) = \{x \mid 1 \leq x < n \ \& \ (x, n) = 1\},$$

then,

$$card(F(n)) = \varphi(n),$$

where (p, q) is the greatest common divisor of the natural numbers p and q and $card(X)$ is the cardinality of set X.

Let us define

$$\varphi_\mu(n) = card(\{x \mid 1 < x \le n \ \& \ x/n\}),$$

where x/n denotes that x divides n, and

$$\varphi_\nu(n) = \varphi(n).$$

Then, we can define the function V that juxtaposes to each natural number $n \ge 2$ the couple

$$V(n) = \langle \frac{\varphi_\mu(n)}{n}, \frac{\varphi_\nu(n)}{n} \rangle.$$

It can be easily seen that this couple is an intuitionistic fuzzy couple, i.e.,

$$0 \le \frac{\varphi_\mu(n)}{n} + \frac{\varphi_\nu(n)}{n} \le 1.$$

Now, we can also define

$$\varphi_\pi(n) = n - \varphi_\mu(n) - \varphi_\nu(n)$$

$$= n - card(F(n)) - card(\{x \mid 1 \le x \le n \ \& \ x/n\})$$

$$= card(\{x \mid 1 < x < n \ \& \ 1 < (x, n) < n\}).$$

For example,

$$V(3) = \langle \frac{1}{3}, \frac{2}{3} \rangle, \ V(12) = \langle \frac{4}{12}, \frac{4}{12} \rangle, \ V(15) = \langle \frac{3}{15}, \frac{8}{15} \rangle.$$

Function V determines the degrees of divisibility and of non-divisibility of each natural number, element of the universe $\mathcal{N}$ of all natural numbers. Therefore, we can construct the following IFS:

$$\mathcal{N}^* = \{\langle n, \frac{\varphi_\mu(n)}{n}, \frac{\varphi_\nu(n)}{n} \rangle \mid n \in \mathcal{N}\}.$$

1.5   An Example Related to the Game of Chess

In this Section, an example for IF-estimations in chess-game is given. For this estimation we have different cases. For instance, the first case is to make an estimation of the static composition of the chess pieces on the chessboard. Next case that is used in all contemporary programs for chess playing is to make estimations on the basis of the estimations of the sub-tree of the current node. It is assumed that better quality of the game is achieved when deeper investigation of the decision tree is made.

First, we must note that the Rook (Ξ), placed on an arbitrary square of the chessboard, in any case without exception can go from its square to 14 squares (different from the square where it resides) and that it cannot go to the rest 49 squares of the board. Therefore, if our evaluation of the value of the rook equals the number $\frac{m}{n}$, where $m = 14$ is the number of all possible squares where the rook can go, to the number of all free squares on the board $n = 63$, then its negation will be $\frac{n - m}{n}$. Hence,

$$\frac{m}{n} + \frac{n - m}{n} = 1.$$

i.e., their evaluation is a standard fuzzy evaluation, because there is no element of indeterminacy (uncertainty) for it. If we wish to give an intuitionistic fuzzy form of the evaluation, it will look like

$$V(\Xi) = \langle \frac{14}{63}, \frac{49}{63} \rangle.$$

The rook is the only chess piece with the property that the number of its possibly reachable squares does not depend on the square where it is.

The Pawn ($\triangle$) has a similar property, if it is neither a white pawn on the second or eighth ranks, nor a black pawn on the first or seventh ranks. Upon this condition, the following formula holds:

$$V(\triangle) = \langle \frac{1}{63}, \frac{62}{63} \rangle.$$

On the other hand, if the pawn is either a white pawn on the second rank, or a black pawn on the seventh rank, then there are two possible squares for it to go.

Finally, if the pawn is either a white pawn on the eight rank, or a black pawn on the first rank, it terminates its existence, generating a new piece with new travelling regularities, and therefore we are not going to treat this case.

With the exception of the 2nd/7th ranks, if we exclude the two boundary cases discussed above, we can state that for every pawn there is at least one possible square and 61 impossible squares, i.e., its evaluation in an intuitionistic fuzzy form will be

$$V(\triangle) = \langle \frac{1}{63}, \frac{61}{63} \rangle.$$

Of course, here we do not consider the case when a pawn captures an opponent's piece (as it is now the only piece on the board); this is planned to be included in a more detailed model.

The cases with the other pieces are more interesting.

We can easily see that if the Bishop (♗) is placed on square $a1$ as in Fig. 1.1, then it can go to 7 separate squares; if it is placed on square $b2$, then it can go to 9 separate squares; if it is placed on square $c3$, then it can go to 11 separate squares; if it is placed on square $d4$, then it can go to 13 separate squares. Therefore, its intuitionistic fuzzy evaluation will be:

$$V(♗) = \langle \frac{7}{63}, \frac{50}{63} \rangle.$$

More complex is the situation with the Queen (♕). If it is placed on square $a1$ as in Fig. 1.2, then it can go to 21 separate squares; if it is placed on square $b2$, then it can go to 23 separate squares; if it is placed on square $c3$, then it can go to 25 separate squares; if it is placed on square $d4$, then it can go to 27 separate squares. Therefore, its intuitionistic fuzzy evaluation will be:

$$V(♕) = \langle \frac{21}{63}, \frac{36}{63} \rangle.$$

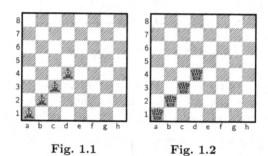

Fig. 1.1   Fig. 1.2

For the King (♔) there are three different situations: if it is placed on square $a1$ from Fig. 1.3, then it can go to 3 separate squares; if it is placed on square $b1$, then it can go to 5 separate squares; if it is placed on square $b2$, then it can go to 8 separate squares. Therefore, its intuitionistic fuzzy evaluation will be:

$$V(♔) = \langle \frac{3}{63}, \frac{55}{63} \rangle.$$

Finally, the situation with the Knight (♘) is the most complex one. If it is placed on square $a1$ as in Fig. 1.4, then it can go to 2 separate squares; if it is placed on square $b1$, then it can go to 3 separate squares; if it is placed on square $b2$ or $c1$ (i.e., there are two different squares with the same property), then it can go to 4 separate squares; if it is placed on square $c2$, then it can go to 6 separate squares; if it is placed on square $c3$, then it can go to 8 separate squares. Therefore, its intuitionistic fuzzy evaluation will be:

$$V(♘) = \langle \frac{2}{63}, \frac{55}{63} \rangle$$

The so constructed chess-examples of intuitionistic fuzziness illustrate also
the possibility for evaluations of the elements of a set with respect to universes
which are intuitionistic fuzzy sets with respect to other universes.

For example, if we have the situation as in Fig. 1.5, then for Bishop there
are 5 different squares where it can go, instead of the 7 possibilities which we
discussed above for the similar square.

In a subsequent study we shall discuss some more complex (and actual)
chess situations and their intuitionistic fuzzy evaluations.

Now we shall introduce the definitions of three new intuitionistic fuzzy
degrees, related to the estimation of the chess game.

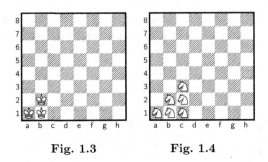

Fig. 1.3   Fig. 1.4

Every move can be estimated as *useful, harmful* or *indifferent (indefinite)*.
If we have the situation as in Fig. 1.6, then examples of these three types of
moves are given in Figs. 1.7 and 1.8, in Fig. 1.9 and on Fig. 1.10, respectively.

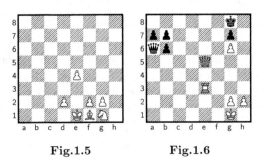

Fig.1.5   Fig.1.6

The moves as in Figs. 1.7 and 1.8 are useful for the white figures, because
they finish the game to their advantage. The move as in Fig. 1.9 is harmful
for the white figures, because the next move of the black figures can be ♛
$f1 \times$ (mate). Finally, there are indifferent moves gaining no direct results,
such one being the move of the whites $h2 - h3$ because neither the whites
lose their possibility to give unpreventable mate with their Queen, nor the
black figures may already on their turn give mate with their Queen.

Now, we define the following numbers, corresponding to the three
situations.

On the board, let there be s_w white figures $f_1^w, f_2^w, ..., f_{s_w}^w$ and s_b black figures $f_1^b, f_2^b, ..., f_{s_b}^b$, and let $\varphi(f^x, t)$ be the number of possible (for the moment t) moves of figures f^x, where $x \in \{w, b\}$. Let $\varphi_u(f^x, t)$, $\varphi_h(f^x, t)$ and $\varphi_i(f^x, t)$ be the number of possible moves (in the same moment) of the figure. Obviously,

$$\varphi_u(f^x, t) + \varphi_h(f^x, t) + \varphi_i(f^x, t) = \varphi(f^x, t).$$

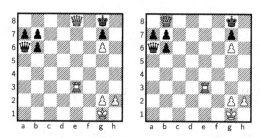

Fig.1.7   Fig.1.8

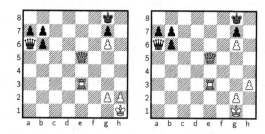

Fig.1.9   Fig.1.10

Therefore, we define

$$\mu_x(t) = \frac{\sum_{i=1}^{s_x} \varphi_u(f^x, t)}{\sum_{i=1}^{s_x} \varphi(f^x, t)},$$

$$\nu_x(t) = \frac{\sum_{i=1}^{s_x} \varphi_h(f^x, t)}{\sum_{i=1}^{s_x} \varphi(f^x, t)},$$

$$\pi_x(t) = \frac{\displaystyle\sum_{i=1}^{s_x} \varphi_i(f^x, t)}{\displaystyle\sum_{i=1}^{s_x} \varphi(f^x, t)}.$$

Obviously,

$$\mu_x(t) + \nu_x(t) + \pi_x(t) = 1,$$

i.e., $\langle \mu_x(t), \nu_x(t) \rangle$ is an intuitionistic fuzzy estimation of the situation on the chessboard at time moment t.

1.6   Some Ways for Altering the Experts' Estimations

In some cases, μ- and ν-function values are calculated on the basis of expert knowledge. Some problems may arise in relation to the correctness of the expert estimations. No such problems arise when we deal with ordinary fuzzy sets. Using the above notations, we discuss some ways for altering the experts' estimations of given events (or objects; for brevity, below we use only the word "event") in case the experts have made unacceptable errors in their Intuitionistic Fuzzy Estimations (IFEs).

While in an everyday conversation the expression "I am 1000% sure" is allowed, this is not the case when considering expert estimations. Clearly it does not matter much whether the estimations will be expressed in percentage or in numbers from the $[0, 1]$ range.

Following the terminology of the fuzzy set theory, we will assume that for a given estimated event or object the experts' estimations are expressed as numbers belonging to $[0, 1]$ range, producing simultaneously two degrees (of truth and falsity, of membership and non-membership, etc.).

If we use fuzzy estimations and if an expert states that he is 101% convinced in something, then we could just assume that he is, for instance, joking or exaggerating, and feel free to adjust his estimation to 100%.

The situation that arises when working with IFS is essentially different. Now it is not only a matter of consideration whether the expert was joking or not, but rather whether his estimations are precise enough.

Let us call the expert E_i ($i \in I$; I is an index set, related to the experts) *unconscientious* if among his estimations $\{\langle \mu_{i,j}, \nu_{i,j} \rangle / j \in J_i\}$, where

$$J = \bigcup_{i \in I} J_i$$

is an index set, related to the events estimated by the experts indexed by the elements of the set I, there are estimations for which $\mu_{i,j} \leq 1$, $\nu_{i,j} \leq 1$, but $\mu_{i,j} + \nu_{i,j} > 1$.

In this case, different ways for a post-adjustment of his incorrect estimations can be introduced in order to convert them to a correct IF-form.

It is obvious that if the estimations are fuzzy, then if an expert happens to assert a given event is 101% possible, we can simply assume his estimation is 1. Unfortunately, in the case of an IF-estimation the procedure of post-adjustment is very complicated. Some different ways for such post-adjustments of the estimations of the unconscientious experts are shown below.

Among the estimations $\{\langle \mu_{i,j}, \nu_{i,j}\rangle / j \in J_i\}$ of the expert E_i, let there be a subset $\{\langle \mu_{i,k}, \nu_{i,k}\rangle / k \in K_i\}$, where $K_i \subset J_i$, for which $\mu_{i,k} \leq 1$, $\nu_{i,k} \leq 1$, but $\mu_{i,k} + \nu_{i,k} > 1$.

Let $\overline{K}_i = J_i - K_i$.

Way 1 (trivial): We increase both IF-degrees of the estimation, removing the degree of non-determinacy, as follows:

$$\overline{\mu}_{i,k} = \frac{\mu_{i,k}}{\mu_{i,k} + \nu_{i,k}},$$

$$\overline{\nu}_{i,k} = \frac{\nu_{i,k}}{\mu_{i,k} + \nu_{i,k}}.$$

In this case, the new values do not leave the expert any possibility of hesitation.

A modification of this way is the following:

$$\overline{\mu}_{i,k} = \frac{\mu_{i,k}}{2},$$

$$\overline{\nu}_{i,k} = \frac{\nu_{i,k}}{2}$$

(the expert use $\mu_{i,k}$ and $\nu_{i,k}$ as independent elements of $[0,1]$).

Way 2: We replace μ- and ν-values with:

$$\overline{\mu}_{i,k} = \mu_{i,k} - \frac{\min(\mu_{i,k}, \nu_{i,k})}{2},$$

$$\overline{\nu}_{i,k} = \nu_{i,k} - \frac{\min(\mu_{i,k}, \nu_{i,k})}{2}.$$

Obviously,

$$\begin{aligned}
&\overline{\mu}_{i,k} + \overline{\nu}_{i,k} \\
&= \mu_{i,k} - \min(\mu_{i,k}, \nu_{i,k})/2 + \nu_{i,k} - \min(\mu_{i,k}, \nu_{i,k})/2 \\
&= \mu_{i,k} + \nu_{i,k} - \min(\mu_{i,k}, \nu_{i,k}) \\
&= \max(\mu_{i,k}, \nu_{i,k}) \leq 1.
\end{aligned}$$

These two ways admit only the values of the expert estimation for the j-th event.

A modification of this way is the following:

$$\overline{\mu}_{i,k} = \frac{(1 + \mu_{i,k} - \nu_{i,k})}{2},$$

$$\overline{\nu}_{i,k} = \frac{(1 - \mu_{i,k} + \nu_{i,k})}{2}.$$

These two ways allow only to alter the expert estimation of the j-th event. In this case, however, the balance between the IF estimations of the various events might be spoiled.

For example, the expert E_i has given the above (erroneous) estimations responding to the question "Is it sunny today?", at the same time having answered the question "Is it warm today?" with the values $\langle 0.5, 0.3 \rangle$, meaning that for 50% of the time the temperature has been above $20°C$, and for 30% of the time it has been below $15°C$.

If we calculate the answer to the first question (cf. above: $\langle 0.7, 0.8 \rangle$) in, say, the first (trivial) way, we would obtain approximately the values $\langle 0.47, 0.53 \rangle$.

Therefore, the expert in his erroneous statement is more convinced that today the weather is sunny (0.7) rather than warm (0.5); but, on the other hand, after adjusting his estimation to a correctly formed one, we get just the opposite – that he is more convinced that the weather is warm (0.5) rather than sunny (0.47).

As for his degree of negation of the statement, things are similar.

The ways considered below are more complex, but take into account more parameters that influence the expert's answers.

Way 3: When $card(K_i)$ is large enough, we determine the number

$$p_i = \frac{1}{card(\overline{K}_i)} \sum_{k \in \overline{K}_i} \pi_{i,k},$$

which corresponds to the average degree of indeterminacy of the correct expert's estimations (in the IF-sense) of the events, where

$$\pi_{i,k} = 1 - \mu_{i,k} - \nu_{i,k}.$$

Then, we can adjust his incorrect estimations as follows:

$$\overline{\mu}_{i,k} = \frac{(1 - p_i)\mu_{i,k}}{\mu_{i,k} + \nu_{i,k}},$$

$$\overline{\nu}_{i,k} = \frac{(1 - p_i)\nu_{i,k}}{\mu_{i,k} + \nu_{i,k}}.$$

Obviously,

$$\overline{\mu}_{i,k} + \overline{\nu}_{i,k} = 1 - p_i$$

i.e. $\overline{\pi}_{i,k} = p_i$.

In this case, the "hesitation" of the expert is simulated, but the hesitations are more homogeneous for all elements of the set K_i.

Thus, we can alter the individual estimations of the experts, but this is not influenced by the opinions of the other experts about the particular event. When $card(\overline{K}_i)$ is not large enough, i.e., when the majority of the experts are unconscious (at least with respect to the given event), we must use one of the first two ways.

Let $L \subset I$ and let for $l \in L$ the l-th expert give the following estimations $\{\langle \mu_{l,k}, \nu_{l,k}\rangle / k \in J_l\}$ where $\mu_{l,k} \le 1$, $\nu_{l,k} \le 1$ and $\mu_{k,l} + \nu_{k,l} > 1$. Let $\overline{L} = I - L$.

If the majority of the experts are unconscious for the estimation of k-th event ($k \in K_l$), we must again use one of the first two ways. If the number of the unconscious experts is small, we can use one of the following ways.

Way 4: Analogous to Way 3, we determine the numbers p_i for all $i \in I - L$ and the number

$$\overline{p}_l = \frac{1}{card(I - L)} \sum_{i \in I-L} p_i.$$

After this, we apply the formulae

$$\overline{\mu}_{l,k} = \frac{(1 - \overline{p}_l)\mu_{l,k}}{\mu_{l,k} + \nu_{l,k}},$$

$$\overline{\nu}_{l,k} = \frac{(1 - \overline{p}_l)\nu_{l,k}}{\mu_{l,k} + \nu_{l,k}}.$$

It must be immediately noted that none of the above considerations make use of information about the experts' scores we might have at our disposal. Below we are going to use that too.

Let the score of the i-th expert ($i \in L$) be specified by the ordered pair of positive real numbers $\langle \delta_i, \varepsilon_i \rangle$ where $\delta_i + \varepsilon_i \le 1$, δ_i is interpreted as his "degree of competence", and ε_i as his "degree of incompetence" (below we present a discussion on possible ways of both calculating and updating the experts' scores).

Then we could apply

Way 5: The numbers p_l are calculated as in Way 4, and then the numbers

$$p_l^{\mu} = \frac{1}{card(I - L)} \sum_{i \in I-L} \delta_i . p_i,$$

$$p_l^{\nu} = \frac{1}{card(I - L)} \sum_{i \in I-L} \varepsilon_i . p_i$$

are calculated. Now, apply the formulae from Way 4 with the values p_l^{μ} and p_l^{ν} instead of $\overline{p}_l$ on the corresponding places:

$$\bar{\mu}_{l,k} = \frac{\mu_{l,k}}{\mu_{l,k} + \nu_{l,k}} \cdot (1 - \frac{1}{card(I - L)} \sum_{i \in I - L} \delta_i . p_i),$$

$$\bar{\mu}_{l,k} = \frac{\nu_{l,k}}{\mu_{l,k} + \nu_{l,k}} \cdot (1 - \frac{1}{card(I - L)} \sum_{i \in I - L} \varepsilon_i . p_i).$$

The suggested techniques are, of course, just a part of all possible ways for post-adjustment of the experts' estimations.

Finally, we will discuss a way for constructing of the experts' scores. Let i-th expert ($i \in L$) have score $\langle \delta_i, \varepsilon_i \rangle$ and let he/she participated in n procedures, on the basis of which this score is determined, then after his/her participation in $(n+1)$-th procedure his/her score will be determined by

$$\langle \delta_i', \varepsilon_i' \rangle = \begin{cases} \langle \frac{\delta n + 1}{n+1}, \frac{\varepsilon n}{n+1} \rangle, & \text{if the expert's estimation is correct} \\ \langle \frac{\delta n}{n+1}, \frac{\varepsilon n}{n+1} \rangle, & \text{if the expert had not given any estimation.} \\ \langle \frac{\delta n}{n+1}, \frac{\varepsilon n + 1}{n+1} \rangle, & \text{if the expert's estimation is incorrect} \end{cases}$$

This text of this Section is based on [24, 29, 39].

In Section **3.6**, new ways for altering the experts' estimations are discussed. In Section **8.5**, more detailed formulae for the expert's score are given. In Section **8.6**, some ways of determining of IFS-membership and non-membership functions are discussed.

Operations and Relations over IFSs

2.1 Basic Operations and Relations over IFSs

In this Section, operations and relations over IFSs extending the definitions of the relations and operations over fuzzy sets (see e.g. [218, 297, 301, 612]) are introduced. Conversely, the fuzzy sets relations and operations will turn out to be particular cases of these new definitions.

Following [11, 14, 28], for every two IFSs A and B the following relations and operations can be defined (everywhere below "iff" means "if and only if"):

$$A \subseteq B \text{ iff } (\forall x \in E)(\mu_A(x) \le \mu_B(x) \ \& \ \nu_A(x) \ge \nu_B(x)); \qquad (2.1)$$

$$A \supseteq B \text{ iff } B \subseteq A; \qquad (2.2)$$

$$A = B \text{ iff } (\forall x \in E)(\mu_A(x) = \mu_B(x) \ \& \ \nu_A(x) = \nu_B(x)); \qquad (2.3)$$

$$A \cap B = \{\langle x, \min(\mu_A(x), \mu_B(x)), \max(\nu_A(x), \nu_B(x))\rangle | x \in E\}; \qquad (2.4)$$

$$A \cup B = \{\langle x, \max(\mu_A(x), \mu_B(x)), \min(\nu_A(x), \nu_B(x))\rangle | x \in E\}; \qquad (2.5)$$

$$A + B = \{\langle x, \mu_A(x) + \mu_B(x) - \mu_A(x)\mu_B(x), \nu_A(x)\nu_B(x)\rangle \mid x \in E\}; \qquad (2.6)$$

$$A.B = \{\langle x, \mu_A(x)\mu_B(x), \nu_A(x) + \nu_B(x) - \nu_A(x)\nu_B(x)\rangle \mid x \in E\}; \qquad (2.7)$$

$$A@B = \{\langle x, \frac{\mu_A(x)+\mu_B(x)}{2}, \frac{\nu_A(x)+\nu_B(x)}{2}\rangle | x \in E\}. \qquad (2.8)$$

Note that the following three operations were defined over the IFSs and included in [39]

$$A \bowtie B = \{\langle x, 2\frac{\mu_A(x).\mu_B(x)}{\mu_A(x) + \mu_B(x)}, 2.\frac{\nu_A(x).\nu_B(x)}{(\nu_A(x) + \nu_B(x))}\rangle | x \subset E\},$$

$$\text{for which if } \mu_A(x) = \mu_B(x) = 0, \text{then } \frac{\mu_A(x).\mu_B(x)}{\mu_A(x) + \mu_B(x)} = 0$$

$$\text{and if } \nu_A(x) = \nu_B(x) = 0, \text{then } \frac{\nu_A(x).\nu_B(x)}{\nu_A(x) + \nu_B(x)} = 0,$$

K.T. Atanassov: On Intuitionistic Fuzzy Sets Theory, STUDFUZZ 283, pp. 17–36.
springerlink.com   © Springer-Verlag Berlin Heidelberg 2012

$$A\$B = \{\langle x, \sqrt{\mu_A(x).\mu_B(x)}, \sqrt{\nu_A(x).\nu_B(x)}\rangle | x \in E\};$$

$$A * B = \{\langle x, \frac{\mu_A(x) + \mu_B(x)}{2.(\mu_A(x).\mu_B(x) + 1)}, \frac{\nu_A(x) + \nu_B(x)}{2.(\nu_A(x).\nu_B(x) + 1)}\rangle | x \in E\},$$

but so far they have not found any application and no interesting properties have been found for them. Hence, they have not been objects for discussion in the present book.

The most interesting was the fate of operations "negation" and "implication":

$$\overline{A} = \{\langle x, \nu_A(x), \mu_A(x)\rangle | x \in E\} \tag{2.9}$$

$$A \mapsto B = \{\langle x, \max(\nu_A(x), \mu_B(x)), \min(\mu_A(x), \nu_B(x))\rangle | x \in E\}. \tag{2.10}$$

Their present forms became the "apple of discord" related to the name of IFSs. For this reason during the last 6 years, I constructed a lot of new negations and implications, and studied their properties. They are discussed in Chapter **9**.

It is easy to demonstrate the correctness of the defined operations and relations. For example, for the "+" operation it is enough to see that the following inequalities hold:

$$0 \le \nu_A(x).\nu_B(x) \le \mu_A(x) + \mu_B(x) - \mu_A(x).\mu_B(x) + \nu_A(x).\nu_B(x)$$

$$\le \mu_A(x) + \mu_B(x) - \mu_A(x).\mu_B(x) + (1 - \mu_A(x)).(1 - \mu_B(x)) = 1.$$

The equality

$$A \mapsto B = \overline{A} \cup B$$

is valid.

The three relations defined above are analogous to the relations of inclusion and equality in the ordinary fuzzy set theory. It also holds that, for every two IFSs A and B:

$$A \subseteq B \text{ and } B \subseteq A \text{ iff } A = B.$$

The first four of the above operations were defined in the present form by the author in 1983 and this was his first step towards a research on the IFS properties.

It must be noted that the definition of the operation "@" is introduced by Toader Buhaescu in [148] independently from the author.

Operations "$", "$*$" and "$\bowtie$" were defined in 1992-93 for the sake of completeness (see [28]). Analogous operations are defined in the theory of fuzzy sets and logics (see, e.g. [301, 364, 612]).

In Stefan Danchev's paper [189], an operation is defined as an extension of all the operations "∩", "∪", "@", "$", "$\bowtie$". It has the form:

$$A \bigotimes_{m,n} B = \{\langle x, f(m, \mu_A(x), \mu_B(x)), f(n, \nu_A(x), \nu_B(x))\rangle | x \in E\}, \tag{2.11}$$

where

$$f(k, a, b) = \begin{cases} (\frac{1}{2} \cdot (a^k + b^k))^{1/k}, & \text{if } k > 0 \ \text{ or } \ (k < 0 \text{ and } a, b > 0) \\ \sqrt{ab}, & \text{if } k = 0 \\ 0, & \text{otherwise .} \end{cases}$$

The following equalities hold for every two IFSs A and B:

$$A \cap B = A \otimes_{-\infty, +\infty} B,$$

$$A \cup B = A \otimes_{+\infty, -\infty} B,$$

$$A @ B = A \otimes_{1,1} B,$$

$$A \$ B = A \otimes_{0,0} B,$$

$$A \bowtie B = A \otimes_{-1, -1} B.$$

At least for the moment, operation $\otimes_{0,0}$ is only of theoretical importance. So far, concrete practical applications exist only for the first five operations and some of these applications are discussed here. The last operation illustrates the relationships that exist between the different operations, defined over the IFSs, despite the considerable differences in their definitions.

The following definitions are well-known. Let M be a fixed set, let $e_* \in M$ be an unitary element of M, and let $*$ be an operation.

$\langle M, * \rangle$ is a *groupoid* iff $(\forall a, b \in M)(a * b \in M)$;

$\langle M, * \rangle$ is a *semi-group* iff $\langle M, * \rangle$ is a groupoid and

$$(\forall a, b, c \in M)((a * b) * c = a * (b * c));$$

$\langle M, *, e_* \rangle$ is a *monoid* iff $\langle M, * \rangle$ is a semi-group and

$$(\forall a \in M)(a * e_* = a = e_* * a).$$

$\langle M, *, e_* \rangle$ is a *commutative monoid* iff $\langle M, *, e_* \rangle$ is a monoid and

$$(\forall a, b \in M)(a * b = b * a).$$

$\langle M, *, e_* \rangle$ is a *group* iff $\langle M, *, e_* \rangle$ is a monoid and

$$(\forall a \in M)(\exists a_* \in M)(a * a_* = e_* = a_* * a).$$

$\langle M, *, e_* \rangle$ is a *commutative group* iff $\langle M, *, e_* \rangle$ is a group and

$$(\forall a, b \in M)(a * b = b * a).$$

A modification of the set-theoretical operator "power-set" is defined for a given set X by

$$P(X) = \{Y|Y \subseteq X\}, \tag{2.12}$$

where relation $\subseteq$ is given by (2.1).

We must note that each $Y \subseteq X$ has equal cardinality with set X. Therefore, $\emptyset \notin P(X)$ for the present modification of this operator.

Let us define the following *special IFSs*

$$O^* = \{\langle x, 0, 1\rangle | x \in E\}, \tag{2.13}$$

$$E^* = \{\langle x, 1, 0\rangle | x \in E\}, \tag{2.14}$$

$$U^* = \{\langle x, 0, 0\rangle | x \in E\}. \tag{2.15}$$

Obviously,

$$P(E^*) = \{A|A = \{\langle x, \mu_A(x), \nu_A(x)\rangle | x \in E\}\},$$

$$P(U^*) = \{B|B = \{\langle x, 0, \nu_B(x)\rangle | x \in E\}\},$$

$$P(O^*) = \{O^*\}.$$

Therefore,

$$\emptyset \notin P(O^*) \cup P(E^*) \cup P(U^*).$$

Theorem 2.1. For a fixed universe E,

1. $\langle P(E^*), \cap, E^*\rangle$ is a commutative monoid;
2. $\langle P(E^*), \cup, O^*\rangle$ is a commutative monoid;
3. $\langle P(E^*), +, O^*\rangle$ is a commutative monoid;
4. $\langle P(E^*), ., E^*\rangle$ is a commutative monoid;
5. $\langle P(E^*), @\rangle$ is a groupoid;
6. $\langle P(U^*), \cap, U^*\rangle$ is a commutative monoid;
7. $\langle P(U^*), \cup, O^*\rangle$ is a commutative monoid;
8. $\langle P(U^*), +, O^*\rangle$ is a commutative monoid;
9. $\langle P(U^*), ., U^*\rangle$ is a commutative monoid;
10. $\langle P(U^*), @\rangle$ is a groupoid;
11. None of these objects is a (commutative) group.

The following assertions are valid for three IFSs A, B and C:

1. $(A \cap B) \cup C = (A \cup C) \cap (B \cup C)$,
2. $(A \cap B) + C = (A + C) \cap (B + C)$,
3. $(A \cap B).C = (A.C) \cap (B.C)$,
4. $(A \cap B)@C = (A@C) \cap (B@C)$,

5. $(A \cup B) \cap C = (A \cap C) \cup (B \cap C)$,

6. $(A \cup B) + C = (A + C) \cup (B + C)$,

7. $(A \cup B).C = (A.C) \cup (B.C)$,

8. $(A \cup B)@C = (A@C) \cup (B@C)$,

9. $(A + B).C \subseteq (A.C) + (B.C)$,

10. $(A + B)@C \subseteq (A@C) + (B@C)$,

11. $(A.B) + C \supseteq (A + C).(B + C)$,

12. $(A.B)@C \supseteq (A@C).(B@C)$,

13. $(A@B) + C = (A + C)@(B + C)$,

14. $(A@B).C = (A.C)@(B.C)$,

15. $(A \cap B) \mapsto C \supseteq (A \mapsto C) \cap (B \mapsto C)$,

16. $(A \cup B) \mapsto C \subseteq (A \mapsto C) \cup (B \mapsto C)$,

17. $(A + B) \mapsto C \supseteq (A \mapsto C) + (B \mapsto C)$,

18. $(A.B) \mapsto C \subseteq (A \mapsto C).(B \mapsto C)$,

19. $A \mapsto (B + C) \subseteq (A \mapsto B) + (A \mapsto C)$,

20. $A \mapsto (B.C) \supseteq (A \mapsto B).(A \mapsto C)$,

21. $A \cap A = A$,

22. $A \cup A = A$,

23. $A@A = A$,

24. $\overline{A \cap B} = A \cup B$,

25. $\overline{A \cup B} = A \cap B$,

26. $\overline{A + B} = A.B$,

27. $\overline{A.B} = A + B$,

28. $\overline{A@B} = A@B$,

29. $E^* \mapsto A = A$,

30. $A \mapsto O^* = \overline{A}$,

31. $A \mapsto E^* = E^*$,

32. $O^* \mapsto A = E^*$,

33. $U^* \mapsto A = E^*$.

While the relations concerning equality and inclusion were defined soon after the definition of the concept of an IFS, the systematic research on the concept of intuitionistic fuzzy relation began later (see Chapter 8).

Now, we formulate one elementary but unusual equality between IFSs.

Theorem 2.2 [34, 39]. For every two IFSs A and B:

$$((A \cap B) + (A \cup B))@((A \cap B).(A \cup B)) = A@B. \qquad (2.16)$$

Proof: We shall use the fact that for every two real numbers a and b it follows that $\max(a,b) + \min(a,b) = a + b$. Now, we obtain

$$((A \cap B) + (A \cup B))@((A \cap B).(A \cup B))$$

$$= (\{\langle x, \min(\mu_A(x), \mu_B(x)), \max(\nu_A(x), \nu_B(x))\rangle | x \in E\}$$
$$+ \{\langle x, \max(\mu_A(x), \mu_B(x)), \min(\nu_A(x), \nu_B(x))\rangle | x \in E\})$$
$$@(\{\langle x, \min(\mu_A(x), \mu_B(x)), \max(\nu_A(x), \nu_B(x))\rangle | x \in E\}$$
$$.\{\langle x, \max(\mu_A(x), \mu_B(x)), \min(\nu_A(x), \nu_B(x))\rangle | x \in E\})$$

$$= \{\langle x, \min(\mu_A(x), \mu_B(x)) + \max(\mu_A(x), \mu_B(x))$$
$$- \min(\mu_A(x), \mu_B(x)). \max(\mu_A(x), \mu_B(x)),$$
$$\max(\nu_A(x), \nu_B(x)). \min(\nu_A(x), \nu_B(x))\rangle | x \in E\}$$
$$@\{\langle x, \min(\mu_A(x), \mu_B(x)). \max(\mu_A(x), \mu_B(x)),$$
$$\max(\nu_A(x), \nu_B(x)) + \min(\nu_A(x), \nu_B(x))$$
$$- \max(\nu_A(x), \nu_B(x)). \min(\nu_A(x), \nu_B(x))\rangle | x \in E\}$$

$$= \{\langle x, \mu_A(x) + \mu_B(x) - \mu_A(x).\mu_B(x), \nu_A(x).\nu_B(x)\rangle | x \in E\}$$
$$@\{\langle x, \mu_A(x).\mu_B(x), \nu_A(x) + \nu_B(x) - \nu_A(x).\nu_B(x)\rangle | x \in E\}$$

$$= \{\langle x, \frac{\mu_A(x) + \mu_B(x) - \mu_A(x).\mu_B(x) + \mu_A(x).\mu_B(x)}{2},$$
$$\frac{\nu_A(x).\nu_B(x) + \nu_A(x) + \nu_B(x) - \nu_A(x).\nu_B(x)}{2}\rangle | x \in E\}$$

$$= \{\langle x, \frac{\mu_A(x) + \mu_B(x)}{2}, \frac{\nu_A(x) + \nu_B(x)}{2}\rangle | x \in E\}$$

$$= A@B.$$

The following equalities also hold:

$$(A \cap B)@(A \cup B) = (A + B)@(A.B),$$
$$(A \cap B) + (A \cup B) = A + B,$$
$$(A \cap B).(A \cup B) = A.B,$$
$$(A \cap B)@(A \cup B) = A@B,$$
$$(A + B)@(A.B) = A@B.$$

In [551], my former student Tsvetan Vasilev proved that

$$((A + B) \cap (A.B))@((A + B) \cup (A.B)) = A@B$$

These equalities can be interpreted as analogues to the idempotency in Boolean algebra.

Theorem 2.3. For every two IFSs A and B

$$A.B \subseteq A \cap B \subseteq A@B \subseteq A \cup B \subseteq A + B. \tag{2.17}$$

The following three equalities are given in [78], for every three IFSs A, B and C,

$$(A \cap B) \to (A \cup B) = (A \to B) \cup (B \to A),\,^1$$
$$(A \to B) \cup (B \to C) \cup (C \to A) = (A \to C) \cup (C \to B) \cup (B \to A),$$
$$(A \to B) \cap (B \to C) \cap (C \to A) = (A \to C) \cap (C \to B) \cap (B \to A).$$

These equalities and inequalities can be regarded as curious facts in the theories of fuzzy sets and IFSs.

2.2 Short Remarks on Intuitionistic Fuzzy Logics. Part 1

The *intuitionistic fuzzy propositional calculus* has been introduced more than 20 years ago (see, e.g., [17, 39]). In it, if x is a variable, then its truth-value is represented by the ordered couple

$$V(x) = \langle a, b \rangle,$$

so that $a, b, a + b \in [0, 1]$, where a and b are the degrees of validity and of non-validity of x and the following definitions are given.

Assume that for the two variables x and y, the equalities $V(x) = \langle a, b \rangle, V(y) = \langle c, d \rangle$ $(a, b, c, d, a + b, c + d \in [0, 1])$ hold.

For two variables x and y operations "conjunction" (&), "disjunction" ($\vee$), "implication" ($\to$), and "(standard) negation" ($\neg$) are defined by:

$$V(x \& y) = \langle \min(a, c), \max(b, d) \rangle,$$

$$V(x \vee y) = \langle \max(a, c), \min(b, d) \rangle,$$

$$V(x \to y) = \langle \max(b, c), \min(a, d) \rangle,$$

$$V(\neg x) = \langle b, a \rangle.$$

All other logical operations also have intuitionistic fuzzy interpretations. The most interesting operators are the intuitionistic fuzzy implications and the respective negations generated by them. Currently, there are about 140 different such implications and about 35 different negations.

During the last 20 years, intuitionistic fuzzy predicative, intuitionistic fuzzy modal and intuitionistic fuzzy temporal logics have been developed. They are the objects of the author's future research.

2.3 Second Example from Area of Number Theory

The second example from the area of number theory is related to two arithmetic functions.

---

[1] In this formula in [78] there is a misprint - instead of the first "$\cup$" it is written "$\cap$".

Every natural number n has a canonical representation in the form $n = \prod_{i=1}^{k} p_i^{\alpha_i}$, where $p_1, p_2, ..., p_k$ are distinct prime numbers and $\alpha_1, \alpha_2, ..., \alpha_k \geq 1$ are natural numbers. Let $\underline{dim}(n) = \sum_{i=1}^{k} \alpha_i$.

Firstly, following [41], we introduce the concept of *converse factor*. The converse factor of a natural number is a natural number, defined as

$$CF(n) = \prod_{i=1}^{k} \alpha_i^{p_i}.$$

It can be easily seen that $CF(n) = 1$ iff for every i $(1 \leq i \leq k)$ $\alpha_i = 1$.
On the other hand, if there is at least one $\alpha_i > 1$, then $CF(n) > 1$.
CF is obviously a multiplicative function.
CF is not a monotonic function. For example,

$$8 < 9 \text{ and } CF(8) = 9 > 8 = CF(9);$$

$$4 < 16 \text{ and } CF(4) = 4 < 16 = CF(16).$$

On the other hand, if for the prime numbers a, b, c, $m = a.b$ and $n = b.c$, then

$$CF(m.n) = CF(a.b^2.c) = 2^b > 1 = CF(ab).CF(bc) = CF(m).CF(n).$$

Obviously, $CF(n) \geq 1$ for every natural number $n > 1$; and for every two natural numbers n and m:

$$CF(n^m) = CF(\prod_{i=1}^{k} p_i^{m\alpha_i}) = \prod_{i=1}^{k} (m\alpha_i)^{p_i} = m^{\sum_{i=1}^{k} p_i} .CF(n).$$

For the well-known Möbius function μ (see e.g., [378]) for every natural number n, the following equality is valid:

$$\mu(n) = (-1)^{cas(n)}.[\frac{1}{CF(n)}], \tag{2.18}$$

where for every integer number n, the function $\underline{cas}(n)$ is the number of the prime divisors of n and $[x]$ is the integer part of real number $x \geq 0$.
    If n is a prime number, then

$$(-1)^{cas(n)}.[\frac{1}{CF(n)}] = -[\frac{1}{1}] = -1 = \mu(n);$$

if $n = p_1 p_2 ... p_s$, where $p_1, p_2, ..., p_s$ are different prime numbers, then $\underline{cas}(n) = s$ and

$$(-1)^{\underline{cas}(n)}.[\frac{1}{CF(n)}] = (-1)^s[\frac{1}{CF(p_1 p_2 ... p_s)}] = (-1)^s[\frac{1}{1}] = (-1)^s = \mu(n);$$

if there exists such a prime p that $n = p^s m$, where s, m are natural numbers, p does not divide m and $s \geq 2$, then

$$(-1)^{\underline{cas}(n)}.[\frac{1}{CF(n)}] = (-1)^{\underline{cas}(n)}[\frac{1}{CF(p^s m)}] = (-1)^{\underline{cas}(n)}[\frac{1}{CF(p^s).CF(m)}]$$

$$= (-1)^{\underline{cas}(n)}[\frac{1}{s^p.CF(m)}] = 0 = \mu(n),$$

because at least $s^p > 1$.

Practically, the most interesting information from the Möbius function is obtained from the last case. Therefore, having in mind (2.18), we can generalize this function to the form:

$$M(n) = (-1)^{\underline{cas}(n)}.\frac{1}{CF(n)}.$$

Now, the new function satisfies inclusion $M(n) \in [-1, 1]$ and it determines, in more detail the form of the number n.

Let N be a fixed natural number and let n be its divisor (i.e., $n' = \frac{N}{n}$ is a natural number, that is a divisor of N). Then, we can define the following function[2]

$$IFF_N(n) \equiv \langle \mu_N(n), \nu_N(n) \rangle$$
$$= \langle \frac{dim(n).CF(n)}{dim(N).CF(N)}, \frac{dim(n').CF(n')}{dim(N).CF(N)} \rangle.$$

It can be easily seen that

$$0 \leq \frac{dim(n).CF(n)}{dim(N).CF(N)} + \frac{dim(n').CF(n')}{dim(N).CF(N)} \leq 1,$$

i.e., IFF has the behaviour of an intuitionistic fuzzy object. For example, we can see that

$$IFF_N(n') \equiv \langle \mu_N(n'), \nu_N(n') \rangle$$
$$= \langle \frac{dim(n').CF(n')}{dim(N).CF(N)}, \frac{dim(n).CF(n)}{dim(N).CF(N)} \rangle = \neg \langle \mu_N(n), \nu_N(n) \rangle.$$

We must immediately note that there are natural numbers for which $\mu_N(n) + \nu_N(n) < 1$, and others such that $\mu_N(n) + \nu_N(n) = 1$. For example, if N is a prime number, then

---

[2] Here and below, the symbol $\equiv$ is used to mention that X is equal to the right side by definition.

$$IFF_N(N) = \langle 1, 0 \rangle$$

$$IFF_N(1) = \langle 0, 1 \rangle$$

(having in mind that $\underline{dim}(1) = 0$), while, if $N = a.b^2$ for prime numbers a, b and if $n = a$, then

$$IFF_N(n) = \langle \frac{1}{3.2^b}, \frac{2^{b+1}}{3.2^b} \rangle$$

and obviously,

$$\frac{1}{3.2^b} + \frac{2^{b+1}}{3.2^b} < 1.$$

If m and n are different divisors of N, $m' = \frac{N}{m}$, $n' = \frac{N}{n}$ and m is a divisor of n, then

$$IFF_N(m) = \langle \frac{dim(m).CF(m)}{dim(N).CF(N)}, \frac{dim(m').CF(m')}{dim(N).CF(N)} \rangle$$

$$\leq \langle \frac{dim(n).CF(n)}{dim(N).CF(N)}, \frac{dim(m').CF(m')}{dim(N).CF(N)} \rangle = IFF_N(n),$$

where for $a, b, c, d \in [0, 1]$ and $a + b \leq 1, c + d \leq 1$: $\langle a, b \rangle \leq \langle c, d \rangle$ iff $a \leq c$ and $b \geq d$.

Therefore, the new function carries more information than Möbius function.

Of course, we can restrict function IFF to the ordinary fuzzy set form, using only its first component.

2.4   Other Operations over IFSs

First, following [99], we introduce two operations, defined over IFSs by Beloslav Riečan and the author. They are analogous of operations *"subtraction"* and *"division"* and for every two given IFSs A and B have the forms:

$$A - B = \{\langle x, \mu_{A-B}(x), \nu_{A-B}(x) \rangle | x \in E\}, \qquad (2.19)$$

where

$$\mu_{A-B}(x) = \begin{cases} \dfrac{\mu_A(x) - \mu_B(x)}{1 - \mu_B(x)}, & \text{if } \mu_A(x) \geq \mu_B(x) \text{ and } \nu_A(x) \leq \nu_B(x) \\ & \text{and } \nu_B(x) > 0 \\ & \text{and } \nu_A(x)\pi_B(x) \leq \pi_A(x)\nu_B(x) \\ 0, & \text{otherwise} \end{cases}$$

and

$$\nu_{A-B}(x) = \begin{cases} \dfrac{\nu_A(x)}{\nu_B(x)}, & \text{if } \mu_A(x) \geq \mu_B(x) \text{ and } \nu_A(x) \leq \nu_B(x) \\ & \text{and } \nu_B(x) > 0 \\ & \text{and } \nu_A(x)\pi_B(x) \leq \pi_A(x)\nu_B(x) \\ 1, & \text{otherwise} \end{cases} ;$$

and

$$A : B = \{\langle x, \mu_{A:B}(x), \nu_{A:B}(x)\rangle | x \in E\}, \qquad (2.20)$$

where

$$\mu_{A:B}(x) = \begin{cases} \dfrac{\mu_A(x)}{\mu_B(x)}, & \text{if } \mu_A(x) \leq \mu_B(x) \text{ and } \nu_A(x) \geq \nu_B(x) \\ & \text{and } \mu_B(x) > 0 \\ & \text{and } \mu_A(x)\pi_B(x) \leq \pi_A(x)\mu_B(x) \\ 0, & \text{otherwise} \end{cases}$$

and

$$\nu_{A:B}(x) = \begin{cases} \dfrac{\nu_A(x) - \nu_B(x)}{1 - \nu_B(x)}, & \text{if } \mu_A(x) \leq \mu_B(x) \text{ and } \nu_A(x) \geq \nu_B(x) \\ & \text{and } \mu_B(x) > 0 \\ & \text{and } \mu_A(x)\pi_B(x) \leq \pi_A(x)\mu_B(x) \\ 1, & \text{otherwise} \end{cases} .$$

We check easily that

$$0 \leq \mu_{A-B}(x) + \nu_{A-B}(x) \leq 1$$

and

$$0 \leq \mu_{A:B}(x) + \nu_{A:B}(x) \leq 1.$$

For every two IFSs A and B:

(a) $A - A = O^*$,

(b) $A : A = E^*$,

(c) $A - O^* = A$,

(d) $A : E^* = A$,

(e) $A - U^* = O^*$,

(f) $A : U^* = O^*$,

(g) $(A - B) + B = A$,

(h) $(A : B).B = A$,

(i) $(A - B) - C = (A - C) - B$,

(j) $(A : B) : C = (A : C) : B$,

(k) $(A@B) - C = (A - C)@(B - C)$,

(l) $(A@B) : C = (A : C)@(B : C)$,

(m) $\overline{A - B} = \overline{A} : \overline{B}$,

(n) $\overline{A : B} = \overline{A} - \overline{B}$.

Obviously,

$$E^* - E^* = O^*, \ E^* - U^* = O^*, \ E^* - O^* = E^*,$$

$$U^* - E^* = O^*, \ U^* - U^* = O^*, \ U^* - O^* = U^*,$$

$$O^* - E^* = O^*, \ O^* - U^* = O^*, \ O^* - O^* = O^*,$$

and

$$E^* : E^* = E^*, \ E^* : U^* = O^*, \ E^* : O^* = O^*,$$

$$U^* : E^* = U^*, \ U^* : U^* = O^*, \ U^* : O^* = O^*,$$

$$O^* : E^* = O^*, \ O^* : U^* = O^*, \ O^* : O^* = O^*.$$

Now, we see that for the two IFSs A and B, so that $B \subset A$ and
(a) for each $x \in E$, if

$$\nu_B(x) > 0,$$

$$\nu_A(x)\pi_B(x) \leq \pi_A(x)\nu_B(x),$$

then

$$A - B = \{\langle x, \frac{\mu_A(x) - \mu_B(x)}{1 - \mu_B(x)}, \frac{\nu_A(x)}{\nu_B(x)}\rangle | x \in E\},$$

(b) for each $x \in E$, if

$$\mu_B(x) > 0,$$

$$\mu_A(x)\pi_B(x) \leq \pi_A(x)\mu_B(x),$$

then

$$A : B = \{\langle x, \frac{\mu_A(x)}{\mu_B(x)}, \frac{\nu_A(x) - \nu_B(x)}{1 - \nu_B(x)}\rangle | x \in E\}.$$

Now, the following problems are open.

Open problem 1. *Are there relations between operations " $-$ " and " $:$ "
from one side and operations " $\cup$ " and " $\cap$ " from the other?*

Open problem 2. *Are there other relations between operations " $-$ " and
" $:$ " from one side and operations " $+$ " and "." from the other?*

Following [410], we introduce a new operation, defined over IFSs by Beloslav Riečan and the author. It is an analogous to operation "subtraction" and for every two given IFSs A and B has the form:

$$A|B = \{\langle \min(\mu_A(x), \nu_B(x)), \max(\mu_B(x), \nu_A(x)) \rangle | x \in E\}. \qquad (2.21)$$

We call the new operation *"set-theoretical subtraction"*.

First, we must check that as a result of the operation we obtain an IFS. For two given IFSs A and B and for each $x \in E$, if $\mu_B(x) \leq \nu_A(x)$, then

$$\min(\mu_A(x), \nu_B(x)) + \max(\mu_B(x), \nu_A(x))$$

$$= \min(\mu_A(x), \nu_B(x)) + \nu_A(x) \leq \mu_A(x) + \nu_A(x) \leq 1;$$

if $\mu_B(x) > \nu_A(x)$, then

$$\min(\mu_A(x), \nu_B(x)) + \max(\mu_B(x), \nu_A(x))$$

$$= \min(\mu_A(x), \nu_B(x)) + \mu_B(x) \leq \nu_B(x) + \nu_A(x) \leq 1.$$

In a similar way, we can prove the following assertions for every two IFSs A and B:

(a) $A|E^* = O^*$,

(b) $A|O^* = A$,

(c) $E^*|A = \overline{A}$,

(d) $O^*|A = O^*$,

(e) $(A|B) \cap C = (A \cap C)|B = A \cap (C|B)$,

(f) $(A|B) \cup C = (A \cup C) \cap \overline{B|C} = (A \cup C)|(B|C)$,

(g) $(A \cap B)|C = (A|C) \cap (B|C)$,

(h) $(A \cup B)|C = (A|C) \cup (B|C)$,

(i) $\overline{A|B} = \overline{A} \cap B = \overline{A} \cup B$.

Obviously,

$$E^*|E^* = O^*, \quad E^*|U^* = O^*, \quad E^*|O^* = E^*,$$

$$U^*|E^* = O^*, \quad U^*|U^* = U^*, \quad U^*|O^* = U^*,$$

$$O^*|E^* = O^*, \quad O^*|U^* = O^*, \quad O^*|O^* = O^*.$$

Now, the following problems are open.

Open problem 3. *Are there relations between operations " $-$ " and " $:$ " from one side and operation "|" from the other?*

Open problem 4. *Are there relations between operations "$+$" and ".". from one side and operation "|" from the other?*

In [194], Supriya Kumar De, Ranjit Biswas and Akhil Ranjan Roy introduced the following two operations:

$$n.A = \{\langle x, 1 - (1 - \mu_A(x))^n, (\nu_A(x))^n \rangle \mid x \in E\}, \tag{2.22}$$

$$A^n = \{\langle x, (\mu_A(x))^n, 1 - (1 - \nu_A(x))^n \rangle \mid x \in E\}, \tag{2.23}$$

where n is a natural number. They are called *multiplication of an IFS with n* and *n-th power of an IFS*.

Two other operations are defined in [411, 413], but we will not discuss them here.

Finally, let us define a new version of operation @:

$$\overset{n}{\underset{i=1}{@}} A_i = \{\langle x, \sum_{i=1}^{n} \frac{\mu_{A_i}(x)}{n}, \sum_{i=1}^{n} \frac{\nu_{A_i}(x)}{n} \rangle \mid x \in E\}.$$

In particular,

$$A_1 @ A_2 = \overset{2}{\underset{i=1}{@}} A_i.$$

Therefore, the following equalities hold:

$$\overline{\overset{n}{\underset{i=1}{@}} \overline{A_i}} = \overset{n}{\underset{i=1}{@}} A_i;$$

$$(\overset{n}{\underset{i=1}{@}} A_i) + B = \overset{n}{\underset{i=1}{@}} (A_i + B);$$

$$(\overset{n}{\underset{i=1}{@}} A_i).B = \overset{n}{\underset{i=1}{@}} (A_i.B).$$

2.5   On the Difference between Crisp Sets and IFSs

It is well known that for each two ordinary (crisp) sets the equality

$$(A \cup B) \cap (\overline{A} \cup \overline{B}) = (\overline{A} \cap B) \cup (A \cap \overline{B}) \tag{2.24}$$

is valid. In [530], it is noted that (2.24) is not always valid for the case of fuzzy sets. Here, we study the cases, when (2.24) is valid and the cases when it is not valid for IFSs.

Let for a fixed universe E, A and B be two IFSs such that

$$A = \{\langle x, \mu_A(x), \nu_A(x) \rangle \mid x \in E\},$$

$$B = \{\langle x, \mu_B(x), \nu_B(x) \rangle \mid x \in E\},$$

where functions $\mu_A, \mu_B : E \to [0,1]$ and $\nu_A, \nu_B : E \to [0,1]$ define the respective degrees of membership and non-membership of the element $x \in E$. Let for every $x \in E$,

$$0 \leq \mu_A(x) + \nu_A(x) \leq 1,$$

$$0 \leq \mu_B(x) + \nu_B(x) \leq 1.$$

Then,

$$(A \cup B) \cap (\overline{A} \cup \overline{B}) = \{\langle x, \min(\max(\mu_A(x), \mu_B(x)), \max(\nu_A(x), \nu_B(x))),$$

$$\max(\min(\nu_A(x), \nu_B(x)), \min(\mu_A(x), \mu_B(x)))\rangle | x \in E\} \qquad (2.25)$$

and

$$(\overline{A} \cap B) \cup (A \cap \overline{B}) = \{\langle x, \max(\min(\nu_A(x), \mu_B(x)), \min(\mu_A(x), \nu_B(x))),$$

$$\min(\max(\mu_A(x), \nu_B(x)), \max(\nu_A(x), \mu_B(x)))\rangle | x \in E\} \qquad (2.26)$$

Now, we shall study the cases when the IFSs in (2.25) and (2.26) coincide; the cases when one of these sets is included in the other, and the cases when these sets are not in any of the two above relations, having in mind (2.1), (2.2) and (2.3). Therefore, we must check the relations between expressions

$$Z_1 = \min(\max(\mu_A(x), \mu_B(x)), \max(\nu_A(x), \nu_B(x))),$$

$$Z_2 = \max(\min(\nu_A(x), \mu_B(x)), \min(\mu_A(x), \nu_B(x))),$$

$$Z_3 = \max(\min(\nu_A(x), \nu_B(x)), \min(\mu_A(x), \mu_B(x))),$$

$$Z_4 = \min(\max(\mu_A(x), \nu_B(x)), \max(\nu_A(x), \mu_B(x))).$$

We onstruct the following Table

Orderings of $\mu_A(x), \mu_B(x), \nu_A(x)$ and $\nu_B(x)$	Relations between Z_1 and Z_2	Relations between Z_3 and Z_4
$\mu_A(x) \leq \mu_B(x) \leq \nu_A(x) \leq \nu_B(x)$	$Z_1 = Z_2$	$Z_3 = Z_4$
$\mu_A(x) \leq \mu_B(x) \leq \nu_B(x) \leq \nu_A(x)$	$Z_1 = Z_2$	$Z_3 = Z_4$
$\mu_A(x) \leq \nu_A(x) \leq \mu_B(x) \leq \nu_B(x)$	$Z_1 \geq Z_2$	$Z_3 \leq Z_4$
$\mu_A(x) \leq \nu_A(x) \leq \nu_B(x) \leq \mu_B(x)$	$Z_1 \geq Z_2$	$Z_3 \leq Z_4$
$\mu_A(x) \leq \nu_B(x) \leq \mu_B(x) \leq \nu_A(x)$	$Z_1 = Z_2$	$Z_3 = Z_4$
$\mu_A(x) \leq \nu_B(x) \leq \nu_A(x) \leq \mu_B(x)$	$Z_1 = Z_2$	$Z_3 = Z_4$
$\mu_B(x) \leq \mu_A(x) \leq \nu_A(x) \leq \nu_B(x)$	$Z_1 = Z_2$	$Z_3 = Z_4$
$\mu_B(x) \leq \mu_A(x) \leq \nu_B(x) \leq \nu_A(x)$	$Z_1 = Z_2$	$Z_3 = Z_4$
$\mu_B(x) \leq \nu_A(x) \leq \mu_A(x) \leq \nu_B(x)$	$Z_1 = Z_2$	$Z_3 = Z_4$
$\mu_B(x) \leq \nu_A(x) \leq \nu_B(x) \leq \mu_A(x)$	$Z_1 = Z_2$	$Z_3 = Z_4$
$\mu_B(x) \leq \nu_B(x) \leq \mu_A(x) \leq \nu_A(x)$	$Z_1 \geq Z_2$	$Z_3 \leq Z_4$
$\mu_B(x) \leq \nu_B(x) \leq \nu_A(x) \leq \mu_A(x)$	$Z_1 \geq Z_2$	$Z_3 \leq Z_4$
$\nu_A(x) \leq \mu_B(x) \leq \mu_A(x) \leq \nu_B(x)$	$Z_1 = Z_2$	$Z_3 = Z_4$
$\nu_A(x) \leq \mu_B(x) \leq \nu_B(x) \leq \mu_A(x)$	$Z_1 = Z_2$	$Z_3 = Z_4$
$\nu_A(x) \leq \mu_A(x) \leq \mu_B(x) \leq \nu_B(x)$	$Z_1 \geq Z_2$	$Z_3 \leq Z_4$
$\nu_A(x) \leq \mu_A(x) \leq \nu_B(x) \leq \mu_B(x)$	$Z_1 \geq Z_2$	$Z_3 \leq Z_4$
$\nu_A(x) \leq \nu_B(x) \leq \mu_B(x) \leq \mu_A(x)$	$Z_1 = Z_2$	$Z_3 = Z_4$
$\nu_A(x) \leq \nu_B(x) \leq \mu_A(x) \leq \mu_B(x)$	$Z_1 = Z_2$	$Z_3 = Z_4$
$\nu_B(x) \leq \mu_B(x) \leq \mu_A(x) \leq \nu_A(x)$	$Z_1 \geq Z_2$	$Z_3 \leq Z_4$

Orderings of $\mu_A(x), \mu_B(x), \nu_A(x)$ and $\nu_B(x)$	Relations between Z_1 and Z_2	Relations between Z_3 and Z_4
$\nu_B(x) \leq \mu_B(x) \leq \nu_A(x) \leq \mu_A(x)$	$Z_1 \geq Z_2$	$Z_3 \leq Z_4$
$\nu_B(x) \leq \nu_A(x) \leq \mu_B(x) \leq \mu_A(x)$	$Z_1 = Z_2$	$Z_3 = Z_4$
$\nu_B(x) \leq \nu_A(x) \leq \mu_A(x) \leq \mu_B(x)$	$Z_1 = Z_2$	$Z_3 = Z_4$
$\nu_B(x) \leq \mu_A(x) \leq \mu_B(x) \leq \nu_A(x)$	$Z_1 = Z_2$	$Z_3 = Z_4$
$\nu_B(x) \leq \mu_A(x) \leq \nu_A(x) \leq \mu_B(x)$	$Z_1 = Z_2$	$Z_3 = Z_4$

From the Table we directly obtain the validity of the following inclusion

$$(\overline{A} \cap B) \cup (A \cap \overline{B}) \subseteq (A \cup B) \cap (\overline{A} \cup \overline{B}) \qquad (2.27)$$

The above Table shows the orderings for which the equality is valid, as well. It is obvious that inclusion (2.27) is valid for fuzzy sets.

Inclusion (2.27) is an example of a difference between the ordinary (crisp) sets on one hand and both fuzzy sets and IFSs on another.

2.6   Intuitionistic Fuzzy Tautological Sets

An IFS A is called *Intuitionistic Fuzzy Tautological Set (IFTS)* iff for every $x \in E$

$$\mu_A(x) \geq \nu_A(x). \qquad (2.30)$$

It can be easily checked that the following properties hold.
Theorem 2.4. For every three IFSs A, B, C :

(a) $A \mapsto A$,

(b) $A \mapsto (B \mapsto A)$,

(c) $A \cap B \mapsto A$,

(d) $A \cap B \mapsto B$,

(e) $A \mapsto (A \cup B)$,

(f) $B \mapsto (A \cup B)$,

(g) $A \mapsto (B \mapsto (A \cap B))$,

(h) $(A \mapsto C) \mapsto ((B \mapsto C) \mapsto ((A \cup B) \mapsto C))$,

(i) $\overline{\overline{A}} \mapsto A$,

(j) $(A \mapsto (B \mapsto C)) \mapsto ((A \mapsto B) \mapsto (A \mapsto C))$,

(k) $(\overline{A} \mapsto \overline{B}) \mapsto ((\overline{A} \mapsto B) \mapsto A)$

are IFTS.
Proof: Let us prove (h)

$$(A \mapsto C) \mapsto ((B \mapsto C) \mapsto ((A \cup B) \mapsto C))$$

$$= \{\langle x, \max(\nu_A(x), \mu_C(x)), \min(\mu_A(x), \nu_C(x))\rangle | x \in E\}$$
$$\mapsto (\{\langle x, \max(\nu_B(x), \mu_C(x)), \min(\mu_B(x), \nu_C(x))\rangle | x \in E\}$$
$$\mapsto (\{\langle x, \max(\mu_A(x), \mu_B(x)), \min(\nu_A(x), \nu_B(x))\rangle | x \in E\}$$
$$\mapsto \{\langle x, \mu_C(x), \nu_C(x)\rangle | x \in E\}))$$
$$= \{\langle x, \max(\nu_A(x), \mu_C(x)), \min(\mu_A(x), \nu_C(x))\rangle | x \in E\}$$
$$\mapsto (\{\langle x, \max(\nu_B(x), \mu_C(x)), \min(\mu_B(x), \nu_C(x))\rangle | x \in E\}$$
$$\mapsto \{\langle x, \max(\mu_C(x), \min(\nu_A(x), \nu_B(x)),$$
$$\min(\nu_C(x), \max(\mu_A(x), \mu_B(x)))\rangle | x \in E\})$$

$$= \{\langle x, \max(\nu_A(x), \mu_C(x)), \min(\mu_A(x), \nu_C(x))\rangle | x \in E\}$$
$$\mapsto \{\langle x, \max(\min(\mu_B(x), \nu_C(x)), \mu_C(x),$$
$$\min(\nu_A(x), \nu_B(x))), \min(\max(\nu_B(x), \mu_C(x)),$$
$$\nu_C(x), \max(\mu_A(x), \mu_B(x)))\rangle | x \in E\}$$
$$= \{\langle x, \max(\min(\mu_A(x), \nu_C(x)), \min(\mu_B(x), \nu_C(x)), \mu_C(x),$$
$$\min(\nu_A(x), \nu_B(x))), \min(\max(\nu_A(x), \mu_C(x)),$$
$$\max(\nu_B(x), \mu_C(X)), \nu_C(x), \max(\mu_A(x), \mu_B(x)))\rangle | x \in E\}$$

From

$$\max(\min(\mu_A(x), \nu_C(x)), \min(\mu_B(x), \nu_C(x)), \mu_C(x),$$
$$\min(\nu_A(x), \nu_B(x)))$$

$$\geq \max(\min(\mu_A(x), \nu_C(x)), \min(\mu_B(x), \nu_C(x)))$$
$$= \min(\nu_C(x), \max(\mu_A(x), \mu_B(x)))$$
$$\geq \min(\max(\nu_A(x), \mu_C(x)), \max(\nu_B(x), \mu_C(x)),$$
$$\nu_C(x), \max(\mu_A(x), \mu_B(x)))$$

it follows that

$$(A \mapsto (B \mapsto C)) \mapsto ((A \mapsto B) \mapsto (A \mapsto C))$$

is an IFTS.

Let us also discuss (k).

$$(\overline{A} \mapsto \overline{B}) \mapsto (\overline{A} \mapsto B) \mapsto A)$$
$$= (\{\langle x, \nu_A(x), \mu_A(x)\rangle | x \in E\} \mapsto \{\langle x, \nu_B(x), \mu_B(x)\rangle | x \in E\})$$
$$\mapsto (\{\langle x, \max(\mu_A(x), \mu_B(x)), \min(\nu_A(x), \nu_B(x))\rangle | x \in E\}$$
$$\mapsto \{\langle x, \mu_A(x), \nu_A(x)\rangle | x \in E\})$$

$$= \{\langle x, \max(\mu_A(x), \nu_B(x)), \min(\nu_A(x), \mu_B(x))\rangle | x \in E\}$$
$$\mapsto \{\langle x, \max(\mu_A(x), \min(\nu_A(x), \nu_B(x))),$$
$$\min(\nu_A(x), \max(\mu_A(x), \mu_B(x)))\rangle | x \in E\}$$
$$= \{\langle x, \max(\min(\nu_A(x), \mu_B(x)), \mu_A(x), \min(\nu_A(x), \nu_B(x))),$$
$$\min(\nu_A(x), \max(\mu_A(x), \mu_B(x)), \max(\mu_A(x), \nu_B(x)))\rangle | x \in E\}$$

From

$$\max(\min(\nu_A(x), \mu_B(x)), \mu_A(x), \min(\nu_A(x), \nu_B(x)))$$
$$\geq \max(\mu_A(x), \min(\nu_A(x), \mu_B(x)))$$
$$\geq \min(\nu_A(x), \max(\mu_A(x), \mu_B(x)))$$
$$\geq \min(\nu_A(x), \max(\mu_A(x), \mu_B(x)), \max(\mu_A(x), \nu_B(x)))$$

it follows that $(\overline{A} \mapsto \overline{B}) \mapsto ((\overline{A} \mapsto B) \mapsto A)$ is an IFTS.

Thus, it is shown that assertions analogous to the axioms of the propositional calculus (see e.g. [372]) hold for every two IFSs A and B :

(a) If $A \subseteq B$, then $A \mapsto B$ is an IFTS,
(b) $(A \cap (A \mapsto B)) \mapsto B$ is an IFTS,
(c) $((A \mapsto B) \cap \overline{B}) \mapsto \overline{A}$ is an IFTS,
(d) $((A \mapsto B) \cap (B \mapsto C)) \mapsto (A \mapsto C)$ is an IFTS.

Finally, we introduce one more theorem dealing with IFTS.

Theorem 2.5. For every five IFSs A, B, C, D and E,

$$(((((A \mapsto B) \mapsto (\overline{C} \mapsto \overline{D})) \mapsto C) \mapsto E) \mapsto ((E \mapsto A) \mapsto (D \mapsto A))$$

is an IFTS.

Proof: $((((A \mapsto B) \mapsto (\overline{C} \mapsto \overline{D})) \mapsto C) \mapsto E) \mapsto ((E \mapsto A) \mapsto (D \mapsto A))$

$= (((\{\langle x, \max(\nu_A(x), \mu_B(x)), \min(\mu_A(x), \nu_B(x))\rangle | x \in E\}$
$\mapsto \{\langle x, \max(\mu_C(x), \nu_D(x)), \min(\nu_C(x), \mu_D(x))\rangle | x \in E\})$
$\mapsto C) \mapsto E) \mapsto ((E \mapsto A) \mapsto (D \mapsto A))$

$= ((\{\langle x, \max(\mu_C(x), \nu_D(x), \min(\mu_A(x), \nu_B(x))),$
$\min(\nu_C(x), \mu_D(x), \max(\nu_A(x), \mu_B(x)))\rangle | x \in E\}$
$\mapsto C) \mapsto E) \mapsto ((E \mapsto A) \mapsto (D \mapsto A))$

$= (\{\langle x, \max(\min(\nu_C(x), \mu_D(x), \max(\nu_A(x), \mu_B(x))), \mu_C(x)),$
$\min(\max(\mu_C(x), \nu_D(x), \min(\mu_A(x), \nu_B(x))), \nu_C(x))\rangle | x \in E\}$
$\mapsto E) \mapsto (\{\langle x, \max(\nu_E(x), \mu_A(x)), \min(\mu_E(x), \nu_A(x))\rangle | x \in E\}$
$\mapsto \{\langle x, \max(\nu_D(x), \mu_A(x)), \min(\mu_D(x), \nu_A(x))\rangle | x \in E\})$

$= \{\langle x, \max(\min(\max(\mu_C(x), \nu_D(x), \min(\mu_A(x), \nu_B(x))), \nu_C(x)), \mu_E(x)),$
$\min(\max(\min(\nu_C(x), \mu_D(x), \max(\nu_A(x), \mu_B(x))), \mu_C(x)), \nu_E(x))\rangle$
$| x \in E\} \mapsto \{\langle x, \max(\nu_D(x), \mu_A(x), \min(\mu_E(x), \nu_A(x))),$
$\min(\mu_D(x), \nu_A(x), \max(\nu_E(x), \mu_A(x)))\rangle | x \in E\}$

$= \{\langle x, \max(\nu_D(x), \mu_A(x), \min(\mu_E(x), \nu_A(x)),$
$\min(\max(\min(\nu_C(x), \mu_D(x), \max(\nu_A(x), \mu_B(x))), \mu_C(x)), \nu_E(x))),$
$\min(\mu_D(x), \nu_A(x), \max(\nu_E(x), \mu_A(x)),$
$\max(\min(\max(\mu_C(x), \nu_D(x), \min(\mu_A(x), \nu_B(x))), \nu_C(x)), \mu_E(x)))\rangle$
$| x \in E\}$

Let for every $x \in E$,

$a(x) = \max(\nu_D(x), \mu_A(x), \min(\mu_E(x), \nu_A(x)),$
$\quad \min(\max(\min(\nu_C(x), \mu_D(x), \max(\nu_A(x), \mu_B(x))), \mu_C(x)), \nu_E(x)))$
$\quad - \min(\mu_D(x), \nu_A(x), \max(\nu_E(x), \mu_A(x))),$
$\quad \max(\min(\max(\mu_C(x), \nu_D(x), \min(\mu_A(x), \nu_B(x))), \nu_C(x)), \mu_E(x)))$

Obviously,

$$\max(\mu_A(x), \nu_A(x)) \geq \min(\mu_A(x), \nu_A(x)),$$

$$\max(\mu_A(x), \nu_A(x)) \geq \min(\nu_A(x), \nu_E(x)),$$

$$\max(\mu_A(x), \mu_E(x)) \geq \min(\mu_A(x), \nu_A(x)).$$

Let $\max(\mu_A(x), \mu_E(x)) \geq \min(\nu_A(x), \nu_E(x))$. Then,

$a(x)$
$\geq \max(\mu_A(x), \min(\mu_E(x), \nu_A(x))) - \min(\nu_A(x), \max(\nu_E(x), \mu_A(x)))$
$= \min(\max(\mu_A(x), \nu_A(x)), \max(\mu_A(x), \mu_E(x)))$
$\quad - \max(\min(\mu_A(x), \nu_A(x)), \min(\nu_A(x), \nu_E(x)))$
$\geq 0.$

Let $\max(\mu_A(x), \mu_E(x)) < \min(\nu_A(x), \nu_E(x))$. Then,

$$\mu_A(x) < \nu_A(x), \; \mu_A(x) < \nu_E(x), \; \mu_E(x) < \nu_A(x), \; \mu_E(x) < \nu_E(x).$$

If $\nu_E(x) \leq \max(\mu_C(x), \min(\nu_C(x), \mu_D(x), \max(\nu_A(x), \mu_B(x))))$, then,

$a(x)$
$= \max(\nu_D(x), \mu_A(x), \mu_E(x), \min(\max(\min(\nu_C(x), \mu_D(x), \max(\nu_A\mu_D(x),$
$\quad \max(\nu_A(x), \mu_B(x))), \mu_C(x)), \nu_E(x))) - \min(\mu_D(x), \nu_A(x), \nu_E(x),$
$\quad \max(\min(\max(\mu_C(x), \nu_D(x), \min(\mu_A(x), \nu_B(x))), \nu_C(x)), \mu_E(x)))$
$\geq \max(\nu_D(x), \mu_A(x), \mu_E(x), \nu_E(x)) - \min(\mu_D(x), \nu_A(x), \nu_E(x))$
$\geq 0.$

If $\mu_E(x) \geq \min(\max(\mu_C(x), \nu_D(x), \min(\mu_A(x), \nu_B(x))), \nu_C(x))$, then,

$a(x)$
$= \max(\nu_D(x), \mu_A(x), \mu_E(x), \min(\max(\min(\nu_C(x), \mu_D(x),$
$\quad \max(\nu_A(x), \mu_B(x))), \mu_C(x)), \nu_E(x))) - \min(\mu_D(x), \nu_A(x), \nu_E(x),$
$\quad \max(\min(\max(\mu_C(x), \nu_D(x), \min(\mu_A(x), \nu_B(x))), \nu_C(x)), \mu_E(x)))$
$\geq \max(\nu_D(x), \mu_A(x), \mu_E(x)) - \min(\mu_D(x), \nu_A(x), \mu_E(x), \nu_E(x))$
$\geq 0.$

Finally, if $\nu_E(x) > \max(\mu_C(x), \min(\nu_C(x), \mu_D(x), \max(\nu_A(x), \mu_B(x))))$ and $\mu_E(x) < \min(\max(\mu_C(x), \nu_D(x), \min(\mu_A(x), \nu_B(x))), \nu_C(x))$, then

$$\nu_E(x) > \mu_C(x), \; \nu_E(x) > \min(\nu_C(x), \mu_D(x), \max(\nu_A(x), \mu_B(x))),$$

$$\mu_E(x) < \nu_C(x), \ \mu_E(x) < \max(\mu_C(x), \nu_D(x), \min(\mu_A(x), \nu_B(x))).$$

Therefore,

$$a(x)$$
$$= \max(\nu_D(x), \mu_A(x), \mu_E(x), \mu_C(x), \min(\nu_C(x), \mu_D(x),$$
$$\max(\nu_A(x), \mu_B(x))))$$
$$- \min(\mu_D(x), \nu_A(x), \nu_E(x), \nu_C(x), \max(\mu_C(x), \nu_D(x),$$
$$\min(\mu_A(x), \nu_B(x))))$$

$$\geq \min(\nu_C(x), \mu_D(x), \max(\nu_A(x), \mu_B(x)))$$
$$- \min(\mu_D(x), \nu_A(x), \nu_E(x), \nu_C(x), \max(\mu_C(x), \nu_D(x),$$
$$\min(\mu_A(x), \nu_B(x))))$$

$$\geq \min(\nu_C(x), \mu_D(x), \nu_A(x))$$
$$- \min(\mu_D(x), \nu_A(x), \nu_E(x), \nu_C(x), \max(\mu_C(x), \nu_D(x),$$
$$\min(\mu_A(x), \nu_B(x))))$$

$$\geq 0.$$

Thus, it has been shown that an analogue to the Meredith's axiom (see [372]) holds.

3

Geometrical Interpretations of IFSs

3.1 Geometrical Interpretations of an IFS

In this Chapter, several geometrical interpretations of the IFSs are discussed.

The ordinary fuzzy sets have only one geometrical interpretation, while in this Section two interpretations of IFSs are given, following [20]. In Section **3.3**, we discuss even more. There exist seven different geometrical interpretations so far. The most relevant of them are discussed below.

The most widely accepted standard geometrical interpretation of the IFSs is shown in Fig. 3.1.

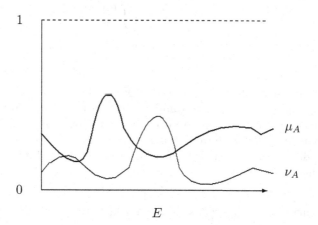

Fig. 3.1

Its analogue is given in Fig. 3.2[1]

---

[1] I will mention the following short story. Some time ago I received a paper for refereeing. In it, it was written that (here, for clearness, I use the present enu-

K.T. Atanassov: On Intuitionistic Fuzzy Sets Theory, STUDFUZZ 283, pp. 37–52.
springerlink.com   © Springer-Verlag Berlin Heidelberg 2012

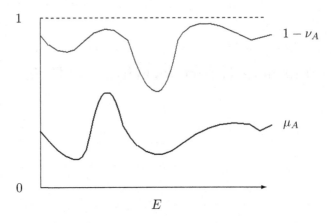

Fig. 3.2

Therefore, to every element $x \in E$ we can map a unit segment of the form:

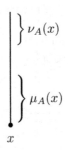

On the other hand, the situation in Fig. 3.3 is impossible.

Let a universe E be given and let us consider the figure F in the Euclidean plane with a Cartesian coordinate system (see Fig. 3.4). Let us call the triangle F *"IFS-interpretational triangle"*.

Let $A \subseteq E$ be a fixed set. Then we can construct a function $f_A : E \to F$ such that if $x \in E$, then

$$p = f_A(x) \in F,$$

and point p has coordinates $\langle a, b \rangle$ for which, $0 \le a + b \le 1$ and equalities $a = \mu_A(x), b = \nu_A(x)$ hold.

---

meration of the Figures instead of these from the paper) vague sets are better than IFSs, because the IFSs have the geometrical interpretation from Fig. 3.1, while, the vague sets have the interpretation from Fig. 3.2! In my report I wrote that it seems the authors had read my book [39] up to page 3, where Fig. 3.1 is placed, but they had obviously not turned over to the next page, where is Fig. 3.2, and by this reason I could not recommend their paper.

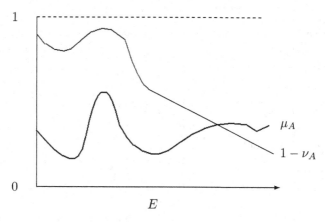

Fig. 3.3

Note that if there exist two different elements $x, y \in E$, $x \neq y$, for which $\mu_A(x) = \mu_A(y)$ and $\nu_A(x) = \nu_A(y)$ with respect to some set $A \subseteq E$, then $f_A(x) = f_A(y)$.

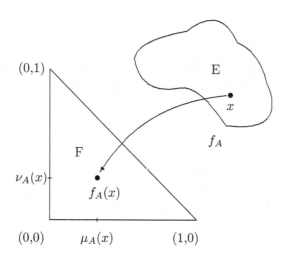

Fig. 3.4

3.2 Geometrical Interpretations of Operations over IFSs

In this section, we discuss the geometrical interpretations of the first five operations defined in Section **2.2**.

If A and B are two IFSs over E, then a function $f_{A \cap B}$ assigns to $x \in E$, a point $f_{A \cap B}(x) \in F$ with coordinates

$$\langle \min(\mu_A(x), \mu_B(x)), \max(\nu_A(x), \nu_B(y)) \rangle.$$

There exist three geometrical interpretations (see Figs. 3.5 (a) – 3.5 (c)) from which one general case (Fig. 3.5 (a)) and two particular cases (Figs. 3.5 (b) and 3.5 (c)).

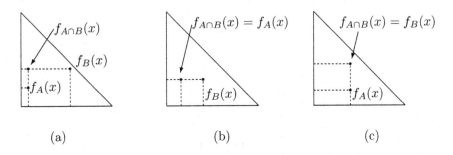

(a)   (b)   (c)

Fig. 3.5

If A and B are two IFSs over E, then a function $f_{A \cup B}$ assigns to $x \in E$, a point $f_{A \cup B}(x) \in F$ with coordinates

$$\langle \max(\mu_A(x), \mu_B(x)), \min(\nu_A(x), \nu_B(y)) \rangle$$

There exist also three geometrical cases as above, again one general case (Fig 3.6 (a)) and two particular cases (Fig. 3.6 (b), 3.6 (c)).

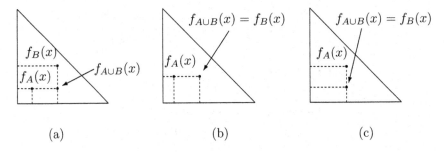

(a)   (b)   (c)

Fig. 3.6

If A and B are two IFSs over E, then a function f_{A+B} assigns to $x \in E$, a point $f_{A+B}(x) \in F$ with coordinates

$$\langle \mu_A(x) + \mu_B(x) - \mu_A(x).\mu_B(x), \nu_A(x).\nu_B(x) \rangle.$$

There exists only one form of geometrical interpretation of this operation (see Fig. 3.7 (a)).

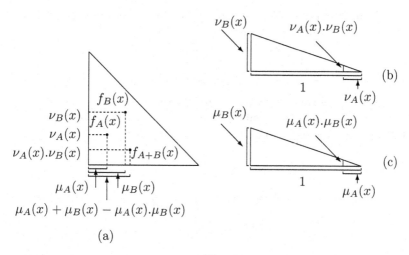

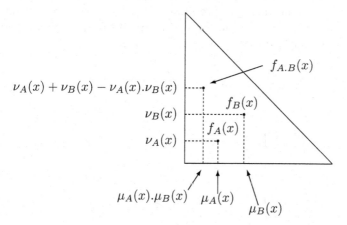

Fig. 3.7

The construction of $\mu_A(x).\mu_B(x)$ and $\nu_A(x).\nu_B(x)$ are shown in Fig. 3.7 (b) and 3.7 (c), respectively.

If A and B are two IFSs over E, then a function $f_{A.B}$ assigns to $x \in E$, a point $f_{A.B}(x) \in F$ with coordinates

$$\langle \mu_A(x).\mu_B(x), \nu_A(x) + \nu_B(x) - \nu_A(x).\nu_B(x) \rangle.$$

There exists only one form of geometrical interpretation of this operation (see Fig. 3.8; the constructions of $\mu_A(x).\mu_B(x)$ and $\nu_A(x).\nu_B(x)$ are as in Fig. 3.7).

Fig. 3.8

If A and B are two IFSs over E, then a function $f_{A@B}$ assigns to $x \in E$, a point $f_{A@B}(x) \in F$ with coordinates

$$\langle \frac{\mu_A(x) + \mu_B(x)}{2}, \frac{\nu_A(x) + \nu_B(x)}{2} \rangle$$

There exists only one form of geometrical interpretation of this operation (see Fig. 3.9). It must be noted that point $f_{A@B}(x)$ is the middle point of section between points $f_A(x)$ and $f_B(x)$, and so, is the name averaging operator.

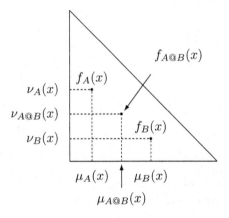

Fig. 3.9

3.3   Other Geometrical Interpretations of the IFSs

The author constructed two other geometrical interpretations (see Fig. 3.10 and 3.11). Two more geometrical interpretations have been invented by Ivan Antonov in [9] and Stefan Danchev in [188]: the first one over a sphere and the second one in an equilateral triangle (see Fig. 3.12).

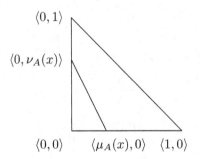

Fig. 3.10

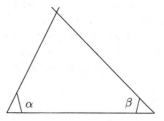

where $\alpha = \pi.\mu_A(x), \beta = \pi.\nu_A(x)$ and here $\pi = 3.14\ldots$

Fig. 3.11

Eulalia Szmidt and Janusz Kacprzyk constructed another three-dimensional geometrical interpretation [455, 474, 475] (see Fig. 3.13).

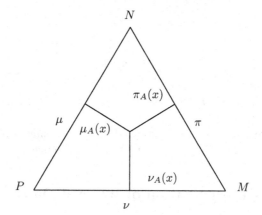

Fig. 3.12

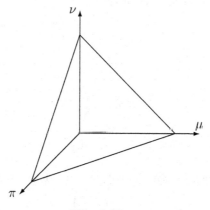

Fig. 3.13

In [122], a geometrical interpretation based on radar chart is proposed by
Vassia Atanassova. In Fig. 3.14, the innermost zone corresponds to the membership degree, the outermost zone to the non-membership degree and the region between both zones to the degree of uncertainty. This IFS-interpretation
can be especially useful for data in time series, multivaried data sets and other
data with cyclic trait.

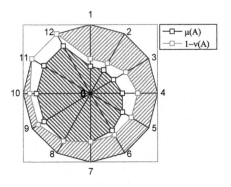

Fig. 3.14

Open problem 5. *What other useful geometrical interpretations of the IFSs
can be constructed?*
Open problem 6. *Is it true that all the above geometrical interpretations
can be constructed only by a ruler and a pair of compasses?*

3.4   On IF-Interpretation of Interval Data

When working with interval data, we can transform them to intuitionistic
fuzzy form. Then interpret them as points of the IFS-interpretation triangle.
For example, let us have the set of intervals

$$[a_1, b_1],$$

$$[a_2, b_2],$$

$$...$$

$$[a_n, b_n].$$

Let $A = \min\limits_{1 \le i \le n} a_i < \max\limits_{1 \le i \le n} b_i = B$. Of course, $A < B$, because otherwise for
all i: $a_i = b_i$. Now, for interval $[a_i, b_i]$ we can construct numbers

$$\mu_i = \frac{a_i - A}{B - A},$$

$$\nu_i = \frac{B - b_i}{B - A}$$

that satisfy the condition $0 \leq \mu_i + \nu_i \leq 1$ and have the geometrical interpretation from Fig. 3.4. This idea is introduced for a first time in [10] and it is used in a series of joint research of V. Kreinovich, M. Mukaidono, H. Nguyen, B. Wu, M. Koshelev, B. Rachamreddy, H. Yasemis and the author (see, [69, 315, 320, 321]).

3.5   On Some Transformation Formulae

There is a transformation between the equilateral triangle (Fig. 3.12) and the rectangular triangle (Fig 3.4).

It is seen that every point $X(x, y)$ in Fig. 3.12 is represented by exactly one corresponding point in the rectangular triangle in Fig 3.4.

It is well-known from elementary geometry that for every point $X(x, y)$ in the equilateral triangle, the equality

$$\mu(x) + \nu(x) + \pi(x) = 1$$

holds, where $\mu(x), \nu(x)$ and $\pi(x)$ are the distances to sections marked by μ, ν and π in the equilateral triangle in Fig. 3.12, respectively.

Now, the transformation function F has the form:

$$\langle u, v \rangle = F(x, y) = \langle \frac{x\sqrt{3} - y}{2}, y \rangle.$$

Therefore,

$$F(0, 0) = \langle 0, 0 \rangle,$$

$$F(\frac{2\sqrt{3}}{3}, 0) = \langle 1, 0 \rangle,$$

$$F(\frac{\sqrt{3}}{3}, 1) = \langle 0, 1 \rangle.$$

The inverse transformation is

$$F^{-1}(u, v) = \langle x, y \rangle \equiv \langle \frac{2u + v}{\sqrt{3}}, v \rangle.$$

It can be easily proved that F is bijective.

The transformation between the three-dimensional equilateral triangle in Fig 3.13 and the rectangular triangle in Fig. 3.4 is very simple and is given by

$$\langle u, v \rangle = G(x, y, z) = \langle x, y \rangle,$$

where $x + y + z = 1$. It is also bijective and its inverse function has the form

$$\langle x, y, z \rangle = G^{-1}(u, v) = \langle u, v, 1 - u - v \rangle.$$

Combining the above two formulae, we obtain the bijective transformation formulae between the three-dimensional equilateral triangle in Fig. 3.13 and the two-dimensional equilateral triangle triangle in Fig. 3.12 (again $x + y + z = 1$):

$$\langle u, v \rangle = H(x, y, z) = \langle \frac{2x + y}{\sqrt{3}}, y \rangle$$

$$\langle x, y, z \rangle = H^{-1}(u, v) = \langle \frac{x\sqrt{3} - y}{2}, y, \frac{2 - x\sqrt{3} - y}{2} \rangle.$$

3.6 Property of the IFS-Interpretational Triangle

In this Section, following [44], the following problem is discussed:
To find a continuous bijective transformation that transforms the unit square ABCD to the IFS interpretation triangle ABD (Fig. 3.15).

The solution to this question gives the possibility:
• to prove that each bi-lattice (see e.g., [234]) can be interpreted by an IFS;
• to prove that each intuitionistic L-fuzzy set with universe L of the same power as the continuum (see [39]) can be interpreted by an IFS;
• to construct a new algorithm for modifying incorrect expert estimations (see [39]).

Theorem 3.1. The transformation

$$F(x, y) = \begin{cases} \langle 0, 0 \rangle & \text{if } x = y = 0 \\ \\ \langle \frac{x^2}{x + y}, \frac{xy}{x + y} \rangle & \text{if } x, y \in [0, 1] \text{ and } x \geq y \\ & \text{and } x + y > 0 \\ \\ \langle \frac{xy}{x + y}, \frac{y^2}{x + y} \rangle & \text{if } x, y \in [0, 1] \text{ and } x \leq y \\ & \text{and } x + y > 0 \end{cases} \tag{3.1}$$

presents a solution to the Problem.

Proof. Let us first we shall prove that F is a bijective transformation.

Case 1: Let $x \geq y$. Consider the points $\langle x, y \rangle$ and $\langle u, v \rangle$ in Fig. 3.16, such that $x, y, u, v \in [0, 1]$ and they are not collinear with the point A $\langle 0, 0 \rangle$. Obviously, points $F(x, y)$ and $F(u, v)$ do not lie on a line containing point A. If $\langle x, y \rangle$ and $\langle u, v \rangle$ lie on a line containing point A, then

$$\frac{x}{y} = \frac{u}{v} = k > 0,$$

$y \neq v$ and then

$$\frac{x^2}{x + y} = \frac{k^2 y}{1 + k} \neq \frac{k^2 v}{1 + k} = \frac{u^2}{u + v}.$$

D (0,1)   C (1,1)

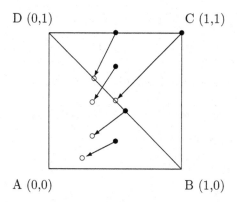

A (0,0)   B (1,0)

Fig. 3.15

Let

$$
\begin{cases}
\dfrac{x^2}{x+y} = \dfrac{u^2}{u+v} \\[2ex]
\dfrac{xy}{x+y} = \dfrac{uv}{u+v}
\end{cases}
.
$$

Let us assume that $x < u$. Of course, if $x = u$, then $y = v$. Therefore, $u = x + a$ and $v = y + b$ for $a > 0$ and $b \in [-1, 1]$. Therefore,

$$
\begin{cases}
\dfrac{x^2}{x+y} = \dfrac{x^2 + 2ax + a^2}{x+y+a+b} \\[2ex]
\dfrac{xy}{x+y} = \dfrac{(x+a)(y+b)}{x+y+a+b}
\end{cases}
.
$$

Hence,

$$
\begin{cases}
a(x^2 + x + 2xy + y) - bx^2 = 0 \\[2ex]
bx(x+a) + ay(y+b) = 0
\end{cases}
. \tag{3.2}
$$

From the first equality of (3.2), it is seen that $a, b \geq 0$ simultaneously, or $a, b \leq 0$ simultaneously. If the first case is valid, then from the second equality of (3.2), it follows that $a \geq 0$ and $b \leq 0$ or $a \leq 0$ and $b \geq 0$, that is possible only for $a = b = 0$. In the second case, having in mind our assumption, we see that it reduces to $a = b = 0$, i.e., $x = u$ and $y = v$.

Therefore, F is a bijective transformation. It can be easily seen that F is a continuous transformation and that

$$
\begin{cases}
\lim_{x \to 0,\ y \to 0} \dfrac{x^2}{x+y} = 0 \\[3ex]
\lim_{x \to 0,\ y \to 0} \dfrac{xy}{x+y} = 0
\end{cases}
.
$$

For every $u, v \in (0, 1]$, such that $u + v \leq 1$, we have

$$F^{-1}(u, v) = \begin{cases} \langle 0, 0 \rangle & \text{if } u = v = 0 \\ \langle u + v, \frac{v}{u}(u + v) \rangle & \text{if } u \geq v \text{ and } u > 0 \\ \langle \frac{u}{v}(u + v), u + v \rangle & \text{if } u \leq v \text{ and } v > 0 \end{cases}$$

Case 2: The proof is similar for $x < y$.

Corollary 3.1. For every $x, y \in [0, 1]$,

$$F(1, y) = \langle \frac{1}{1+y}, \frac{y}{1+y} \rangle, \quad F(x, 1) = \langle \frac{x}{x+1}, \frac{1}{x+1} \rangle,$$

$$F(1, 1) = \langle \frac{1}{2}, \frac{1}{2} \rangle, \quad F(x, 1 - x) = \langle x^2, x - x^2 \rangle,$$

$$F(\frac{1}{2}, \frac{1}{2}) = \langle \frac{1}{4}, \frac{1}{4} \rangle, \quad F(0, 0) = \langle 0, 0 \rangle.$$

The following theorem can be proved in a similar manner,

Theorem 3.2. The transformation

$$G(x, y) = \begin{cases} \langle x - \frac{y}{2}, \frac{y}{2} \rangle & \text{if } x, y \in [0, 1] \text{ and } x \geq y \\ \langle \frac{x}{2}, y - \frac{x}{2} \rangle & \text{if } x, y \in [0, 1] \text{ and } x \leq y \end{cases} \tag{3.3}$$

presents a solution to the Problem (see Fig. 3.16).

D (0,1)   C (1,1)

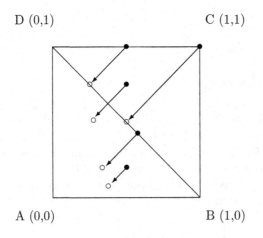

A (0,0)   B (1,0)

Fig. 3.16

For every $u, v \in (0, 1]$ such that $u + v \leq 1$,

$$G^{-1}(u,v) = \begin{cases} \langle u+v, 2v \rangle & \text{if } u \geq v \\[2mm] \langle 2u, u+v \rangle & \text{if } u \leq v \end{cases}.$$

Corollary 3.2. For every $x, y \in [0,1]$,

$$G(1,y) = \langle 1 - \frac{y}{2}, \frac{y}{2} \rangle, \quad G(x,1) = \langle \frac{x}{2}, 1 - \frac{x}{2} \rangle,$$

$$G(1,1) = \langle \frac{1}{2}, \frac{1}{2} \rangle, \quad G(x, 1-x) = \langle \frac{3x-1}{2}, \frac{1-x}{2} \rangle, \text{ if } x \geq \frac{1}{2},$$

$$G(x, 1-x) = \langle \frac{x}{2}, 1 - \frac{3x}{2} \rangle, \text{ if } x \leq \frac{1}{2}, \quad G(\frac{1}{2}, \frac{1}{2}) = \langle \frac{1}{4}, \frac{1}{4} \rangle,$$

$$G(0,0) = \langle 0,0 \rangle.$$

3.7 Other Ways for Altering Experts' Estimations

In this Section, three new ways for altering IF-experts' estimations, introduced in [74], are discussed. The notations are the same as in Section **1.7**.

Let, among the estimations $\{\langle \mu_{i,j}, \nu_{i,j} \rangle / j \in J_i\}$ made by expert E_i, there be a subset $\{\langle \mu_{i,k}, \nu_{i,k} \rangle / k \in K_i\}$, where $K_i \subset J_i$, for which $\mu_{i,k} \leq 1$, $\nu_{i,k} \leq 1$, but $\mu_{i,k} + \nu_{i,k} > 1$.

The first two ways for solving the problem by correcting of the incorrect estimations involve using the two transformation formulae above. Obviously, they will transform the experts estimations to new correct ones. In both cases, we change the expert's estimations only on the basis of his/her estimations, i.e., without using any additional information, as it was done for some of the five algorithms in Section **1.7**.

These two new algorithms are simple from computational point of view. The third algorithm that we discuss here decreases the original expert's estimations.

Let us determine the pair $\langle p_i, q_i \rangle$ for the i-th expert, so that $p_i = \mu_{i,m}$, $q_i = \nu_{i,m}$ and

$$\mu_{i,m}^2 + \nu_{i,m}^2 = \max_{k \in K_i}(\mu_{i,k}^2 + \nu_{i,k}^2).$$

The new transformation has the form

$$H(\mu_{i,k}, \nu_{i,k}) = \langle \frac{\mu_{i,k}p_i}{p_i + q_i}, \frac{\nu_{i,k}q_i}{p_i + q_i} \rangle.$$

Theorem 3.3. The transformation H is bijective.
Proof. Let the expert's estimations have geometrical interpretation by points with coordinates $\langle \mu_{i,k}, \nu_{i,k} \rangle$ and $\langle \mu_{i,j}, \nu_{i,j} \rangle$ (for $k, j \in J_i$). Then, from

$$H(\mu_{i,k}, \nu_{i,k}) = H(\mu_{i,j}, \nu_{i,j})$$

it follows that

$$\langle \frac{\mu_{i,k}p_i}{p_i+q_i}, \frac{\nu_{i,k}q_i}{p_i+q_i} \rangle = \langle \frac{\mu_{i,j}p_i}{p_i+q_i}, \frac{\nu_{i,j}q_i}{p_i+q_i} \rangle,$$

i.e.

$$\frac{\mu_{i,k}p_i}{p_i+q_i} = \frac{\mu_{i,j}p_i}{p_i+q_i}$$

and

$$\frac{\nu_{i,k}q_i}{p_i+q_i} = \frac{\nu_{i,j}q_i}{p_i+q_i},$$

and hence it follows directly that

$$\langle \mu_{i,k}, \nu_{i,k} \rangle = \langle \mu_{i,j}, \nu_{i,j} \rangle.$$

On the other hand, if

$$\langle \mu_{i,k}, \nu_{i,k} \rangle \neq \langle \mu_{i,j}, \nu_{i,j} \rangle,$$

then

$$\mu_{i,k} \neq \mu_{i,j} \text{ or } \nu_{i,k} \neq \nu_{i,j}.$$

Hence,

$$\frac{\mu_{i,k}p_i}{p_i+q_i} \neq \frac{\mu_{i,j}p_i}{p_i+q_i} \text{ or } \frac{\nu_{i,k}q_i}{p_i+q_i} \neq \frac{\nu_{i,j}q_i}{p_i+q_i},$$

i.e., $H(\mu_{i,k}, \nu_{i,k}) \neq H(\mu_{i,j}, \nu_{i,j})$.

Therefore, H is a bijective transformation. It can be easily seen that H is a continuous transformation and that

$$\begin{cases} \lim\limits_{x \to 0} \frac{xp_i}{p_i+q_i} = 0 \\[2ex] \lim\limits_{y \to 0} \frac{yq_i}{p_i+q_i} = 0 \end{cases} .$$

In [553], Peter Vassilev introduced a new approach for correcting expert's estimations through finding the smallest p for which the corresponding p-IFS contains all expert's evaluation and retransforming it to an IFS with new degrees of membership and non-membership.

3.8   An Example: IFSs and Complex Analysis

Following [63], a relation between the geometrical interpretation of the IFSs and element of complex analysis is discussed hereunder.

Two elements x and y of E are said to be in relation *negation* if

$$\mu_A(x) = \nu_A(y) \text{ and } \nu_A(x) = \mu_A(y).$$

Their geometrical interpretation in the intuitionistic fuzzy interpretation triangle is shown in Fig. 3.17.

It is well-known that numbers $a+ib$ and $a-ib$ are conjugate in the complex plane, where i is the imaginary unit. Let $a, b \in [0, 1]$ and let the geometrical representation of both points is given in Fig. 3.18. Let sections OA, OB and OC in Fig. 3.18 have unit length.

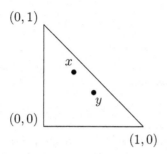

Fig. 3.17

We introduce formulae that transform the points of triangle ABC into triangle ABO. One of the possible forms of these formulae is:

$$f(a, b) = \begin{cases} (\frac{a}{2}, \frac{a}{2} + b), & \text{if } b \geq 0 \\ (\frac{a}{2} - b, \frac{a}{2}), & \text{if } b \leq 0 \end{cases}, \tag{3.4}$$

where (a, b) are the coordinates of the complex number $a+ib$ and $a \in [0, 1], b \in [-1, 1]$.

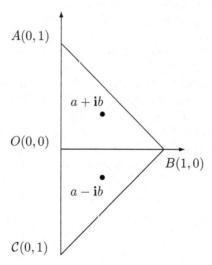

Fig. 3.18

It can immediately be seen, that

$$f(0,0) = (0,0),$$

$$f(0,1) = (0,1),$$

$$f(1,0) = (\frac{1}{2}, \frac{1}{2}),$$

$$f(0,-1) = (1,0).$$

It can be easily checked that function f is continuous and bijective.

From (3.4) we directly see that

$$f^{-1}(a,b) = (2a, b - a).$$

Now, we see that for the arbitrary complex conjugate numbers $a + ib$ and $a - ib$ in triangle ABC (here $a, b, a + b \in [0,1]$):

$$f(a,b) = (\frac{a}{2}, \frac{a}{2} + b),$$

$$f(a,-b) = (\frac{a}{2} + b, \frac{a}{2}).$$

Therefore, if IFS-element x has degrees of membership and non-membership $\frac{a}{2}$ and $\frac{a}{2} + b$, respectively, then IFS-element y that has degrees of membership and non-membership $\frac{a}{2} + b$ and $\frac{a}{2}$, respectively, shares with x, a relation *negation*.

The above construction generates a lot of interesting (and currently open) problems. Some of them are the following:

Open Problem 7. *To develop complex analysis for the intuitionistic fuzzy interpretation triangle.*

Open Problem 8. *To find transformations from the complex plane to the intuitionistic fuzzy interpretation triangle and vice versa in order to acquire more convenient methods for studying different properties of given IFSs or of given objects in the complex plane.*

Open Problem 9. *To find interpretations of the other intuitionistic fuzzy negations (see Section* **9.3***) in the complex plane and study their behaviour.*

Open Problem 10. *To construct complex analysis interpretations of the IFS-operators from modal, topological and level types (see Chapters* **4, 5** *and* **6***).*

4

Modal and Topological Operators Defined over IFSs

4.1 "Necessity" and "Possibility" Operators Defined over IFSs

This Chapter includes several operators over IFSs which have no counterparts in the ordinary fuzzy set theory.

Initially, following [11, 39], we introduce two operators over IFSs that transform an IFS into a fuzzy set (i.e., a particular case of an IFS). They are similar to the operators *"necessity"* and *"possibility"* defined in some modal logics. Their properties resemble these of the modal logic (see e.g. [227]).

Let, for every IFS A,

$$\Box A = \{\langle x, \mu_A(x), 1 - \mu_A(x)\rangle | x \in E\}, \tag{4.1}$$

$$\Diamond A = \{\langle x, 1 - \nu_A(x), \nu_A(x)\rangle | x \in E\}. \tag{4.2}$$

Obviously, if A is an ordinary fuzzy set, then

$$\Box A = A = \Diamond A. \tag{4.3}$$

The equalities (4.3) show that these operators do not have analogues in the case of fuzzy sets; and therefore, this is a new demonstration of the fact that IFSs are proper extensions of the ordinary fuzzy sets.

Most of the results described so far were, to some extent, similar to these of the ordinary fuzzy set theory. It was the definition of the two new operators that made it clear that IFSs are objects of different nature.

Both operators (4.1) and (4.2) were defined in March 1983 and this stimulated the author to continue his research on IFSs. At that moment George Gargov named the new sets "Intuitionistic Fuzzy Sets".

The following paragraphs consider the properties, modifications and extensions of these new operators.

For every IFS A,

K.T. Atanassov: On Intuitionistic Fuzzy Sets Theory, STUDFUZZ 283, pp. 53–76.
springerlink.com

(a) $\overline{\Box \overline{A}} = \Diamond A,$

(b) $\overline{\Diamond \overline{A}} = \Box A,$

(c) $\Box A \subseteq A \subseteq \Diamond A,$

(d) $\Box \Box A = \Box A,$

(e) $\Box \Diamond A = \Diamond A,$

(f) $\Diamond \Box A = \Box A,$

(g) $\Diamond \Diamond A = \Diamond A.$

For example, the validity of (a) is checked as follows:

$$\overline{\Box \overline{A}} = \overline{\Box \{\langle x, \nu_A(x), \mu_A(x)\rangle | x \in E\}}$$

$$= \overline{\{\langle x, \nu_A(x), 1 - \nu_A(x)\rangle | x \in E\}} = \{\langle x, 1 - \nu_A(x), \nu_A(x)\rangle | x \in E\} = \Diamond A.$$

For every two IFSs A and B,

(a) $\Box (A \cap B) = \Box A \cap \Box B,$

(b) $\Box (A \cup B) = \Box A \cup \Box B,$

(c) $\Box (A + B) = \Box A + \Box B,$

(d) $\Box (A.B) = \Box A. \Box B,$

(e) $\Box (\overline{\overline{A} + \overline{B}}) = \Diamond A. \Diamond B,$

(f) $\Box (\overline{\overline{A}.\overline{B}}) = \Diamond A + \Diamond B,$

(g) $\Box (A@B) = \Box A@ \Box B,$

(h) $\Diamond (A \cap B) = \Diamond A \cap \Diamond B,$

(i) $\Diamond (A \cup B) = \Diamond A \cup \Diamond B,$

(j) $\Diamond (A + B) = \Diamond A + \Diamond B,$

(k) $\Diamond (A.B) = \Diamond A. \Diamond B,$

(l) $\Diamond (\overline{\overline{A} + \overline{B}}) = \Box A. \Box B,$

(m) $\Diamond (\overline{\overline{A}.\overline{B}}) = \Box A + \Box B,$

(n) $\Diamond (A@B) = \Diamond A@ \Diamond B,$

(o) $\Box (\underset{i=1}{\overset{n}{@}} A_i) = \underset{i=1}{\overset{n}{@}} (\Box A_i),$

(p) $\Diamond (\underset{i=1}{\overset{n}{@}} A_i) = \underset{i=1}{\overset{n}{@}} (\Diamond A_i).$

Two new relations are defined as follows:

$$A \subseteq_\Box B \quad \text{iff} \quad (\forall x \in E)(\mu_A(x) \le \mu_B(x)), \tag{4.4}$$

$$A \subseteq_\Diamond B \quad \text{iff} \quad (\forall x \in E)(\nu_A(x) \ge \nu_B(x)). \tag{4.5}$$

The following statement, for every two IFSs A and B, justifies the choice of the denotations $\subseteq_\square$ and $\subseteq_\Diamond$:

(a) $A \subseteq_\square B$ iff $\square A \subseteq \square B$,
(b) $A \subseteq_\Diamond B$ iff $\Diamond A \subseteq \Diamond B$.
(c) $A \subseteq_\square B$ and $A \subseteq_\Diamond B$ iff $A \subseteq B$.

We introduce one more relation:

$$A \sqsubset B \text{ iff } (\forall x \in E)(\pi_A(x) \le \pi_B(x)). \tag{4.6}$$

For this relation, it holds that for every two IFSs A and B,

(a) if $A \subseteq_\square B$ and $A \sqsubset B$, then $A \subseteq B$,
(b) if $A \subseteq_\Diamond B$ and $B \sqsubset A$, then $A \subseteq B$.

Two different geometrical interpretations of both operators are given in Figs. 4.1, 4.2, 4.3 and 4.4 respectively.

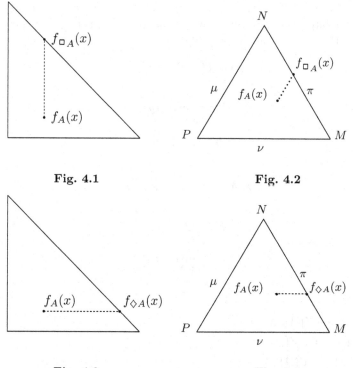

Fig. 4.1

Fig. 4.2

Fig. 4.3

Fig. 4.4

The relations "$\subseteq_\square$" and "$\subseteq_\Diamond$" defined over the elements of the set $E' = E \times [0,1] \times [0,1]$ for a fixed universe E, are quasi-orderings in the sense of [401], since for every three IFSs $A, B, C \in E^*$,

$$A \le A$$

$$\text{if } A \le B \text{ and } B \le C, \text{ then } A \le C$$

where $\le \in \{\subseteq_\square, \subseteq_\diamond\}$.

It is easily seen that $\subseteq_\square$ and $\subseteq_\diamond$ are not relations of linear ordering, because there may exist two IFSs $A, B \in E$ such that neither $A \le B$, nor $B \le A$ will hold, where $\le$ is one of the above three relations. For example, if $E = \{x, y\}$ and

$$A = \{\langle x, 0.3, 0.5 \rangle, \langle y, 0.4, 0.4 \rangle\},$$

$$B = \{\langle x, 0.4, 0.4 \rangle, \langle y, 0.3, 0.5 \rangle\},$$

then, obviously, neither $A \le B$ nor $B \le A$.

4.2   Topological Operators over IFSs

Following [14], we introduce two operators which are analogous to the topological operators of closure and interior (see, e.g. [328, 590]). They were defined in October 1983 by the author and their basic properties were studied. Three years later, the relations between the modal and the topological operators over IFSs were studied (see [39]).

For every IFS A,
$$\mathcal{C}(A) = \{\langle x, K, L \rangle | x \in E\}, \tag{4.7}$$

where
$$K = \sup_{y \in E} \mu_A(y), \tag{4.8}$$

$$L = \inf_{y \in E} \nu_A(y) \tag{4.9}$$

and
$$\mathcal{I}(A) = \{\langle x, k, l \rangle | x \in E\}, \tag{4.10}$$

where
$$k = \inf_{y \in E} \mu_A(y), \tag{4.11}$$

$$l = \sup_{y \in E} \nu_A(y). \tag{4.12}$$

For every two IFSs A and B,

(a) $\mathcal{C}(A)$ and $\mathcal{I}(A)$ are IFSs;
(b) $\mathcal{I}(A) \subseteq A \subseteq \mathcal{C}(A)$;
(c) $\mathcal{C}(\mathcal{C}(A)) = \mathcal{C}(A)$;
(d) $\mathcal{C}(\mathcal{I}(A)) = \mathcal{I}(A)$;
(e) $\mathcal{I}(\mathcal{C}(A)) = \mathcal{C}(A)$;

(f) $\mathcal{I}(\mathcal{I}(A)) = \mathcal{I}(A)$;

(g) $\mathcal{C}(A \cup B) = \mathcal{C}(A) \cup \mathcal{C}(B)$;

(h) $\mathcal{C}(A \cap B) \subseteq \mathcal{C}(A) \cup \mathcal{C}(B)$;

(i) $\mathcal{C}(A@B) \subseteq \mathcal{C}(A)@\mathcal{C}(B)$;

(j) $\mathcal{I}(A \cup B) \supseteq \mathcal{I}(A) \cup \mathcal{I}(B)$;

(k) $\mathcal{I}(A \cap B) = \mathcal{I}(A) \cap \mathcal{I}(B)$;

(l) $\mathcal{I}(A@B) \supseteq \mathcal{I}(A)@\mathcal{I}(B)$;

(m) $\overline{\mathcal{I}(\overline{A})} = \mathcal{C}(A)$;

(n) $\overline{\mathcal{C}(\overline{A})} = \mathcal{I}(A)$;

(o) $\mathcal{C}(O^*) = O^*$;

(p) $\mathcal{C}(E^*) = E^*$;

(q) $\mathcal{C}(U^*) = U^*$;

(r) $\mathcal{I}(O^*) = O^*$;

(s) $\mathcal{I}(E^*) = E^*$,

(t) $\mathcal{I}(U^*) = U^*$,

(u) $\square\,(\mathcal{C}(A)) = \mathcal{C}(\square\,(A))$,

(v) $\square\,(\mathcal{I}(A)) = \mathcal{I}(\square\,(A))$,

(w) $\Diamond(\mathcal{C}(A)) = \mathcal{C}(\Diamond(A))$,

(x) $\Diamond(I(A)) = \mathcal{I}(\Diamond(A))$,

where O^*, E^* and U^* are defined by (2.13), (2.14) and (2.15).

For example, we prove property (m).

$$\overline{\mathcal{I}(\overline{A})} = \overline{\mathcal{I}(\{\langle x, \nu_A(y), \mu_A(y)\rangle | x \in E\})}$$
$$= \overline{\{\langle x, L, K\rangle | x \in \overline{E}\}}$$
$$= \{\langle x, K, L\rangle | x \in E\} = \mathcal{C}(A).$$

It must be noted that $\mathcal{I}(A) \subseteq \mathcal{C}(B)$ follows from the inclusion $A \subseteq B$, but in the general case, $\mathcal{C}(A) \subseteq \mathcal{I}(B)$ does not hold.

For brevity, below we use $\mathcal{C}A$ and $\mathcal{I}A$ instead of $\mathcal{C}(A)$ and $\mathcal{I}(A)$, respectively, throughout this chapter. For every IFS A:

(a) $\square\mathcal{C}\square A = \Diamond\mathcal{C}\square A = \overline{\square\mathcal{I}\Diamond\overline{A}} = \overline{\Diamond\mathcal{I}\Diamond\overline{A}} = \{\langle x, K, 1 - K\rangle | x \in E\}$,

(b) $\square\mathcal{C}\Diamond A = \Diamond\mathcal{C}\Diamond A = \overline{\square\mathcal{I}\square\overline{A}} = \overline{\Diamond\mathcal{I}\square\overline{A}} = \{\langle x, 1 - L, L\rangle | x \in E\}$,

(c) $\square\mathcal{I}\square A = \Diamond\mathcal{I}\square A = \overline{\square\mathcal{C}\Diamond\overline{A}} = \overline{\Diamond\mathcal{C}\Diamond\overline{A}} = \{\langle x, k, 1 - k\rangle | x \in E\}$,

(d) $\square\mathcal{I}\Diamond A = \Diamond\mathcal{I}\Diamond A = \overline{\square\mathcal{C}\square\overline{A}} = \overline{\Diamond\mathcal{C}\square\overline{A}} = \{\langle x, 1 - l, l\rangle | x \in E\}$,

(e) $\square\mathcal{C}\square\overline{A} = \Diamond\mathcal{C}\square\overline{A} = \overline{\square\mathcal{I}\Diamond A} = \overline{\Diamond\mathcal{I}\Diamond A} = \{\langle x, l, 1 - l\rangle | x \in E\}$,

(f) $\Box C \Diamond \overline{A} = \Diamond C \Diamond \overline{A} = \Box \overline{\mathcal{I} \Box A} = \Diamond \overline{\mathcal{I} \Box A} = \{\langle x, 1-k, k\rangle | x \in E\}$,

(g) $\Box \mathcal{I} \Box \overline{A} = \Diamond \mathcal{I} \Box \overline{A} = \Box C \Diamond A = \overline{\Diamond C \Diamond A} = \{\langle x, L, 1-L\rangle | x \in E\}$,

(h) $\Box \mathcal{I} \Diamond \overline{A} = \Diamond \mathcal{I} \Diamond \overline{A} = \Box C \Box A = \Diamond C \Box A = \{\langle x, 1-K, K\rangle | x \in E\}$.

Let for a fixed IFS A:

$$s(A) = \{\Box C \Box A, \Diamond C \Box A, \Box \mathcal{I} \Diamond \overline{A}, \overline{\Diamond \mathcal{I} \Diamond \overline{A}}\},$$

$$t(A) = \{\Box C \Diamond A, \Diamond C \Diamond A, \overline{\Box \mathcal{I} \Box \overline{A}}, \Diamond \mathcal{I} \Box \overline{A}\},$$

$$u(A) = \{\Box \mathcal{I} \Box A, \Diamond \mathcal{I} \Box A, \Box C \Diamond \overline{A}, \overline{\Diamond C \Diamond \overline{A}}\},$$

$$v(A) = \{\Box \mathcal{I} \Diamond A, \Diamond \mathcal{I} \Diamond A, \overline{\Box C \Box \overline{A}}, \Diamond C \Box \overline{A}\},$$

$$w(A) = \{\Box C \Box \overline{A}, \Diamond C \Box \overline{A}, \Box \mathcal{I} \Diamond A, \overline{\Diamond \mathcal{I} \Diamond A}\},$$

$$x(A) = \{\Box C \Diamond \overline{A}, \Diamond C \Diamond \overline{A}, \overline{\Box \mathcal{I} \Box A}, \Diamond \mathcal{I} \Box A\},$$

$$y(A) = \{\Box \mathcal{I} \Box \overline{A}, \Diamond \mathcal{I} \Box \overline{A}, \Box C \Diamond A, \overline{\Diamond C \Diamond A}\},$$

$$z(A) = \{\Box \mathcal{I} \Diamond \overline{A}, \Diamond \mathcal{I} \Diamond \overline{A}, \overline{\Box C \Box A}, \Diamond C \Box A\}.$$

It can be directly seen that for every two IFSs P and Q:

(a) If $P \in s(A)$ and $Q \in t(A)$, then $P \subseteq CA \subseteq Q$;

(b) If $P \in u(A)$ and $Q \in v(A)$, then $P \subseteq \mathcal{I}A \subseteq Q$;

(c) If $P \in w(A)$ and $Q \in x(A)$, then $P \subseteq \mathcal{I}A \subseteq Q$;

(d) If $P \in y(A)$ and $Q \in z(A)$, then $P \subseteq CA \subseteq Q$.

The geometrical interpretations of both operators are given in Figs. 4.5 and 4.6, respectively.

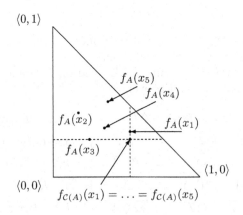

Fig. 4.5

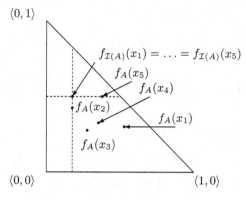

Fig. 4.6

An IFS A is called "*s-Normal*" (s-NIFS), if

$$\sup_{x \in E} \mu_A(x) = 1 \text{ and } \inf_{x \in E} \nu_A(x) = 0.$$

An IFS A is called "*i-Normal*" (i-NIFS) if

$$\inf_{x \in E} \mu_A(x) = 0 \text{ and } \sup_{x \in E} \nu_A(x) = 1.$$

Therefore, we have,

(a) A is a s-NIFS iff $\mathcal{C}(A) = E^*$,

(b) A is an i-NIFS iff $\mathcal{I}(A) = O^*$.

For every natural number n, IFSs $A_1, A_2, \ldots, A_n$, and for every two numbers $\alpha, \beta \in [0, 1]$:

(a) $\mathcal{C}(\underset{i=1}{\overset{n}{@}} A_i) \subseteq \underset{i=1}{\overset{n}{@}} \mathcal{C}(A_i);$

(b) $\mathcal{I}(\underset{i=1}{\overset{n}{@}} A_i) \supseteq \underset{i=1}{\overset{n}{@}} \mathcal{I}(A_i).$

To prove (a),

$$\mathcal{C}(\underset{i=1}{\overset{n}{@}} A_i) = \mathcal{C}(\{\langle x, \frac{\sum\limits_{i=1}^{n} \mu_{A_i}(x)}{n}, \frac{\sum\limits_{i=1}^{n} \nu_{A_i}(x)}{n} \rangle | x \in E\})$$

$$= \{\langle x, \sup_{y \in E} \frac{\sum\limits_{i=1}^{n} \mu_{A_i}(y)}{n}, \inf_{y \in E} \frac{\sum\limits_{i=1}^{n} \nu_{A_i}(y)}{n} \rangle | x \in E\}$$

$$\subseteq \{\langle x, \frac{\sum\limits_{i=1}^{n} \sup\limits_{y \in E} \mu_{A_i}(y)}{n}, \frac{\sum\limits_{i=1}^{n} \inf\limits_{y \in E} (\nu_{A_i}(y)}{n} \rangle | x \in E\} = \underset{i=1}{\overset{n}{@}} \mathcal{C}(A_i).$$

4.3   Extended Topological Operators

The following operators are defined in [40], as extensions of the two topological operators C and I:

$$C_\mu(A) = \{\langle x, K, \min(1 - K, \nu_A(x))\rangle | x \in E\}; \tag{4.13}$$

$$C_\nu(A) = \{\langle x, \mu_A(x), L\rangle | x \in E\}; \tag{4.14}$$

$$\mathcal{I}_\mu(A) = \{\langle x, k, \nu_A(x)\rangle | x \in E\}; \tag{4.15}$$

$$\mathcal{I}_\nu(A) = \{\langle x, \min(1 - l, \mu_A(x)), l\rangle | x \in E\}, \tag{4.16}$$

where K, L, k, l have the forms (4.8), (4.9), (4.11) and (4.12), respectively.

The geometrical interpretations of these operators applied on the IFS A in Fig. 4.7 are shown in Figs. 4.8-4.11.

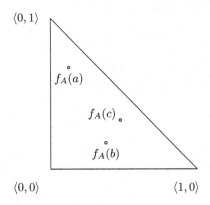

$\langle 0, 1 \rangle$

$\overset{\circ}{f_A(a)}$

$f_A(c)_{\,\circ}$

$\overset{\circ}{f_A(b)}$

$\langle 0, 0 \rangle$ $\langle 1, 0 \rangle$

Fig. 4.7

Obviously, for every IFS A,

$$\mathcal{I}(A) \subseteq \mathcal{I}_\mu(A) \subseteq \mathcal{I}_\nu(A) \subseteq A \subseteq C_\nu(A) \subseteq C_\mu(A) \subseteq C(A). \tag{4.17}$$

For every IFS A,
(a) $C_\mu(C_\nu(A)) = C_\nu(C_\mu(A)) = C(A)$,

(b) $\mathcal{I}_\mu(\mathcal{I}_\nu(A)) = \mathcal{I}_\nu(\mathcal{I}_\mu(A)) = \mathcal{I}(A)$,

(c) $C_\mu(\mathcal{I}_\mu(A)) = \mathcal{I}_\mu(C_\mu(A))$,

(d) $C_\nu(\mathcal{I}_\nu(A)) = \mathcal{I}_\nu(C_\nu(A))$,

(e) $\Box C_\mu(A) = C_\mu(\Box A)$,

(f) $\Diamond C_\mu(A) \subseteq C_\mu(\Diamond A)$,

(g) $\Box C_\nu(A) \subseteq C_\nu(\Box A)$,

(h) $\Diamond C_\nu(A) \supseteq C_\nu(\Diamond A)$,

(i) $\Box \mathcal{I}_\mu(A) \subseteq \mathcal{I}_\mu(\Box A)$,

(j) $\Diamond \mathcal{I}_\mu(A) \supseteq \mathcal{I}_\mu(\Diamond A)$,

(k) $\Box \mathcal{I}_\nu(A) \supseteq \mathcal{I}_\nu(\Box A)$,

(l) $\Diamond \mathcal{I}_\nu(A) = \mathcal{I}_\nu(\Diamond A)$,

(m) $\overline{\mathcal{C}_\mu(\overline{A})} = \mathcal{I}_\nu(A)$,

(n) $\overline{\mathcal{I}_\mu(\overline{A})} = \mathcal{C}_\nu(A)$.

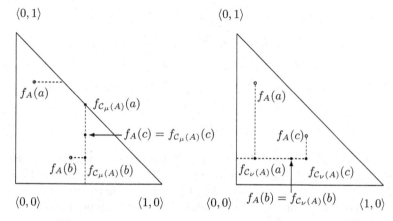

Fig. 4.8   Fig. 4.9

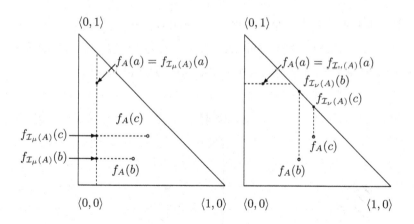

Fig. 4.10   Fig. 4.11

For every two IFSs A and B:

(a) $\mathcal{C}_\mu(A \cup B) = \mathcal{C}_\mu(A) \cup \mathcal{C}_\mu(B)$,

(b) $\mathcal{C}_\nu(A \cup B) = \mathcal{C}_\nu(A) \cup \mathcal{C}_\nu(B)$,

(c) $\mathcal{I}_\mu(A \cap B) = \mathcal{I}_\mu(A) \cap \mathcal{I}_\mu(B)$,

(d) $\mathcal{I}_\nu(A \cap B) = \mathcal{I}_\nu(A) \cap \mathcal{I}_\nu(B)$.

Another example of the action of operator $\mathcal{C}_\mu(A)$ is illustrated in Fig. 4.12

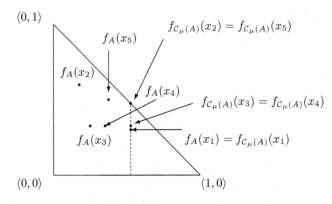

Fig. 4.12

It is seen from Fig. 4.12 that the operator $\mathcal{C}_\mu$ transforms points x_2 and x_5 to one point on the hypotenuse. More generally, all points from the hatching trapezoid can be transformed to point A in Fig. 4.13. Similar is the situation in Fig. 4.14, where all points from the hatching trapezoid can be transformed to point B by the operator $\mathcal{I}_\nu$.

In [79], the two new topological operators $\mathcal{C}_\mu^*$ and $\mathcal{I}_\nu^*$ are introduced as

$$\mathcal{C}_\mu^*(A) = \{\langle x, \min(K, 1 - \nu_A(x)), \min(1 - K, \nu_A(x))\rangle | x \in E\}; \qquad (4.18)$$

$$\mathcal{I}_\nu^*(A) = \{\langle x, \min(1 - l, \mu_A(x)), \min(l, 1 - \mu_A(x))\rangle | x \in E\}, \qquad (4.19)$$

where K, L, k, l are as defined by (4.8), (4.9), (4.11) and (4.12), respectively.

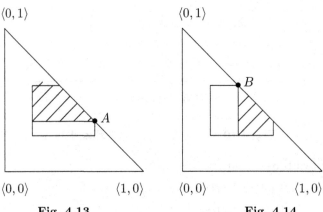

Fig. 4.13　　　　　　　　　　　**Fig. 4.14**

Let us first prove that $\mathcal{C}_\mu^*(A)$ and $\mathcal{I}_\nu^*(A)$ are IFSs.

Let, for $\mathcal{C}_\mu^*(A)$ and for a fixed element $x \in E$,

$$Z \equiv \min(K, 1 - \nu_A(x)) + \min(1 - K, \nu_A(x)).$$

If $K \geq 1 - \nu_A(x)$, then,

$$Z = 1 - \nu_A(x) + \min(1 - K, \nu_A(x)) = \min(2 - K - \nu_A(x), 1) \leq 1.$$

If $K < 1 - \nu_A(x)$, then,

$$Z = K + \min(1 - K, \nu_A(x)) = \min(1, K + \nu_A(x)) \leq 1.$$

Therefore, set $\mathcal{C}_\mu^*(A)$ is an IFS. Proving $\mathcal{I}_\nu^*(A)$ is also an IFS is analogous.

The geometrical interpretations of the new operators applied on the IFS A having the form in Fig. 4.15 are given in Fig. 4.16 and 4.17.

Obviously, for every IFS A:

$$\mathcal{I}(A) \subseteq \mathcal{I}_\nu(A) \subseteq \mathcal{I}_\nu^*(A) \subseteq A \subseteq \mathcal{C}_\mu^*(A) \subseteq \mathcal{C}_\mu(A) \subseteq \mathcal{C}(A). \tag{4.20}$$

Theorem 4.1: For every IFS A,

(a) $\overline{\mathcal{C}_\mu^*(\overline{A})} = \mathcal{I}_\nu^*(A)$,

(b) $\overline{\mathcal{I}_\mu^*(\overline{A})} = \mathcal{C}_\nu^*(A)$,

(c) $\mathcal{C}_\mu^*(\mathcal{C}_\nu(A)) = \mathcal{C}(A)$,

(d) $\mathcal{I}_\nu^*(\mathcal{I}_\mu(A)) = \mathcal{I}(A)$.

Proof: (a) Let A be an IFS. Then

$$\overline{\mathcal{C}_\mu^*(\overline{A})} = \overline{\mathcal{C}_\mu^*(\{\langle x, \nu_A(x), \mu_A(x)\rangle \mid x \in E\})}$$

$$= \overline{\{\langle x, \min(l, 1 - \mu_A(x)), \min(1 - l, \mu_A(x))\rangle | x \in E\}}$$

$$= \{\langle x, \min(1 - l, \mu_A(x)), \min(l, 1 - \mu_A(x))\rangle | x \in E\} = \mathcal{I}_\nu^*(A).$$

The proofs of the remaining assertions are analogous.

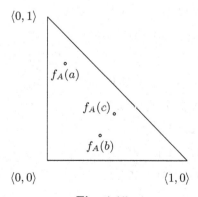

$\langle 0, 1 \rangle$

$\overset{\circ}{f_A(a)}$

$f_A(c)_\circ$

$\overset{\circ}{f_A(b)}$

$\langle 0, 0 \rangle$ $\langle 1, 0 \rangle$

Fig. 4.15

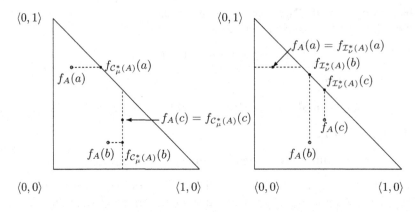

Fig. 4.16   Fig. 4.17

4.4 Weight-Center Operator

In [87], Adrian Ban and the author introduced *"weight-center operator"* over a given IFS A by:

$$W(A) = \{\langle x, \frac{\underset{y \in E}{\Sigma} \mu_A(y)}{card(E)}, \frac{\underset{y \in E}{\Sigma} \nu_A(y)}{card(E)}\rangle | x \in E\}, \tag{4.21}$$

where $card(E)$ is the number of the elements of a finite set E. For the continuous case, the "summation" may be replaced by integrals over E.

Obviously, for every IFS A,

$$\mathcal{I}(A) \subseteq W(A) \subseteq \mathcal{C}(A).$$

Theorem 4.2: For every IFS A:

(a) $W(W(A)) = W(A)$,

(b) $\overline{W(\overline{A})} = W(A)$,

(c) $\mathcal{I}_\mu(W(A)) = \mathcal{I}_\nu(W(A)) = \mathcal{C}_\mu(W(A)) = \mathcal{C}_\nu(W(A))$
$= \mathcal{I}(W(A)) = \mathcal{C}(W(A)) = W(A)$,

(d) $W(\mathcal{C}_\mu(A)) \supseteq W(A)$,

(e) $W(\mathcal{C}_\nu(A)) \supseteq W(A)$,

(f) $W(\mathcal{I}_\mu(A)) \subseteq W(A)$,

(g) $W(\mathcal{I}_\nu(A)) \subseteq W(A)$,

(h) $W(\mathcal{C}(A)) = \mathcal{C}(A)$,

(i) $W(\mathcal{I}(A)) = \mathcal{I}(A)$,

(j) $\square W(A) = W(\square A)$,

(k) $\Diamond W(A) = W(\Diamond A)$.

4.5 Other Extended Topological Operators

A convex hull of a set of n points, say $p_1, p_2, ..., p_n$ may be defined formally as the intersection of all convex sets, containing the given n points. In other words, it is the minimal convex set that contains the points (from the point of view of computational geometry, however, it is more significant that the set is also a convex polygon). A convex hull may be formally defined as the set (see [566]):

$$CH \equiv \sum_{j=1}^{n} \alpha_j p_j$$

for $i = 1, .., n$ and $\alpha_i \geq 0$, where

$$\sum_{j=1}^{n} \alpha_j = 1.$$

For the planar case, there exist various algorithms, which can find the convex hull in reasonable time. The quickest known algorithms are Kirkpatrick and Seidel's and Chan's (see [566]). One of the earliest and most widely used algorithms was designed by R. Graham in 1972. It is an incremental algorithm with complexity $O(n \log n)$. In the first phase of the algorithm, the point with the lowest y coordinate (we shall denote it by P_0) is found and the other points are sorted in increasing order (to avoid possible problems in the future) of the angle between the straight line passing through the respective point and P_0 and the x-axis. Then, the algorithm proceeds with considering for each point whether moving to it from the two points considered previously is a "right turn" (in which case the second to-last point is discarded, and the process is continued as long as the last three points make a right turn) or "left turn" (meaning they form a convex combination, in which case the algorithm moves to the next point) and incrementally updates points entering the convex hull. One of the main disadvantages of the Graham Scan is that it can only be used in the planar case. A more detailed description may be found in ([138]). Another popular algorithm is the Jarvis March (also reffered to as the Gift Wrapping Algorithm). In the Jarvis March Algorithm, we start from one point that we would be on the hull (for example the one with lowest y coordinate). Then find a point at maximal distance such that all other points lie to the left of the line connecting this point to the starting point. We proceed with the found point as a starting point. The Jarvis March Algorithm has a complexity $O(nh)$ and works well with sets that have small number of hull points.

We may use either one of Jarvis March or Graham Scan algorithms with the following modification: In step one, the point with minimal ν (in case this is a set of collinear points, we start with the point with minimal μ) is taken as the starting point. Then, the algorithm proceeds as described above.

Let, for a given IFS A, the region obtained by the above algorithm is marked by $R(A)$. Therefore, $R(A)$ is the minimal convex region containing IFS A.

In [397], Parvathi Rangasamy, Peter Vassilev and the author introduced four new operators acting in IF-interpretation triangle. Let, for $x \in E$, the point $\langle \mu(x), \nu(x) \rangle$ be given. Then,

$$\gamma_{L,\nu}(\mu(x)) = \langle \mu'(x), \nu(x) \rangle,$$

$$\gamma_{R,\nu}(\mu(x)) = \langle \mu''(x), \nu(x) \rangle,$$

$$\gamma_{T,\mu}(\nu(x)) = \langle \mu(x), \nu'(x) \rangle,$$

$$\gamma_{B,\mu}(\nu(x)) = \langle \mu(x), \nu''(x) \rangle,$$

where $\mu'(x)$ and $\mu''(x)$ are the left and the right intersection points of the line

$$l : u = \nu$$

and $R(A)$; ν' and ν'' are the top and the bottom intersection points of the line

$$l : v = \mu$$

and $R(A)$.

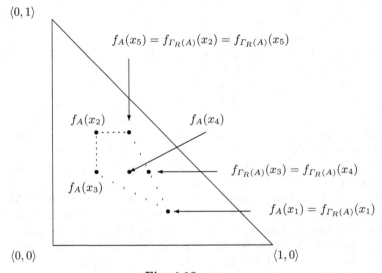

Fig. 4.18

Now, the new operators have the forms:

$$\Gamma_L(A) = \{\langle x, \gamma_{L,\nu}(\mu(x)), \nu(x)\rangle | x \in E\}, \tag{4.22}$$

$$\Gamma_R(A) = \{\langle x, \gamma_{R,\nu}(\mu(x)), \nu(x)\rangle | x \in E\}, \tag{4.23}$$

$$\Gamma_T(A) = \{\langle x, \mu(x), \gamma_{T,\mu}(\nu(x))\rangle | x \in E\}, \tag{4.24}$$

$$\Gamma_B(A) = \{\langle x, \mu(x), \gamma_{B,\mu}(\nu(x))\rangle | x \in E\}. \tag{4.25}$$

These operators have the geometrical interpretations as in Figs. 4.18 - 4.21, respectively.

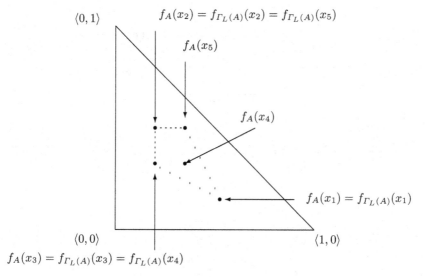

$$f_A(x_2) = f_{\Gamma_L(A)}(x_2) = f_{\Gamma_L(A)}(x_5)$$

$$f_A(x_3) = f_{\Gamma_L(A)}(x_3) = f_{\Gamma_L(A)}(x_4)$$

Fig. 4.19

For the new operators the following assertion is checked directly.

Theorem 4.3: For any IFS A,

$$\text{(a)} \quad \mathcal{I}(A) \subseteq \left\{ \begin{array}{c} \mathcal{I}_\mu(A) \subseteq \Gamma_L(A) \\ \mathcal{I}_\nu(A) \subseteq \Gamma_T(A) \end{array} \right\} \subseteq A \subseteq \left\{ \begin{array}{c} \Gamma_R(A) \subseteq \mathcal{C}_\mu(A) \\ \Gamma_B(A) \subseteq \mathcal{C}_\nu(A) \end{array} \right\} \subseteq \mathcal{C}(A).$$

(b) $\mathcal{I}(A) \subseteq \mathcal{I}_\nu^*(A) \subseteq \Gamma_T(A) \subseteq A \subseteq \Gamma_R(A) \subseteq \mathcal{C}_\mu^*(A) \subseteq \mathcal{C}(A)$.

Theorem 4.4: For any IFS A, the following equalities hold:

(a) $\Gamma_L(\mathcal{I}_\mu(A)) = \mathcal{I}_\mu(A) = \mathcal{I}_\mu(\Gamma_L(A)$,

(b) $\Gamma_L(\mathcal{I}_\nu(A)) = \mathcal{I}(A)$,

(c) $\Gamma_L(\mathcal{C}_\mu(A)) = \mathcal{C}_\mu(A)$,

(d) $\Gamma_L(\mathcal{C}_\nu(A)) = \{\langle x, k, L\rangle | x \in E\}$,

(e) $\Gamma_L(\mathcal{I}(A)) = \mathcal{I}(A) = \mathcal{I}(\Gamma_L(A))$,

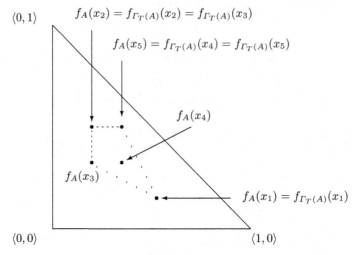

Fig. 4.20

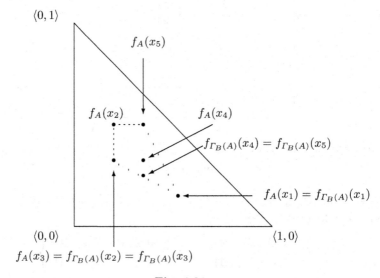

Fig. 4.21

(f) $\Gamma_L(C(A)) = C(A)$,

(g) $\Gamma_L(W(A)) = W(A)$,

(h) $\Gamma_R(\mathcal{I}_\mu(A)) = \mathcal{I}_\mu(A)$,

(i) $\Gamma_R(\mathcal{I}_\nu(A)) = \{\langle x, \sup_{y \in E} \min(1 - l, \mu_A(y)), l \rangle | x \in E\}$,

(j) $\Gamma_R(C_\mu(A)) = C_\mu(A) = C_\mu(\Gamma_R(A)$,

(k) $\Gamma_R(C_\nu(A)) = C(A)$,

(l) $\Gamma_R(\mathcal{I}(A)) = \mathcal{I}(A)$,

(m) $\Gamma_R(\mathcal{C}(A)) = \mathcal{C}(A) = \mathcal{C}(\Gamma_R(A))$,

(n) $\Gamma_R(W(A)) = W(A)$,

(o) $\Gamma_T(\mathcal{I}_\mu(A)) = \mathcal{I}(A)$,

(p) $\Gamma_T(\mathcal{I}_\nu(A)) = \mathcal{I}_\nu(A) = \mathcal{I}_\nu(\Gamma_T(A))$,

(q) $\Gamma_T(\mathcal{C}_\mu(A)) = \{\langle x, K, \sup_{y \in E} \min(1 - K, \nu_A(y))\rangle | x \in E\}$,

(r) $\Gamma_T(\mathcal{C}_\nu(A)) = \mathcal{C}_\nu(A)$,

(s) $\Gamma_T(\mathcal{I}(A)) = \mathcal{I}(A) = \mathcal{I}(\Gamma_T(A))$,

(t) $\Gamma_T(\mathcal{C}(A)) = \mathcal{C}(A)$,

(u) $\Gamma_T(W(A)) = W(A)$,

(v) $\Gamma_B(\mathcal{I}_\mu(A)) = \{\langle x, k, L\rangle | x \in E\}$,

(w) $\Gamma_B(\mathcal{I}_\nu(A)) = \mathcal{I}_\nu(A)$,

(x) $\Gamma_B(\mathcal{C}_\mu(A)) = \mathcal{C}(A)$,

(y) $\Gamma_B(\mathcal{C}_\nu(A)) = \mathcal{C}_\nu(A) = \mathcal{C}_\nu(\Gamma_B(A))$,

(z) $\Gamma_B(\mathcal{I}(A)) = \mathcal{I}(A)$,

(α) $\Gamma_B(\mathcal{C}(A)) = \mathcal{C}(A) = \mathcal{C}(\Gamma_B(A))$,

(β) $\Gamma_B(W(A)) = W(A)$.

The validity of these assertions follows directly from the definitions of the respective operators.

4.6   Short Remarks on Intuitionistic Fuzzy Logics. Part 2

We use the same notations as in Section **2.2**.

Let $x \in E$ be a variable and let $P(x)$ be a predicate. Let

$$V(P(x)) = \langle \mu(P(x)), \nu(P(x))\rangle.$$

The IF-interpretations of the quantifiers *for all* and *there exists* are introduced in [94] by

$$V(\forall x P(x)) = \langle \sup_{y \in E} \mu(P(y)), \inf_{y \in E} \nu(P(y))\rangle,$$

$$V(\exists x P(x)) = \langle \inf_{y \in E} \mu(P(y)), \sup_{y \in E} \nu(P(y))\rangle.$$

It is very interesting to note that these interpretations of both quantifiers coincide, respectively, with the IFS-interpretations of the topological operators "interior" and "closure".

Open problem 11. *Can we construct logical objects, e.g., from quantifier type, that are analogous of the operators $C_\mu, C_\nu, \mathcal{I}_\mu, \mathcal{I}_\nu, C_\mu^*, \mathcal{I}_\nu^*$, discussed in the present Chapter?*

The following version of Craig's interpolation theorem [133] is valid.

Theorem 4.5 [183]: Let F and G be different formulae and let $F \to G$ be an IFT. Then, there exists a formula H different from F and G, such that $F \to H$ and $H \to G$ are IFTs.

4.7 An Example: IF-Interpretations of Conway's Game of Life. Part 1

Conway's Game of Life is devised by John Horton Conway in 1970 (see, [233]), and already 40 years it is an object of research, software implementations and modifications. In [570], there is a list of many papers devoted to Conway's Game of Life. In 1976, the author and his wife, who were then students in Sofia University, also introduced one modification of this game. In [117, 118, 119], another modification was introduced. It is based on the idea of the intuitionistic fuzziness.

The standard Conway's Game of Life has a "universe", which is an infinite two-dimensional orthogonal grid of square cells[1], each of which is in one of two possible states, alive or dead, or as we used to learn the game in the middle of 1970s, in the square there is an asterisk or not.

Every cell interacts with its eight neighbours, namely the cells that are directly adjacent either in horizontal, vertical, or diagonal direction. In a stepwise manner, the state of each cell in the grid preserves or alternates with respect to a given list of rules.

In [117, 118, 119], Lilija Atanassova and the author discussed some versions of the game in which the condition is kept for the necessary number of existing neighbours asterisks for birth or death of an asterisk in some square.

4.7.1 IF-Criteria of Existing, Birth and Death of an Asterisk

Let us have a plane tessellated with squares. Let in some of these squares there be asterisk "*", meaning that the squares are "alive". Now, we will extend this construction of the Game of Life to some new forms.

Let us assume that the square $\langle i, j \rangle$ is assigned a pair of real numbers $\langle \mu_{i,j}, \nu_{i,j} \rangle$, so that $\mu_{i,j} + \nu_{i,j} \leq 1$. We can call the numbers $\mu_{i,j}$ and $\nu_{i,j}$ degree of existence and degree of non-existence of symbol "*" in square $\langle i, j \rangle$. Therefore, $\pi_{i,j} = 1 - \mu_{i,j} - \nu_{i,j} \leq 1$ will correspond to the degree of uncertainty, e.g., lack of information about existence of an asterisk in the respective square.

Here, we formulate a series of different criteria for correctness of the intuitionistic fuzzy interpretations that includes the standard game as a particular case.

---

[1] Below, we use words "cells" and "squares" as synonyms.

Seven Criteria of Existence of an Asterisk

We suppose that there exists an asterisk in square $\langle i, j \rangle$ if:

(1.1) $\mu_{i,j} = 1$ and $\nu_{i,j} = 0$. This is the classical case.

(1.2) $\mu_{i,j} > 0.5$. Therefore, $\nu_{i,j} < 0.5$. In the particular case, when $\mu_{i,j} = 1 > 0.5$ we obtain $\nu_{i,j} = 0 < 0.5$, i.e., the standard existence of the asterisk.

(1.3) $\mu_{i,j} \geq 0.5$. Therefore, $\nu_{i,j} \leq 0.5$. Obviously, if case (1.2) is valid, then case (1.3) will also be valid.

(1.4) $\mu_{i,j} > \nu_{i,j}$. Obviously, case (1.2) is a particular case of the present one, but case (1.3) is not included in the currently discussed case for $\mu_{i,j} = 0.5 = \nu_{i,j}$.

(1.5) $\mu_{i,j} \geq \nu_{i,j}$. Obviously, cases (1.1), (1.2), (1.3) and (1.4) are particular cases of the present one.

(1.6) $\mu_{i,j} > 0$. Obviously, cases (1.1), (1.2), (1.3) and (1.4) are particular cases of the present one, but case (1.5) is not included in the currently discussed case for $\mu_{i,j} = 0.0 = \nu_{i,j}$.

(1.7) $\nu_{i,j} < 1$. Obviously, cases (1.1), (1.2), (1.3) and (1.4) are particular cases of the present one, but case (1.6) is not included in the currently discussed case for $\mu_{i,j} = 0.0$.

From these criteria, it follows that if one is valid – let it be the s-th criterion ($1 \leq s \leq 6$) then we can assert that the asterisk exists with respect to the s-th criterion and, therefore, it will exist with respect to all other criteria, whose validity follows from the validity of the s-th criterion.

On the other hand, if s-th criterion is not valid, then we say that the asterisk does not exist with respect to s-th criterion. It is very important that in this case the square may not be absolutely empty. It is appropriate to tell that the square $\langle i, j \rangle$ is totally empty, if its degrees of existence and non-existence are $\langle 0, 1 \rangle$.

It is suitable to tell that the square is s-full, if it contains an asterisk with respect to the s-th criterion and that the same square is s-empty, if it does not satisfy the s-th criterion.

For the aims of the game-method for modelling, it is suitable to use (with respect to the type of the concrete model) one of the first four criteria for existence of an asterisk. Let us say for each fixed square $\langle i, j \rangle$ that therein is an asterisk by s-th criterion for $1 \leq s \leq 4$, if this criterion confirms the existence of an asterisk.

Four Criteria for the Birth of an Asterisk

In the standard game, the rule for birth of a new asterisk is: the (empty) square has exactly 2 or 3 neighbouring squares containing asterisks. Now, we formulate a series of different rules that include the standard rule as a particular case.

2.1 (extended standard rule): The s-empty square has exactly 2 or 3 neighbouring s-full squares. Obviously, this rule for birth is a direct extension of the standard rule.

2.2 (pessimistic rule): For the natural number $s \geq 2$, the s-empty square has exactly 2 or 3 neighbouring $(s-1)$-full squares.

2.3 (optimistic rule): For the natural number $s \leq 5$, the s-empty square has exactly 2 or 3 neighbouring $(s+1)$-full squares.

2.4 (average rule): Let $M_{i,j}$ and $N_{i,j}$ be, respectively, the sums of the μ-degrees and of the ν-degrees of all neighbours of the s-empty square. Then, the inequality

$$\frac{1}{4}.N_{i,j} \leq M_{i,j} \leq \frac{3}{8}.N_{i,j}$$

holds.

Four Criteria for the Death of an Asterisk

In the standard game, the rule for the death of an existing asterisk is: the (full) square has exactly 2 or 3 neighbouring squares containing asterisks. Now we will formulate a series of different rules that will include the standard rule as a particular case.

3.1 (extended standard rule): The s-full square has less than 2 or more than 3 neighbouring s-full squares. Obviously, this rule for death is a direct extension of the standard rule.

3.2 (pessimistic rule): For the natural number $s \geq 2$, the s-full square has less than 2 or more than 3 neighbouring $(s-1)$-full squares.

3.3 (optimistic rule): For the natural number $s \leq 5$, the s-full square has less than 2 or more than 3 neighbouring $(s+1)$-full squares.

3.4 (average rule): Let $M_{i,j}$ and $N_{i,j}$ be, respectively, the sums of the μ-degrees and of the ν-degrees of all neighbours of the s-full square. Then, one of the inequalities

$$\frac{1}{4}.N_{i,j} > M_{i,j} \ \text{ or } \ M_{i,j} > \frac{3}{8}.N_{i,j}$$

holds.

4.7.2   IF-Rules for Changing the Game-Field

In the standard game, the game-field is changed by the above mentioned rules for birth and death of the asterisks. Now, we discuss some intuitionistic fuzzy rules for changing the game-field. They use the separate forms of operation "negation".

Let us suppose that in a fixed square there is an asterisk if and only if the square is s-full. Therefore, we tell that in the square there is no asterisk if and only if the square is not s-full. In this case we can call this square s-empty.

As it is seen above, the difference between standard and intuitionistic fuzzy form of the game is the existence of values corresponding to the separate squares. In the standard case, they are 1 or 0, or "there is an asterisk", "there is no asterisk". In the intuitionistic fuzzy form of the game, we have pairs of real numbers as in the case when the asterisk exists, as well as in the opposite case. In the classical case, the change of the status of the square is obvious. In the intuitionistic fuzzy we can construct different rules. They are of two types.

The first type contains two modifications of the standard rule:

4.1 (extended standard rule): If an s-full square $\langle i, j \rangle$ has to be changed, then we can use negation $\neg_1$ for pair $\langle \mu_{i,j}, \nu_{i,j} \rangle$ and as a result we will obtain pair $\langle \nu_{i,j}, \mu_{i,j} \rangle$.

4.2 (non-standard, or intuitionistic fuzzy rule): If an s-full square $\langle i, j \rangle$ has to be changed, then we can use any of the other negations $\neg_m$ from Table 9.11 ($2 \leq m \leq 34$).

The second type contains three non-standard modifications. The standard rule and the above two rules for changing the current content of the fixed square (existence or absence of an asterisk) are related only to this content. Now, we include a new parameter, that can be conditionally called *"influence of the environment"*.

5.1 (optimistic (s, m)-rule) If an s-full/empty square $\langle i, j \rangle$ has to be changed, then we can apply m-th negation $\neg_m$ from Table 9.11 to pair $\langle \mu_{i,j}, \nu_{i,j} \rangle$ (before change) and to juxtapose to it the pair $\langle \mu_{i,j}^*, \nu_{i,j}^* \rangle$, so that

$$\mu_{i,j}^* = \max(\mu_{i,j}', \max_{u \in \{i-1,i,i+1\}; v \in \{j-1,j,j+1\}; \langle u,v \rangle \neq \langle i,j \rangle}^{*} \mu_{u,v})$$

$$\nu_{i,j}^* = \min(\nu_{i,j}', \min_{u \in \{i-1,i,i+1\}; v \in \{j-1,j,j+1\}; \langle u,v \rangle \neq \langle i,j \rangle}^{*} \nu_{u,v}),$$

where

$$\langle \mu_{i,j}', \nu_{i,j}' \rangle = \neg_m \langle \mu_{i,j}, \nu_{i,j} \rangle$$

and $\max^*, \min^*$ mean that we use only values that are connected to s-empty/full squares.

5.2 (optimistic-average (s, m)-rule) If an s-full/empty square $\langle i, j \rangle$ has to be changed, then we can apply m-th negation $\neg_m$ to pair (before change) $\langle \mu_{i,j}, \nu_{i,j} \rangle$ and to juxtapose to it the pair $\langle \mu_{i,j}^*, \nu_{i,j}^* \rangle$, so that

$$\mu_{i,j}^* = \max(\mu_{i,j}', \frac{1}{t(i,j)} \sum_{u \in \{i-1,i,i+1\}; v \in \{j-1,j,j+1\}; \langle u,v \rangle \neq \langle i,j \rangle}^{*} \mu_{u,v})$$

$$\nu_{i,j}^* = \min(\nu_{i,j}', \frac{1}{t(i,j)} \sum_{u \in \{i-1,i,i+1\}; v \in \{j-1,j,j+1\}; \langle u,v \rangle \neq \langle i,j \rangle}^{*} \nu_{u,v}),$$

where $\langle \mu'_{i,j}, \nu'_{i,j} \rangle$ is as above, $\sum^*$ mean that we use only values that are connected to s-empty/full squares and $t(i,j)$ is the number of these squares.

5.3 (average (s,m)-rule) If an s-full/empty square $\langle i,j \rangle$ has to be changed, then we can apply m-th negation $\neg_m$ to pair (before change) $\langle \mu_{i,j}, \nu_{i,j} \rangle$ and to juxtapose to it the pair $\langle \mu^*_{i,j}, \nu^*_{i,j} \rangle$, so that

$$\mu^*_{i,j} = \frac{1}{2}(\mu'_{i,j} + \frac{1}{t(i,j)} \sum_{u\in\{i-1,i,i+1\}; v\in\{j-1,j,j+1\}; \langle u,v\rangle\neq\langle i,j\rangle}^* \mu_{u,v})$$

$$\nu^*_{i,j} = \frac{1}{2}(\nu'_{i,j} + \frac{1}{t(i,j)} \sum_{u\in\{i-1,i,i+1\}; v\in\{j-1,j,j+1\}; \langle u,v\rangle\neq\langle i,j\rangle}^* \nu_{u,v}),$$

where $\langle \mu'_{i,j}, \nu'_{i,j} \rangle$, $\sum^*$ and $t(i,j)$ are as in 5.1 and 5.2.

5.4 (pessimistic-average (s,m)-rule) If an s-full/empty square $\langle i,j \rangle$ has to be changed, then we can apply m-th negation $\neg_m$ to pair (before change) $\langle \mu_{i,j}, \nu_{i,j} \rangle$ and to juxtapose to it the pair $\langle \mu^*_{i,j}, \nu^*_{i,j} \rangle$, so that

$$\mu^*_{i,j} = \min(\mu'_{i,j}, \frac{1}{t(i,j)} \sum_{u\in\{i-1,i,i+1\}; v\in\{j-1,j,j+1\}; \langle u,v\rangle\neq\langle i,j\rangle}^* \mu_{u,v})$$

$$\nu^*_{i,j} = \max(\nu'_{i,j}, \frac{1}{t(i,j)} \sum_{u\in\{i-1,i,i+1\}; v\in\{j-1,j,j+1\}; \langle u,v\rangle\neq\langle i,j\rangle}^* \nu_{u,v}),$$

where $\langle \mu'_{i,j}, \nu'_{i,j} \rangle$, $\sum^*$ and $t(i,j)$ are as in 5.1 and 5.2.

5.5 (pessimistic (s,m)-rule) If an s-full/empty square $\langle i,j \rangle$ has to be changed, then we can apply m-th negation $\neg_m$ to pair (before change) $\langle \mu_{i,j}, \nu_{i,j} \rangle$ and to juxtapose to it the pair $\langle \mu^*_{i,j}, \nu^*_{i,j} \rangle$, so that

$$\mu^*_{i,j} = \min(\mu'_{i,j}, \min_{u\in\{i-1,i,i+1\}; v\in\{j-1,j,j+1\}; \langle u,v\rangle\neq\langle i,j\rangle}^* \mu_{u,v})$$

$$\nu^*_{i,j} = \max(\nu'_{i,j}, \max_{u\in\{i-1,i,i+1\}; v\in\{j-1,j,j+1\}; \langle u,v\rangle\neq\langle i,j\rangle}^* \nu_{u,v}),$$

where $\langle \mu'_{i,j}, \nu'_{i,j} \rangle$ and $\max^*, \min^*$ are as in case 5.1.

4.7.3   Transformations of the Game Field by Standard Topological Operators

For brevity, let us work with a finite field F with length of the sides n, i.e., with n^2 squares. Let us construct, as above, the IFS A over the universe F, where the degrees of asterisks' membership and non-membership will be interpreted as the cell estimations.

Using the relation $\geq$ between the IF-estimations of the cells, we can determine those cells, whose liveliness estimations are the highest.

The cells with highest asterisks' estimations can generate some special regions of the field. We discuss two different possibilities for these regions.

The first possibility is: the minimal rectangle, that contains all cells with the highest estimations. In this case, we construct the set $E \subseteq F$ with the elements being the squares of the rectangle. Now, we can construct the set B, that is a restriction of A over set E, i.e., for each $x \in E$:

$$\mu_B(x) = \mu_A(x), \quad \nu_B(x) = \nu_A(x).$$

Obviously, B is an IFS, but on a universe $E \subseteq F$. Now, we can apply over each of the two standard topological operators C and I.

In this way, we modify a part of the region F. Of course, this modification is in either positive or negative direction, depending on whether we use operator C or I.

To fulfil the second possibility, we must have some natural number $s \geq 1$. Around each cell $\langle i, j \rangle \in F$, whose asterisk's estimation is one of the highest, we construct a square with side $2s + 1$, so that the discussed cell is in the center. If the intersection of two bigger squares contains at least one cell of F, then we construct a rectangle that includes the two bigger squares. In a result of this procedure, we can obtain either only one big rectangle, that in some cases may coincide with the rectangle from the first case, or several different rectangles and/or squares. For each of these figures we can apply the procedure of changing the liveliness estimations with operators C and I from the first case.

4.7.4   Transformations of the Game Field by Extended Topological Operators

Here we discuss other ideas for transformation. They are based on the possibility to represent the elements of a given IFS in the interpretation triangle from Fig. 3.4.

Now, we apply one of the extended topological operators $\mathcal{C}_\mu, \mathcal{C}_\nu, \mathcal{I}_\mu, \mathcal{I}_\nu,$ $\mathcal{C}_\mu^*, \mathcal{I}_\nu^*$, defined by (4.13), (4.14), (4.15), (4.16), (4.18) and (4.19).

It is important to note that now we change the values of the asterisks' estimations by one of the above operators in the frames of the IFS-interpretation triangle and after this we transform these changes for the cells in the field.

We will illustrate the present idea with the following example (see Fig. 4.22).

Let the IFS A over universe F have the (descriptive) form

$$A = \{\langle a, 0.7, 0.3 \rangle, \langle b, 0.3, 0.3 \rangle, \langle c, 0.4, 0.4 \rangle, \langle d, 0.6, 0.2 \rangle, \langle e, 0.5, 0.3 \rangle,$$

$$\langle f, 0.4, 0.4 \rangle, \langle g, 0.5, 0.4 \rangle, \langle h, 0.4, 0.6 \rangle, \langle i, 0.2, 0.5 \rangle\}.$$

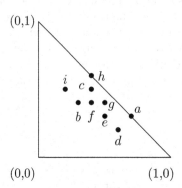

Fig. 4.22

First, we must note that

$$\mathcal{C}(A) = \{\langle a, 0.7, 0.2\rangle, \langle b, 0.7, 0.2\rangle, \langle c, 0.7, 0.2\rangle, \langle d, 0.7, 0.2\rangle, \langle e, 0.7, 0.2\rangle,$$
$$\langle f, 0.7, 0.2\rangle, \langle g, 0.7, 0.2\rangle, \langle h, 0.7, 0.2\rangle, \langle i, 0.7, 0.2\rangle\}.$$
$$\mathcal{I}(A) = \{\langle a, 0.2, 0.6\rangle, \langle b, 0.2, 0.6\rangle, \langle c, 0.2, 0.6\rangle, \langle d, 0.2, 0.6\rangle, \langle e, 0.2, 0.6\rangle,$$
$$\langle f, 0.2, 0.6\rangle, \langle g, 0.2, 0.6\rangle, \langle h, 0.2, 0.6\rangle, \langle i, 0.2, 0.6\rangle\}.$$

Now, the values for the other IFSs that we obtain using the other topological operators are the following.

$$\mathcal{C}_\mu(A) = \{\langle a, 0.7, 0.3\rangle, \langle b, 0.7, 0.3\rangle, \langle c, 0.4, 0.3\rangle, \langle d, 0.7, 0.2\rangle, \langle e, 0.7, 0.3\rangle,$$
$$\langle f, 0.7, 0.3\rangle, \langle g, 0.7, 0.3\rangle, \langle h, 0.7, 0.3\rangle, \langle i, 0.7, 0.3\rangle\},$$
$$\mathcal{C}_\nu(A) = \{\langle a, 0.7, 0.2\rangle, \langle b, 0.3, 0.2\rangle, \langle c, 0.4, 0.2\rangle, \langle d, 0.6, 0.2\rangle, \langle e, 0.5, 0.2\rangle,$$
$$\langle f, 0.4, 0.2\rangle, \langle g, 0.5, 0.2\rangle, \langle h, 0.4, 0.2\rangle, \langle i, 0.2, 0.2\rangle\},$$
$$\mathcal{I}_\mu(A) = \{\langle a, 0.2, 0.3\rangle, \langle b, 0.2, 0.4\rangle, \langle c, 0.2, 0.5\rangle, \langle d, 0.2, 0.2\rangle, \langle e, 0.2, 0.3\rangle,$$
$$\langle f, 0.2, 0.4\rangle, \langle g, 0.2, 0.4\rangle, \langle h, 0.2, 0.6\rangle, \langle i, 0.2, 0.5\rangle\},$$
$$\mathcal{I}_\nu(A) = \{\langle a, 0.4, 0.6\rangle, \langle b, 0.3, 0.6\rangle, \langle c, 0.4, 0.6\rangle, \langle d, 0.4, 0.6\rangle, \langle e, 0.4, 0.6\rangle,$$
$$\langle f, 0.4, 0.6\rangle, \langle g, 0.4, 0.6\rangle, \langle h, 0.4, 0.6\rangle, \langle i, 0.2, 0.6\rangle\},$$
$$\mathcal{C}_\mu^*(A) = \{\langle a, 0.7, 0.3\rangle, \langle b, 0.6, 0.3\rangle, \langle c, 0.5, 0.3\rangle, \langle d, 0.7, 0.2\rangle, \langle e, 0.7, 0.3\rangle,$$
$$\langle f, 0.6, 0.3\rangle, \langle g, 0.6, 0.3\rangle, \langle h, 0.4, 0.3\rangle, \langle i, 0.5, 0.3\rangle\}.$$
$$\mathcal{I}_\nu^*(A) = \{\langle a, 0.4, 0.3\rangle, \langle b, 0.3, 0.6\rangle, \langle c, 0.4, 0.6\rangle, \langle d, 0.4, 0.4\rangle, \langle e, 0.4, 0.5\rangle,$$
$$\langle f, 0.4, 0.6\rangle, \langle g, 0.4, 0.5\rangle, \langle h, 0.4, 0.6\rangle, \langle i, 0.2, 0.6\rangle\}.$$

This Section is written on the basis of [117].

5

Extended Modal Operators

5.1 Operators D_α and $F_{\alpha,\beta}$

Following [18, 19, 39], we construct an operator which represents both operators $\square$ from (4.1) and $\lozenge$ from (4.2). It has no analogue in the ordinary modal logic, but the author hopes that the search for such an analogue in modal logic will be interesting.

Let $\alpha \in [0,1]$ be a fixed number. Given an IFS A, we define an operator D_α as follows:

$$D_\alpha(A) = \{\langle x, \mu_A(x) + \alpha.\pi_A(x), \nu_A(x) + (1-\alpha).\pi_A(x)\rangle | x \in E\}. \qquad (5.1)$$

From this definition it follows that $D_\alpha(A)$ is a fuzzy set, because:

$$\mu_A(x) + \alpha.\pi_A(x) + \nu_A(x) + (1-\alpha).\pi_A(x) = \mu_A(x) + \nu_A(x) + \pi_A(x) = 1.$$

Some of the specific properties of this operator are:

(a) if $\alpha \leq \beta$, then $D_\alpha(A) \subseteq D_\beta(A)$;
(b) $D_0(A) = \square A$;
(c) $D_1(A) = \lozenge A$,

for every IFS A and for every $\alpha, \beta \in [0,1]$.

To every point $x \in E$ the operator $f_{D_\alpha}(A)$ assigns a point of the segment between $f_{\square A}(x)$ and $f_{\lozenge A}(x)$ depending on the value of the argument $\alpha \in [0,1]$ (see Fig. 5.1). As in the case of some of the above operations, this construction needs auxiliary elements which are shown in Fig. 5.1.

As we noted above, the operator D_α is an extension of the operators $\square$ and $\lozenge$, but it can be extended even further.

Let $\alpha, \beta \in [0,1]$ and $\alpha + \beta \leq 1$. Define (see [18, 19, 39]) the operator $F_{\alpha,\beta}$, for the IFS A, by

$$F_{\alpha,\beta}(A) = \{\langle x, \mu_A(x) + \alpha.\pi_A(x), \nu_A(x) + \beta.\pi_A(x)\rangle | x \in E\}. \qquad (5.2)$$

K.T. Atanassov: On Intuitionistic Fuzzy Sets Theory, STUDFUZZ 283, pp. 77–111.
springerlink.com   © Springer-Verlag Berlin Heidelberg 2012

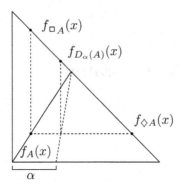

$f_{\Box A}(x)$

$f_{D_\alpha(A)}(x)$

$f_{\Diamond A}(x)$

$f_A(x)$

α

Fig. 5.1

For every IFS A, and for every $\alpha, \beta, \gamma \in [0, 1]$ such that $\alpha + \beta \le 1$,

(a) $F_{\alpha,\beta}(A)$ is an IFS;

(b) if $0 \le \gamma \le \alpha$, then $F_{\gamma,\beta}(A) \subseteq F_{\alpha,\beta}(A)$;

(c) if $0 \le \gamma \le \beta$, then $F_{\alpha,\beta}(A) \subseteq F_{\alpha,\gamma}(A)$;

(d) $D_\alpha(A) = F_{\alpha,1-\alpha}(A)$;

(e) $\Box A = F_{0,1}(A)$;

(f) $\Diamond A = F_{1,0}(A)$;

(g) $\overline{F_{\alpha,\beta}(\overline{A})} = F_{\beta,\alpha}(A)$

(h) $\mathcal{C}(F_{\alpha,\beta}(A)) \subseteq F_{\alpha,\beta}\mathcal{C}(A)$,

(i) $\mathcal{I}(F_{\alpha,\beta}(A)) \supseteq F_{\alpha,\beta}\mathcal{I}(A)$.

Let us prove property (h):

$$\mathcal{C}(F_{\alpha,\beta}(A)) = \mathcal{C}(\{\langle x, \mu_A(x) + \alpha.\pi_A(x), \nu_A(x) + \beta.\pi_A(x)\rangle | x \in E\})$$

$$= \{\langle x, K_1, L_1\rangle | x \in E\},$$

where

$$K_1 = \sup_{y \in E}(\mu_A(y) + \alpha.\pi_A(y)),$$

$$L_1 = \inf_{y \in E}(\nu_A(y) + \beta.\pi_A(y)),$$

and

$$F_{\alpha,\beta}(\mathcal{C}(A)) = F_{\alpha,\beta}(\{\langle x, K, L\rangle | x \in E\})$$
$$= \{\langle x, K + \alpha.(1 - K - L), L + \beta.(1 - K - L)\rangle | x \in E\},$$

where K and L are defined by (4.8) and (4.9).

From

$$K + \alpha.(1 - K - L) - K_1$$

$$= \sup_{y \in E}(\mu_A(y) + \alpha.(1 - \sup_{y \in E} \mu_A(y) - \inf_{y \in E} \nu_A(y))$$

$$\sup_{y \in E}(\mu_A(y) + \alpha.(1 - \mu_A(y) - \nu_A(y))))$$

$$\geq \sup_{y \in E}(\mu_A(y) - \alpha.\sup_{y \in E} \mu_A(y) - \alpha.\inf_{y \in E} \nu_A(y))$$

$$-(1 - \alpha).\sup_{y \in E} \mu_A(y) + \alpha.\inf_{y \in E} \nu_A(y) = 0$$

and from

$$L_1 - L - \beta.(1 - K - L)$$

$$= \inf_{y \in E}(\nu_A(y) + \beta.\pi_A(y)) - \inf_{y \in E} \nu_A(y) - \beta.(1 - \sup_{y \in E} \mu_A(y) - \inf_{y \in E} \nu_A(y)) \geq 0,$$

it follows that

$$\mathcal{C}(F_{\alpha,\beta}(A)) \subseteq F_{\alpha,\beta}(\mathcal{C}(A)).$$

For every two IFSs A and B and for every $\alpha, \beta \in [0,1]$, such that $\alpha + \beta \leq 1$, it holds that

(a) $F_{\alpha,\beta}(A \cap B) \subseteq F_{\alpha,\beta}(A) \cap F_{\alpha,\beta}(B)$;

(b) $F_{\alpha,\beta}(A \cup B) \supseteq F_{\alpha,\beta}(A) \cup F_{\alpha,\beta}(B)$;

(c) $F_{\alpha,\beta}(A + B) \subseteq F_{\alpha,\beta}(A) + F_{\alpha,\beta}(B)$;

(d) $F_{\alpha,\beta}(A.B) \supseteq F_{\alpha,\beta}(A).F_{\alpha,\beta}(B)$;

(e) $F_{\alpha,\beta}(A@B) = F_{\alpha,\beta}(A)@F_{\alpha,\beta}(B)$.

(f) $D_\alpha(\underset{i=1}{\overset{n}{@}} A_i) = \underset{i=1}{\overset{n}{@}} D_\alpha(A_i)$;

(g) $F_{\alpha,\beta}(\underset{i=1}{\overset{n}{@}} A_i) = \underset{i=1}{\overset{n}{@}} F_{\alpha,\beta}(A_i)$, for $\alpha + \beta \leq 1$.

Theorem 5.1: For every IFS A and for every $\alpha, \beta, \gamma, \delta \in [0,1]$ such that, $\alpha + \beta \leq 1$ and $\gamma + \delta \leq 1$, we have

$$F_{\alpha,\beta}(F_{\gamma,\delta}(A)) = F_{\alpha+\gamma-\alpha.\gamma-\alpha.\delta,\beta+\delta-\beta.\gamma-\beta.\delta}(A).$$

Proof: Let for $\alpha, \beta, \gamma, \delta \in [0,1]$,

$$\alpha + \beta \leq 1 \text{ and } \gamma + \delta \leq 1.$$

Then,

$$F_{\alpha,\beta}(F_{\gamma,\delta}(A))$$
$$= F_{\alpha,\beta}(\{\langle x, \mu_A(x) + \gamma.\pi_A(x), \nu_A(x) + \delta.\nu_A(x)\rangle | x \in E\})$$
$$= \{\langle x, \mu_A(x) + \gamma.\pi_A(x) + \alpha.(1 - \mu_A(x) - \gamma.\pi_A(x) - \nu_A(x) - \delta.\pi_A(x)),$$
$$\nu_A(x) + \delta.\pi_A(x) + \beta.(1 - \mu_A(x) - \gamma.\pi_A(x) - \nu_A(x) - \delta.\pi_A(x))\rangle | x \in E\}$$
$$= \{\langle x, \mu_A(x) + (\alpha + \gamma - \alpha.\gamma - \alpha.\delta).\pi_A(x),$$
$$\nu_A(x) + (\beta + \delta - \beta.\gamma - \beta.\delta).\pi_A(x)\rangle | x \in E\}$$
$$= F_{\alpha+\gamma-\alpha.\gamma-\alpha.\delta, \beta+\delta-\beta.\gamma-\beta.\delta}(A).$$

In [454], Alexander Stamenov formulated and proved the following assertion.

Theorem 5.2: Let the sequences $\{\alpha_i\}_{i=0}^n$ and $\{\beta_i\}_{i=0}^n$ be given, where $\alpha_i, \beta_i, \alpha_i + \beta_i \in [0,1]$ and let $\gamma_i = 1 - (\alpha_i + \beta_i)$ for $i = 0, 1, ..., n$. Then, for every IFS A,

$$F_{\alpha_n,\beta_n}(F_{\alpha_{n-1},\beta_{n-1}}(...(F_{\alpha_1,\beta_1}(F_{\alpha_0,\beta_0}(A)))...))$$

$$= F_{\sum_{i=1}^n \alpha_i . \prod_{j=0}^{i-1} \gamma_j, \ \sum_{i=1}^n \beta_i . \prod_{j=0}^{i-1} \gamma_j}(A).$$

Theorem 5.2 generalizes Theorem 5.1. Its modification for the case when $n \to \infty$ is published by Peter Vassilev in [552]. Both authors were my students in the Faculty of Mathematics and Informatics of Sofia University and their papers were prepared for examination on my course on IFSs that I read there.

To every point $x \in E$ the operator $f_{F_{\alpha,\beta}(A)}$ assigns a point of the triangle with vertices $f_A(x)$, $f_{\square A}(x)$ and $f_{\Diamond A}(x)$, depending on the value of the arguments $\alpha, \beta \in [0,1]$ for which $\alpha + \beta \leq 1$ (see Fig. 5.2).

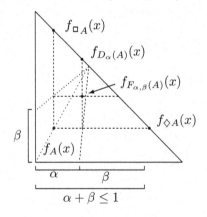

Fig. 5.2

The same results are also valid for the operator D_α as follows:

$$\overline{D_\alpha(\overline{A})} = D_{1-\alpha}(A),$$

$$\mathcal{C}(D_\alpha(A)) \subseteq D_\alpha(\mathcal{C}(A)),$$
$$\mathcal{I}(D_\alpha(A)) \supseteq D_\alpha(\mathcal{I}(A)),$$
$$D_\alpha(D_\beta(A)) = D_\beta(A),$$
$$D_\alpha(F_{\beta,\gamma}(A)) = D_{\alpha+\beta-\alpha.\beta-\alpha.\gamma}(A), \text{ for } \beta + \gamma \le 1,$$
$$F_{\alpha,\beta}(D_\gamma(A)) = D_\gamma(A).$$

A feature that both operators share, together with the first two modal operators, is that each of them changes the degree of uncertainty. While the first three operators make this degree equal to zero, the operator $F_{\alpha,\beta}$ only decreases its value, increasing the degrees of membership and non-membership of the IFS' elements.

5.2 Short Remarks on Intuitionistic Fuzzy Logics. Part 3

We use the notation from Section **2.2**. Let for proposition p,

$$V(p) = \langle a, b \rangle.$$

Then

$$V(\Box p) = \langle a, 1 - a \rangle$$

and

$$V(\Diamond p) = \langle 1 - b, b \rangle.$$

In modal logic (see, e.g., [227]) both operators $\Box$ and $\Diamond$ are related to the two equalities

$$\Box p = \neg \Diamond \neg p$$

and

$$\Diamond p = \neg \Box \neg p,$$

where $\neg$ is the negation of proposition p. Now, define

$$V(D_\alpha p) = \langle a + \alpha(1 - a - b), b + (1 - \alpha)(1 - a - b) \rangle$$

$$V(F_{\alpha,\beta} p) = \langle a + \alpha(1 - a - b), b + \beta(1 - a - b) \rangle$$

and as in Section **5.1** it is seen that operators D_α and $F_{\alpha,\beta}$ ($\alpha, \beta \in [0,1]$ and $\alpha + \beta \le 1$) are direct extensions of the ordinary modal operators. The equalities

$$\Box p = D_0 p = F_{0,1} p,$$

$$\Diamond p = D_1 p = F_{1,0} p$$

show a deeper interconnection between the two ordinary modal logic operators.

Open problem 12. *Can there be constructed logical objects, e.g., from modal operator type, that are analogous of the operators D_α and $F_{\alpha,\beta}$?*

5.3 Operator $G_{\alpha,\beta}$

Let $\alpha,\ \beta \in [0,1]$. Define the operator (see [21])

$$G_{\alpha,\beta}(A) = \{\langle x, \alpha.\mu_A(x), \beta.\nu_A(x)\rangle | x \in E\}. \qquad (5.3)$$

Obviously, $G_{1,1}(A) = A$ and $G_{0,0}(A) = U^*$, where U^* is defined by (2.15).

For every IFS A and for every three real numbers $\alpha, \beta, \gamma \in [0,1]$,

(a) $G_{\alpha,\beta}(A)$ is an IFS;

(b) If $\alpha \leq \gamma$, then $G_{\alpha,\beta}(A) \subseteq G_{\gamma,\beta}(A)$;

(c) If $\beta \leq \gamma$, then $G_{\alpha,\beta}(A) \supseteq G_{\alpha,\gamma}(A)$;

(d) If $\gamma,\ \delta \in [0,1]$, then $G_{\alpha,\beta}(G_{\gamma,\delta}(A)) = G_{\alpha.\gamma,\beta.\delta}(A) = G_{\gamma,\delta}(G_{\alpha,\beta}(A))$;

(e) $G_{\alpha,\beta}(\mathcal{C}(A)) = \mathcal{C}(G_{\alpha,\beta}(A))$;

(f) $G_{\alpha,\beta}(\mathcal{I}(A)) = \mathcal{I}(G_{\alpha,\beta}(A))$;

(g) If $\gamma,\ \delta \in [0,1]$ and $\gamma + \delta \leq 1$, then $G_{\alpha,\beta}(F_{\gamma,\delta}(A)) \subseteq_{\square} F_{\gamma,\delta}(G_{\alpha,\beta}(A))$;

(h) If $\gamma,\ \delta \in [0,1]$ and $\gamma + \delta \leq 1$, then $F_{\gamma,\delta}(G_{\alpha,\beta}(A)) \subseteq_{\diamond} G_{\alpha,\beta}(F_{\gamma,\delta}(A))$;

(i) $\overline{G_{\alpha,\beta}(\overline{A})} = G_{\beta,\alpha}(A)$.

As shown above, the operators $G_{\alpha,\beta}$ and $F_{\gamma,\delta}$ (and hence the operator $G_{\alpha,\beta}$ and operators $\square,\ \diamond,\ D_{\gamma}$) do not commute.

For every two IFSs A and B and for every $\alpha, \beta \in [0,1]$,

(a) $G_{\alpha,\beta}(A \cap B) = G_{\alpha,\beta}(A) \cap G_{\alpha,\beta}(B)$;

(b) $G_{\alpha,\beta}(A \cup B) = G_{\alpha,\beta}(A) \cup G_{\alpha,\beta}(B)$;

(c) $G_{\alpha,\beta}(A + B) \subseteq G_{\alpha,\beta}(A) + G_{\alpha,\beta}(B)$;

(d) $G_{\alpha,\beta}(A.B) \supseteq G_{\alpha,\beta}(A).G_{\alpha,\beta}(B)$;

(e) $G_{\alpha,\beta}(A@B) = G_{\alpha,\beta}(A)@G_{\alpha,\beta}(B)$;

(f) $G_{\alpha,\beta}(\overset{n}{\underset{i=1}{@}} A_i) = \overset{n}{\underset{i=1}{@}} G_{\alpha,\beta}(A_i)$.

The operator f assigns a point $f_{G_{\alpha,\beta}}(x)$ in the rectangle with vertex $f_A(x)$ and vertices with coordinates, $\langle pr_1 f_A(x), 0\rangle, \langle 0, pr_2 f_A(x)\rangle$ and $\langle 0,0\rangle$, where $pr_i p$ is the i-th projection $(i = 1,2)$ of the point p, to every point $x \in E$, depending on the value of the arguments $\alpha, \beta \in [0,1]$ (see Fig. 5.3).

Let $n \geq 1$ be an integer and $\alpha_i, \beta_i \in [0,1], i = 1,\ldots,n$. Then, we can construct the IFS

$$G_{\alpha_n,\beta_n}(\ldots(G_{\alpha_1,\beta_1}(A))\ldots)$$

$$= \{\langle x, \mu_A(x)\prod_{i=1}^{n}\alpha_i, \nu_A(x)\prod_{i=1}^{n}\beta_i\rangle | x \in E\} = G_{\prod_{i=1}^{n}\alpha_i, \prod_{i=1}^{n}\beta_i}(A).$$

In [554] Peter Vassilev studied some properties of this IFS.

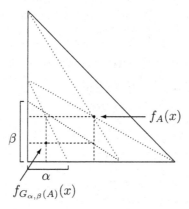

Fig. 5.3

5.4 Operators $H_{\alpha,\beta}$, $H^*_{\alpha,\beta}$, $J_{\alpha,\beta}$, and $J^*_{\alpha,\beta}$

By analogy to the operators D_α, $F_{\alpha,\beta}$ and $G_{\alpha,\beta}$, we define here four other operators over an IFS A, given the fixed numbers α, $\beta \in [0,1]$ (see [21]), as

$$H_{\alpha,\beta}(A) = \{\langle x, \alpha.\mu_A(x), \nu_A(x) + \beta.\pi_A(x)\rangle | x \in E\}, \tag{5.4}$$

$$H^*_{\alpha,\beta}(A) = \{\langle x, \alpha.\mu_A(x), \nu_A(x) + \beta.(1 - \alpha.\mu_A(x) - \nu_A(x))\rangle | x \in E\}, \tag{5.5}$$

$$J_{\alpha,\beta}(A) = \{\langle x, \mu_A(x) + \alpha.\pi_A(x), \beta.\nu_A(x)\rangle | x \in E\}, \tag{5.6}$$

$$J^*_{\alpha,\beta}(A) = \{\langle x, \mu_A(x) + \alpha.(1 - \mu_A(x) - \beta.\nu_A(x)), \beta.\nu_A(x)\rangle | x \in E\}, \tag{5.7}$$

For every IFS A, and tor every $\alpha, \beta \in [0,1]$, the following properties hold:

$$\overline{H_{\alpha,\beta}(\overline{A})} = J_{\beta,\alpha}(A),$$

$$\overline{J_{\alpha,\beta}(\overline{A})} = H_{\beta,\alpha}(A),$$

$$\overline{H^*_{\alpha,\beta}(\overline{A})} = J^*_{\beta,\alpha}(A),$$

$$\overline{J^*_{\alpha,\beta}(\overline{A})} = H^*_{\beta,\alpha}(A).$$

These equalities show that the operators $H_{\alpha,\beta}$ and $J_{\alpha,\beta}$, and the operators $H^*_{\alpha,\beta}$ and $J^*_{\alpha,\beta}$, respectively, are dual. In the same sense, the three last operators D_α, $F_{\alpha,\beta}$ and $G_{\alpha,\beta}$ are autodual.

For every two IFSs A and B, and for every $\alpha, \beta \in [0, 1]$,

(a) $H_{\alpha,\beta}(A \cap B) \subseteq H_{\alpha,\beta}(A) \cap H_{\alpha,\beta}(B)$,

(b) $H_{\alpha,\beta}(A \cup B) \supseteq H_{\alpha,\beta}(A) \cup H_{\alpha,\beta}(B)$,

(c) $H_{\alpha,\beta}(A + B) \subseteq H_{\alpha,\beta}(A) + H_{\alpha,\beta}(B)$,

(d) $H_{\alpha,\beta}(A.B) \supseteq H_{\alpha,\beta}(A).H_{\alpha,\beta}(B)$,

(e) $H_{\alpha,\beta}(A@B) = H_{\alpha,\beta}(A)@H_{\alpha,\beta}(B)$,

(f) $J_{\alpha,\beta}(A \cap B) \subseteq J_{\alpha,\beta}(A) \cap J_{\alpha,\beta}(B)$,

(g) $J_{\alpha,\beta}(A \cup B) \supseteq J_{\alpha,\beta}(A) \cup J_{\alpha,\beta}(B)$,

(h) $J_{\alpha,\beta}(A + B) \subseteq J_{\alpha,\beta}(A) + J_{\alpha,\beta}(B)$,

(i) $J_{\alpha,\beta}(A.B) \supseteq J_{\alpha,\beta}(A).J_{\alpha,\beta}(B)$,

(j) $J_{\alpha,\beta}(A@B) = J_{\alpha,\beta}(A)@J_{\alpha,\beta}(B)$,

(k) $H^*_{\alpha,\beta}(A \cap B) \subseteq H^*_{\alpha,\beta}(A) \cap H^*_{\alpha,\beta}(B)$,

(l) $H^*_{\alpha,\beta}(A \cup B) \supseteq H^*_{\alpha,\beta}(A) \cup H^*_{\alpha,\beta}(B)$,

(m) $H^*_{\alpha,\beta}(A + B) \subseteq H^*_{\alpha,\beta}(A) + H^*_{\alpha,\beta}(B)$,

(n) $H^*_{\alpha,\beta}(A.B) \supseteq H^*_{\alpha,\beta}(A).H^*_{\alpha,\beta}(B)$,

(o) $H^*_{\alpha,\beta}(A@B) = H^*_{\alpha,\beta}(A)@H^*_{\alpha,\beta}(B)$,

(p) $J^*_{\alpha,\beta}(A \cap B) \subseteq J^*_{\alpha,\beta}(A) \cap J^*_{\alpha,\beta}(B)$,

(q) $J^*_{\alpha,\beta}(A \cup B) \supseteq J^*_{\alpha,\beta}(A) \cup J^*_{\alpha,\beta}(B)$,

(r) $J^*_{\alpha,\beta}(A + B) \subseteq J^*_{\alpha,\beta}(A) + J^*_{\alpha,\beta}(B)$,

(s) $J^*_{\alpha,\beta}(A.B) \supseteq J^*_{\alpha,\beta}(A).J^*_{\alpha,\beta}(B)$.

(t) $J^*_{\alpha,\beta}(A@B) = J^*_{\alpha,\beta}(A)@J^*_{\alpha,\beta}(B)$.

For every IFS A, and for every $\alpha, \beta \in [0,1]$,

(a) $\Box H_{\alpha,\beta}(A) \subseteq H_{\alpha,\beta}(\Box A)$,

(b) $\Diamond H_{\alpha,\beta}(A) \subseteq_\Diamond H_{\alpha,\beta}(\Diamond A)$,

(c) $J_{\alpha,\beta}(\Box A) \subseteq \Box J_{\alpha,\beta}(A)$,

(d) $J_{\alpha,\beta}(\Diamond A) \subseteq \Diamond J_{\alpha,\beta}(A)$,

(e) $\Box H^*_{\alpha,\beta}(A) \subseteq H^*_{\alpha,\beta}(\Box A)$,

(f) $\Diamond H^*_{\alpha,\beta}(A) \subseteq_\Diamond H^*_{\alpha,\beta}(\Diamond A)$,

(g) $J^*_{\alpha,\beta}(\Box A) \subseteq_\Box \Box J^*_{\alpha,\beta}(A)$,

(h) $J^*_{\alpha,\beta}(\Diamond A) \subseteq \Diamond J^*_{\alpha,\beta}(A)$.

For every IFS A, for every $\alpha, \beta \in [0,1]$ and for every $\gamma, \delta \in [0,1]$ such that $\gamma + \delta \leq 1$,

(a) $F_{\gamma,\delta}(H_{\alpha,\beta}(A)) \subseteq_\diamond H_{\alpha,\beta}(F_{\gamma,\delta}(A))$,

(b) $J_{\alpha,\beta}(F_{\gamma,\delta}(A)) \subseteq_\square F_{\gamma,\delta}(J_{\alpha,\beta}(A))$.

For every IFS A, for every $\alpha,\ \beta,\ \gamma,\ \delta \in [0,1]$,

(a) $H_{\alpha,\beta}(G_{\gamma,\delta}(A)) \subseteq G_{\gamma,\delta}(H_{\alpha,\beta}(A))$,

(b) $J_{\alpha,\beta}(G_{\gamma,\delta}(A)) \subseteq G_{\gamma,\delta}(J_{\alpha,\beta}(A))$,

(c) $H^*_{\alpha,\beta}(G_{\gamma,\delta}(A)) \subseteq G_{\gamma,\delta}(H^*_{\alpha,\beta}(A))$,

(d) $J^*_{\alpha,\beta}(G_{\gamma,\delta}(A)) \subseteq G_{\gamma,\delta}(J^*_{\alpha,\beta}(A))$.

No other relation of this type holds between the operators $F_{\gamma,\delta}$ (respectively D_α) and $H_{\alpha,\beta}, J_{\alpha.\beta}, H^*_{\alpha,\beta}, J^*_{\alpha.\beta}$.

The operator $f_{H_{\alpha,\beta}(A)}$ assigns to every point $x \in E$ a point $f_{H_{\alpha,\beta}(A)}(x)$ of the rectangle with vertices with coordinates $\langle 0, pr_2 f_A(x) \rangle, \langle 0, pr_2 f_{\square A}(x) \rangle$ and vertices $f_{\square A}(x)$ and $f_A(x)$, depending on the value of the parameters $\alpha, \beta \in [0,1]$ (see Fig. 5.4).

The operator $f_{J_{\alpha,\beta}(A)}$ assigns to every point $x \in E$ a point $f_{J_{\alpha,\beta}(A)}(x)$ of the rectangle with vertices with coordinates $\langle pr_1 f_\diamond A(x), 0 \rangle, \langle pr_1 f_A(x), 0 \rangle$ and vertices $f_A(x)$ and $f_\diamond A(x)$, depending on the value of the parameters $\alpha, \beta \in [0,1]$ (see Fig. 5.5).

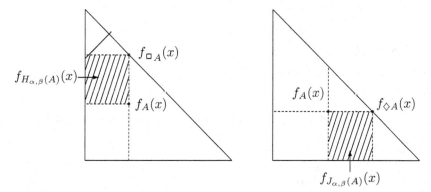

Fig. 5.4   Fig. 5.5

The operator $f_{H^*_{\alpha,\beta}(A)}$ assigns to every point $x \in E$ a point $f_{H^*_{\alpha,\beta}(A)}(x)$ from the figure with vertices with coordinates $\langle 0, pr_2 f_A(x) \rangle$ and $\langle 0, 1 \rangle$ and vertices $f_{\square A}(x)$ and $f_A(x)$, depending on the value of the parameters $\alpha, \beta \in [0,1]$ (see Fig. 5.6).

The operator $f_{J^*_{\alpha,\beta}(A)}$ assigns to every point $x \in E$ a point $f_{J^*_{\alpha,\beta}(A)}(x)$ from the figure with vertices with coordinates $\langle 1, 0 \rangle$ and $\langle pr_1 f_A(x), 0 \rangle$ and vertices $f_A(x)$ and $f_\diamond A(x)$, depending on the value of the parameters $\alpha, \beta \in [0,1]$ (see Fig. 5.7).

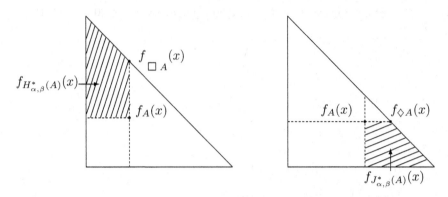

Fig. 5.6   **Fig. 5.7**

The relations shown below are true for the operators $H_{\alpha,\beta}$, $H^*_{\alpha,\beta}$, $J_{\alpha,\beta}$, and $J^*_{\alpha,\beta}$ and operators $\mathcal{C}$ and $\mathcal{I}$.

For every IFS A, and for $\alpha, \beta \in [0,1]$,

(a) $\mathcal{C}(H_{\alpha,\beta}(A)) \subseteq H_{\alpha,\beta}(\mathcal{C}(A))$,

(b) $\mathcal{I}(H_{\alpha,\beta}(A)) \supseteq H_{\alpha,\beta}(\mathcal{I}(A))$,

(c) $\mathcal{C}(H^*_{\alpha,\beta}(A)) \subseteq H^*_{\alpha,\beta}(\mathcal{C}(A))$,

(d) $\mathcal{I}(H^*_{\alpha,\beta}(A)) \supseteq H^*_{\alpha,\beta}(\mathcal{I}(A))$,

(e) $\mathcal{C}(J_{\alpha,\beta}(A)) \subseteq J_{\alpha,\beta}(\mathcal{C}(A))$,

(f) $\mathcal{I}(J_{\alpha,\beta}(A)) \supseteq J_{\alpha,\beta}(\mathcal{I}(A))$,

(g) $\mathcal{C}(J^*_{\alpha,\beta}(A)) \subseteq J^*_{\alpha,\beta}(\mathcal{C}(A))$,

(h) $\mathcal{I}(J^*_{\alpha,\beta}(A)) \supseteq J^*_{\alpha,\beta}(\mathcal{I}(A))$.

For any n IFSs $A_1, A_2, \ldots, A_n$, and for every two numbers $\alpha, \beta \in [0,1]$,

(a) $H_{\alpha,\beta} \underset{i=1}{\overset{n}{@}} A_i = \underset{i=1}{\overset{n}{@}} H_{\alpha,\beta}(A_i)$;

(b) $H^*_{\alpha,\beta}(\underset{i=1}{\overset{n}{@}} A_i) = \underset{i=1}{\overset{n}{@}} H^*_{\alpha,\beta}(A_i)$;

(c) $J_{\alpha,\beta}(\underset{i=1}{\overset{n}{@}} A_i) = \underset{i=1}{\overset{n}{@}} J_{\alpha,\beta}(A_i)$;

(d) $J^*_{\alpha,\beta}(\underset{i=1}{\overset{n}{@}} A_i) = \underset{i=1}{\overset{n}{@}} J^*_{\alpha,\beta}(A_i)$.

It is noteworthy that for any IFS A, and for every two $\alpha, \beta \in [0,1]$,

$$H^*_{\alpha,\beta}(A) \subseteq H_{\alpha,\beta}(A) \subseteq A \subseteq J_{\alpha,\beta}(A) \subseteq J^*_{\alpha,\beta}(A).$$

5.5 An Example: IF-Interpretations of Conway's Game of Life. Part 2

Here we use the notation from Section **4.7**.

5.5.1 Global Transformations of the Game Field by Modal and Extended Modal Operators

Now, following [119], we can apply each of the nine modal operators from Sections **4.1, 5.1, 5.2, 5.3** over the game field F. Having in mind the above discussed IF-interpretations of the asterisk, construct the set

$$A = \{\langle\langle i,j\rangle, \mu_{i,j}, \nu_{i,j}\rangle | \langle i,j\rangle \in F\}.$$

On one hand, it corresponds to the game-field, while, on the other, it is an IFS. Now, if we apply some of the modal operators over IFS A, we change the game-field in this way. For this aim, the values of arguments α and β of the extended modal operators must be determinde in a suitable way. Thus, we globally transform the game field. The first two (standard) modal operators can be used only for this type of transformation.

Let for a fixed cell $\langle i,j\rangle$, the subset of A that contains its neighbouring cells be

$$A_{(i,j)} = \{\langle\langle p,q\rangle, \mu_{p,q}, \nu_{p,q}\rangle | \langle p,q\rangle \in F \,\&\, p \in \{i-1, i, i+1\} \,\&\, q \in \{j-1, j, j+1\}$$

$$\&\, \langle p,q\rangle \neq \langle i,j\rangle\}.$$

We use this set in the discussions below.

5.5.2 Local Transformations of the Game Field by Modal and Extended Modal Operators

These transformations are related to the way of determining arguments α and β of the extended modal operators. In the first case (previous Subsection), they were given independently on the game. Here, they are generated on the basis of the IF-values of the asterisks.

The transformations can be of two types. In the first case, the values of the asterisks that occupy the eight neighbours of a fixed alive cell will determine the values of the parameters α and β. There are three cases for the calculation of these parameters.

Below we continue the enumeration from Section **4.7**.

6.1 (Pessimistic parameter values)

$$\alpha = \min_{\langle p,q\rangle \in A_{(i,j)}} \mu_{p,q},$$

$$\beta = \max_{\langle p,q \rangle \in A_{(i,j)}} \nu_{p,q}.$$

6.2 (Average parameter values)

$$\alpha = \frac{1}{8} \sum_{\langle p,q \rangle \in A_{(i,j)}} \mu_{p,q},$$

$$\beta = \frac{1}{8} \sum_{\langle p,q \rangle \in A_{(i,j)}} \nu_{p,q}.$$

6.3 (Optimistic parameter values)

$$\alpha = \max_{\langle p,q \rangle \in A_{(i,j)}} \mu_{p,q},$$

$$\beta = \min_{\langle p,q \rangle \in A_{(i,j)}} \nu_{p,q}.$$

The values of α and β determined by (6.1) - (6.3) are used as parameters of a previously fixed extended modal operator that is applied over the IF-values of the fixed asterisk.

In the second case, the values of the asterisks that occupy the eight neighbours of a fixed asterisk determine the values of the μ^* and ν^*-parameters for the asterisk in cell $\langle i, j \rangle$, while its original values $\mu_{i,j}$ and $\nu_{i,j}$ are used for α and β-parameters, i.e, the extended modal operators that are used for the transformation of the game field for cell $\langle i, j \rangle$ have values $\alpha = \mu_{i,j}$ and $\beta = \nu_{i,j}$.

The μ^* and ν^* parameters are calculated by the following three ways.

7.1 (Pessimistic parameter values):

$$\mu_{i,j}^* = \min_{\langle p,q \rangle \in A_{(i,j)}} \mu_{p,q},$$

$$\nu_{i,j}^* = \max_{\langle p,q \rangle \in A_{(i,j)}} \nu_{p,q}.$$

7.2 (Average parameter values):

$$\mu_{i,j}^* = \frac{1}{8} \sum_{\langle p,q \rangle \in A_{(i,j)}} \mu_{p,q},$$

$$\nu_{i,j}^* = \frac{1}{8} \sum_{\langle p,q \rangle \in A_{(i,j)}} \nu_{p,q}.$$

7.3 (Optimistic parameter values):

$$\mu_{i,j}^* = \max_{\langle p,q \rangle \in A_{(i,j)}} \mu_{p,q},$$

$$\nu_{i,j}^* = \min_{\langle p,q \rangle \in A_{(i,j)}} \nu_{p,q}.$$

5.6   Relations between Operators Defined over IFSs

In this Section, some basic relations between the operators D_α, $F_{\alpha,\beta}$, $G_{\alpha,\beta}$, $H_{\alpha,\beta}$, $H^*_{\alpha,\beta}$, $J_{\alpha,\beta}$, $J^*_{\alpha,\beta}$ are investigated.

Let $s \geq 2$ be a fixed natural number. The s-tuple $(X_1, \ldots, X_s)$, where $X_1, \ldots, X_s \in S = \{D_\alpha, F_{\alpha,\beta}, G_{\alpha,\beta}, H_{\alpha,\beta}, H^*_{\alpha,\beta}, J_{\alpha,\beta}, J^*_{\alpha,\beta}\}$ will be called a *basic s-tuple of operators* from S if each of the operators of S can be represented by the operators of the s-tuple, using the above operations and the "composition" operation over operators.

Theorem 5.3: $(D, G), (F, G), (H, J), (H, J^*), (H^*, J)$ and (H^*, J^*) are the only possible basic 2-tuples (or couples) of operators.

Proof: The Theorem is proved in two steps.

First, it is shown that each of the operators can be represented by each of the above couples. To end, we construct equalities, connecting every one of the operators with the operators of the particular couples.

Let us assume that A is a fixed IFS over the universe E and let the real numbers α, $\beta \in [0, 1]$ be fixed. For the couple (D, G) we have (let for the first equality below $\alpha + \beta \leq 1$):

$$F_{\alpha,\ \beta}(A)$$
$$= \{\langle x,\ \mu_A(x) + \alpha.\pi_A(x),\ \nu_A(x) + \beta.\pi_A(x)\rangle | x \in E\}$$
$$(\text{ because } \mu_A(x) + \alpha.\pi_A(x) \leq \mu_A(x) + (1 - \beta).\pi_A(x))$$
$$= \{\langle x,\ \mu_A(x) + \alpha.\pi_A(x),\ 0\rangle | x \in E\}$$
$$\cap \{\langle x,\ \mu_A(x) + (1 - \beta).\pi_A(x),\ \nu_A(x) + \beta.\pi_A(x)\rangle | x \in E\}$$
$$= G_{1,\ 0}(\{\langle x,\ \mu_A(x) + \alpha.\pi_A(x),\ \nu_A(x) + (1 - \alpha).\pi_A(x)\rangle | x \in E\})$$
$$\cap \{\langle x,\ \mu_A(x) + (1 - \beta).\pi_A(x),\ \nu_A(x) + \beta.\pi_A(x)\rangle | x \in E\}$$
$$= G_{1,\ 0}(D_\alpha(A)) \cap D_{1-\beta}(A);$$

$$H_{\alpha,\ \beta}(A)$$
$$= \{\langle x,\ \alpha.\mu_A(x),\ \nu_A(x) + \beta.\pi_A(x)\rangle | x \in E\},$$
$$= \{\langle x,\ \alpha.\mu_A(x),\ \beta.\nu_A(x)\rangle | x \in E\}$$
$$\cap \{\langle x,\ \mu_A(x) + (1 - \beta).\pi_A(x),\ \nu_A(x) + \beta.\pi_A(x)\rangle | x \in E\}$$
$$= G_{\alpha,\ \beta}(A) \cap D_{1-\beta}(A);$$

$$H^*_{\alpha,\ \beta}(A)$$
$$= \{\langle x,\ \alpha.\mu_A(x),\ \nu_A(x) + \beta.(1 - \alpha.\mu_A(x) - \nu_A(x))\rangle | x \in E\})$$
$$= \{\langle x,\ \min(\alpha.\mu_A(x) + (1 - \beta).(1 - \alpha.\mu_A(x) - \nu_A(x)),\ \alpha.\mu_A(x)),$$
$$\max(\nu_A(x) + \beta.(1 - \alpha.\mu_A(x) - \nu_A(x)),\ \nu_A(x))\rangle | x \in E\})$$
$$= \{\langle x,\ \alpha.\mu_A(x) + (1 - \beta).(1 - \alpha.\mu_A(x) - \nu_A(x)),$$
$$\nu_A(x) + \beta.(1 - \alpha.\mu_A(x) - \nu_A(x))\rangle | x \in E\})$$
$$\cap \{\langle x,\ \alpha.\mu_A(x),\ \nu_A(x)\rangle | x \in E\}$$
$$= D_{1-\beta}(\{\langle x,\ \alpha.\mu_A(x),\ \nu_A(x)\rangle | x \in E\}) \cap \{\langle x,\ \alpha.\mu_A(x),\ \nu_A(x)\rangle | x \in E\}$$
$$= D_{1-\beta}(G_{\alpha,\ 1}(A)) \cap G_{\alpha,\ 1}(A);$$

$$J_{\alpha,\ \beta}(A)$$
$$= \{\langle x,\ \mu_A(x) + \alpha.\pi_A(x),\ \beta.\nu_A(x)\rangle | x \in E\}$$
$$= \{\langle x,\ \mu_A(x) + \alpha.\pi_A(x),\ \nu_A + (1-\alpha).\pi_A(x)\rangle | x \in E\}$$
$$\cup \{\langle x,\ \mu_A(x),\ \beta.\nu_A(x)\rangle | x \in E\}$$
$$= D_\alpha(A) \cup G_{1,\ \beta}(A);$$

$$J^*_{\alpha,\ \beta}(A)$$
$$= \{\langle x,\ \mu_A(x) + \alpha.(1 - \mu_A(x) - \beta.\nu_A(x)),\ \beta.\nu_A(x)\rangle | x \in E\}$$
$$= \{\langle x,\ \mu_A(x) + \alpha.(1 - \mu_A(x) - \beta.\nu_A(x)),$$
$$\beta.\nu_A(x) + (1 - \alpha).(1 - \mu_A(x) - \beta.\nu_A(x))\rangle | x \in E\}$$
$$\cup \{\langle x,\ \mu_A(x),\ \beta.\nu_A(x)\rangle | x \in E\}$$
$$= D_\alpha(\{\langle x,\ \mu_A(x),\ \beta.\nu_A(x)\rangle | x \in E\}) \cup \{\langle x,\ \mu_A(x),\ \beta.\nu_A(x)\rangle | x \in E\}$$
$$= D_\alpha(G_{1,\ \beta}(A)) \cup G_{1,\ \beta}(A).$$

For the couple $(F,\ G)$, we have (assume $\alpha + \beta \le 1$),

$$D_\alpha(A) = G_{1,\ 1}(F_{\alpha,\ 1-\alpha}(A));$$

$$H_{\alpha,\ \beta}(A)$$
$$= \{\langle x,\ \alpha.\mu_A(x),\ \nu_A(x) + \beta.\pi_A(x)\rangle | x \in E\}$$
$$= \{\langle x,\ \min(\mu_A(x),\ \alpha.\mu_A(x)), \max(\nu_A(x) + \beta.\pi(x),\ \nu_A(x))\rangle | x \in E\}$$
$$= \{\langle x,\ \mu_A(x),\ \nu_A(x) + \beta.\pi_A(x)\rangle | x \in E\} \cap \{\langle x,\ \alpha.\mu_A(x),\ \nu_A(x)\rangle | x \in E\}$$
$$= F_{0,\ \beta}(A) \cap G_{\alpha,\ 1}(A);$$

$$J_{\alpha,\ \beta}(A)$$
$$= \{\langle x,\ \mu_A(x) + \alpha.\pi_A(x),\ \beta.\nu_A(x)\rangle | x \in E\}$$
$$= \{\langle x,\ \mu_A(x) + \alpha.\pi_A(x),\ \nu_A(x)\rangle | x \in E\} \cup \{\langle x,\ \mu_A(x),\ \beta.\nu_A(x)\rangle | x \in E\}$$
$$= F_{\alpha,\ 0}(A) \cup G_{1,\ \beta}(A);$$

$$H^*_{\alpha,\ \beta}(A)$$
$$= \{\langle x,\ \alpha.\mu_A(x),\ \nu_A(x) + \beta.(1 - \alpha.\mu_A(x) - \nu_A(x))\rangle | x \in E\}$$
$$= F_{0,\ \beta}(\{\langle x,\ \alpha.\mu_A(x),\ \nu_A(x)\rangle | x \in E\})$$
$$= F_{0,\ \beta}(G_{\alpha,\ 1}(A));$$

$$J^*_{\alpha,\ \beta}(A)$$
$$= \{\langle x,\ \mu_A(x) + \alpha.(1 - \mu_A(x) - \beta.\nu_A(x)),\ \beta.\nu_A(x)\rangle | x \in E\}$$
$$= F_{\alpha,\ 0}(\{\langle x,\ \mu_A(x),\ \beta.\nu_A(x)\rangle | x \in E\})$$
$$= F_{\alpha,\ 0}(G_{1,\ \beta}(A)).$$

For the couple $(H,\ J)$, we have (let for the second equality below $\alpha + \beta \le 1$),

$$D_\alpha(A)$$
$$= \{\langle x,\ \mu_A(x) + \alpha.\pi_A(x),\ \nu_A(x) + (1-\alpha).\pi_A(x)\rangle | x \in E\}$$
$$= \{\langle x,\ 1 - \nu_A(x) - (1-\alpha).\pi_A(x),\ \ \nu_A(x) + (1-\alpha).\pi_A(x)\rangle | x \in E\}$$
$$= J_{1,\ 1}(\{\langle x,\ 0,\ \nu_A(x) + (1-\alpha).\pi_A(x)\rangle | x \in E\})$$
$$= J_{1,\ 1}(H_{0,\ 1-\alpha}(A));$$

$$F_{\alpha,\,\beta}(A)$$
$$= \{\langle x,\ \mu_A(x) + \alpha.\pi_A(x),\ \nu_A(x) + \beta.\pi_A(x)\rangle | x \in E\}$$
$$= \{\langle x,\ \min(\mu_A(x) + \alpha.\pi_A(x),\ 1 - \nu_A(x) - \beta.\pi_A(x)),$$
$$\max(\nu_A(x) + \beta.\pi_A(x),\ \nu_A(x))\rangle | x \in E\}$$
$$= \{\langle x,\ 1 - \nu_A(x) - \beta.\pi_A(x),\ \nu_A(x) + \beta.\pi_A(x)\rangle | x \in E\}$$
$$\cap \{\langle x,\ \mu_A(x) + \alpha.\pi_A(x),\ \nu_A(x)\rangle | x \in E\}$$
$$= J_{1,\,1}(\{\langle x,\ 0,\ \nu_A(x) + \beta.\pi_A(x)\rangle | x \in E\}) \cap J_{\alpha,\,1}(A)$$
$$= J_{1,\,1}(H_{0,\,\beta}(A)) \cap J_{\alpha,\,1}(A);$$

$$G_{\alpha,\,\beta}(A)$$
$$= \{\langle x,\ \alpha.\mu_A(x),\ \beta.\nu_A(x)\rangle | x \in E\})$$
$$= H_{\alpha,\,0}(\{\langle x,\ \mu_A(x),\ \beta.\nu_A(x)\rangle | x \in E\})$$
$$= H_{\alpha,\,0}(J_{0,\,\beta}(A));$$

$$H^*_{\alpha,\,\beta}(A)$$
$$= \{\langle x,\ \alpha.\mu_A(x),\ \nu_A(x) + \beta.(1 - \alpha.\mu_A(x) - \nu_A(x))\rangle | x \in E\})$$
$$= H_{1,\,\beta}(\{\langle x,\ \alpha.\mu_A(x),\ \nu_A(x)\rangle | x \in E\})$$
$$= H_{1,\,\beta}(H_{\alpha,\,0}(\{\langle x,\ \mu_A(x),\ \nu_A(x)\rangle | x \in E\}))$$
$$= H_{1,\,\beta}(H_{\alpha,\,0}(J_{0,\,1}(A)));$$

$$J^*_{\alpha,\,\beta}(A)$$
$$= \{\langle x,\ \mu_A(x) + \alpha.(1 - \mu_A(x) - \beta.\nu_A(x)),\ \beta.\nu_A(x)\rangle | x \in E\}$$
$$= J_{\alpha,\,1}(\{\langle x,\ \mu_A(x),\ \beta.\nu_A(x)\rangle | x \in E\})$$
$$= J_{\alpha,\,1}(H_{1,\,0}(\{\langle x,\ \mu_A(x),\ \beta.\nu_A(x)\rangle | x \in E\}))$$
$$= J_{\alpha,\,1}(H_{1,\,0}(J_{0,\,\beta}(A))).$$

For the couple $(H,\ J^*)$, we have (let for the second equality below it is valid that $\alpha + \beta \le 1$),

$$D_\alpha(A)$$
$$= \{\langle x,\ \mu_A(x) + \alpha.\pi_A(x),\ \nu_A(x) + (1 - \alpha).\pi_A(x)\rangle | x \in E\}$$
$$= \{\langle x,\ 1 - \nu_A(x) - (1 - \alpha).\pi_A(x),\ \nu_A(x) + (1 - \alpha).\pi_A(x)\rangle | x \in E\}$$
$$= J^*_{1,\,1}(\{\langle x,\ 0,\ \nu_A(x) + (1 - \alpha).\pi_A(x)\rangle | x \in E\})$$
$$= J^*_{1,\,1}(H_{0,\,1-\alpha}(A));$$

$$F_{\alpha,\,\beta}(A)$$
$$= \{\langle x,\ \mu_A(x) + \alpha.\pi_A(x),\ \nu_A(x) + \beta.\pi_A(x)\rangle | x \in E\}$$
$$= \{\langle x,\ 1 - \nu_A(x) - \beta.\pi_A(x),\ \nu_A(x) + \beta.\pi_A(x)\rangle | x \in E\}$$
$$\cap \{\langle x,\ \mu_A(x) + \alpha.\pi_A(x),\ \nu_A(x)\rangle | x \in E\}$$
$$= J^*_{1,\,1}(\{\langle x,\ 0,\ \nu_A(x) + \beta.\pi_A(x)\rangle | x \in E\}) \cap J^*_{\alpha,\,1}(A)$$
$$= J^*_{1,\,1}(H_{0,\,\beta}(A)) \cap J^*_{\alpha,\,1}(A);$$

$$G_{\alpha,\,\beta}(A)$$
$$= \{\langle x,\ \alpha.\mu_A(x),\ \beta.\nu_A(x)\rangle | x \in E\})$$
$$= H_{\alpha,\,0}(\{\langle x,\ \mu_A(x),\ \beta.\nu_A(x)\rangle | x \in E\})$$
$$= H_{\alpha,\,0}(J^*_{0,\,\beta}(A));$$

$$H^*_{\alpha,\ \beta}(A)$$
$$= \{\langle x,\ \alpha.\mu_A(x),\ \nu_A(x) + \beta.(1 - \alpha.\mu_A(x) - \nu_A(x))\rangle | x \in E\})$$
$$= H_{1,\ \beta}(\{\langle x,\ \alpha.\mu_A(x),\ \nu_A(x)\rangle | x \in E\})$$
$$= H_{1,\ \beta}(H_{\alpha,\ 0}(\{\langle x,\ \mu_A(x),\ \nu_A(x)\rangle | x \in E\}))$$
$$= H_{1,\ \beta}(H_{\alpha,\ 0}(J^*_{0,\ 1}(A)));$$

$$J_{\alpha,\ \beta}(A)$$
$$= \{\langle x,\ \mu_A(x) + \alpha.(1 - \mu_A(x) - \nu_A(x)),\ \beta.\nu_A(x)\rangle | x \in E\}$$
$$= \{\langle x,\ \mu_A(x) + \alpha.(1 - \mu_A(x) - \nu_A(x)),\ \nu_A(x)\rangle | x \in E\}$$
$$\cup\{\langle x,\ \alpha.\mu_A(x),\ \beta.\nu_A(x)\rangle | x \in E\})$$
$$= J^*_{\alpha,\ 1}(A) \cup H_{\alpha,\ 0}(\{\langle x,\ \mu_A(x),\ \beta.\nu_A(x)\rangle | x \in E\})$$
$$= J^*_{\alpha,\ 1}(A) \cup H_{\alpha,\ 0}(J^*_{0,\ \beta}(A)).$$

For the couple $(H^*,\ J)$, we have (let for the second equality below $\alpha + \beta \leq 1$),

$$D_\alpha(A)$$
$$= \{\langle x,\ \mu_A(x) + \alpha.\pi_A(x),\ \nu_A(x) + (1 - \alpha).\pi_A(x)\rangle | x \in E\}$$
$$= \{\langle x,\ 1 - \nu_A(x) - (1 - \alpha).\pi_A(x),\ \nu_A(x)) + (1 - \alpha).\pi_A(x)\rangle | x \in E\}$$
$$= J_{1,\ 1}(\{\langle x,\ 0,\ \nu_A(x) + (1 - \alpha).\pi_A(x)\rangle | x \in E\})$$
$$= J_{1,\ 1}(\{\langle x,\ \mu_A(x),\ \nu_A(x) + (1 - \alpha).\pi_A(x)\rangle | x \in E\}$$
$$\cap\{\langle x,\ 0,\ (1 - \alpha).\nu_A(x)\rangle | x \in E\})$$
$$= J_{1,\ 1}(H^*_{1,\ 1-\alpha}(A) \cap H^*_{0,\ 0}(\{\langle x,\ \mu_A(x),\ (1 - \alpha).\nu_A(x)\rangle | x \in E\}))$$
$$= J_{1,\ 1}(H^*_{1,\ 1-\alpha}(A) \cap H^*_{0,\ 0}(J_{0,\ 1-\alpha}(A)))$$

$$F_{\alpha,\ \beta}(A)$$
$$= \{\langle x,\ \mu_A(x) + \alpha.\pi_A(x),\ \nu_A(x) + \beta.\pi_A(x)\rangle | x \in E\}$$
$$= \{\langle x,\ 1 - \nu_A(x) - \beta.\pi_A(x)),\ \nu_A(x) + \beta.\pi_A(x)\rangle | x \in E\}$$
$$\cap\{\langle x,\ \mu_A(x) + \alpha.(1 - \mu_A(x) - \nu_A(x)),\ \nu_A(x)\rangle | x \in E\}$$
$$= J_{1,\ 1}(\{\langle x,\ 0,\ \nu_A(x) + \beta.(1 - \mu_A(x) - \nu_A(x))\rangle | x \in E\} \cap J^*_{\alpha,\ 1}(A)$$
$$= J_{1,\ 1}(\{\langle x,\ \mu_A(x),\ \nu_A(x) + \beta.(1 - \mu_A(x) - \nu_A(x))\rangle | x \in E\}$$
$$\cap\{\langle x,\ 0,\ \beta.\nu_A(x)\rangle | x \in E\}) \cap J_{\alpha,\ 1}(A)$$
$$= J_{1,\ 1}(H^*_{1,\ \beta}(A) \cap H^*_{0,\ 0}(\{\langle x,\ \mu_A(x),\ \beta.\nu_A(x)\rangle | x \in E\})) \cap J_{\alpha,\ 1}(A)$$
$$= J_{1,\ 1}(H^*_{1,\ \beta}(A) \cap H^*_{0,\ 0}(J_{0,\ \beta}(A))) \cap J_{\alpha,\ 1}(A);$$

$$G_{\alpha,\ \beta}(A)$$
$$= \{\langle x,\ \alpha.\mu_A(x),\ \beta.\nu_A(x)\rangle | x \in E\})$$
$$= H^*_{\alpha,\ 0}(\{\langle x,\ \mu_A(x),\ \beta.\nu_A(x)\rangle | x \in E\})$$
$$= H^*_{\alpha,\ 0}(J_{0,\ \beta}(A));$$

$$H_{\alpha,\ \beta}(A)$$
$$= \{\langle x,\ \alpha.\mu_A(x),\ \nu_A(x) + \beta.\pi_A(x)\rangle | x \in E\}$$
$$= \{\langle x,\ \mu_A(x),\ \nu_A(x) + \beta.(1 - \mu_A(x) - \nu_A(x))\rangle | x \in E\}$$
$$\cap\{\langle x,\ \alpha.\mu_A(x),\ \beta.\nu_A(x)\rangle | x \in E\})$$
$$= H^*_{1,\ \beta}(A) \cap H^*_{\alpha,\ 0}(\{\langle x,\ \mu_A(x),\ \beta.\nu_A(x)\rangle | x \in E\})$$
$$= H^*_{1,\ \beta}(A) \cap H^*_{\alpha,\ 0}(J_{0,\ \beta}(A));$$

$$J^*_{\alpha, \, \beta}(A)$$
$$= \{\langle x, \; \mu_A(x) + \alpha.(1 - \mu_A(x) - \beta.\nu_A(x)), \; \beta.\nu_A(x)\rangle | x \in E\}$$
$$= J_{\alpha, \, 1}(\{\langle x, \; \mu_A(x), \; \beta.\nu_A(x)\rangle | x \in E\})$$
$$= J_{\alpha, \, 1}(H^*_{1, \, 0}(\{\langle x, \; \mu_A(x), \; \beta.\nu_A(x)\rangle | x \in E\}))$$
$$= J_{\alpha, \, 1}(H^*_{1, \, 0}(J_{0, \, \beta}(A))).$$

Finally, for the couple $(H^*, \; J^*)$, we have (let for the second equality below $\alpha + \beta \leq 1$),

$$D_\alpha(A)$$
$$= \{\langle x, \; \mu_A(x) + \alpha.\pi_A(x), \; (1 - \alpha).\pi_A(x)\rangle | x \in E\}$$
$$= \{\langle x, \; 1 - \nu_A(x) - (1 - \alpha).\pi_A(x), \; (1 - \alpha).\pi_A(x)\rangle | x \in E\}$$
$$= J^*_{1, \, 1}(\{\langle x, \; 0, \; \nu_A(x) + (1 - \alpha).(1 - \mu_A(x) - \nu_A(x))\rangle | x \in E\})$$
$$= J^*_{1, \, 1}(\{\langle x, \; \mu_A(x), \; \nu_A(x) + (1 - \alpha).(1 - \mu_A(x) - \nu_A(x))\rangle | x \in E\}$$
$$\cap \{\langle x, \; 0, \; (1 - \alpha).\nu_A(x)\rangle | x \in E\})$$
$$= J^*_{1, \, 1}(H^*_{1, \, 1-\alpha}(A) \cap H^*_{0, \, 0}(\{\langle x, \; \mu_A(x), \; (1 - \alpha).\nu_A(x)\rangle | x \in E\}))$$
$$= J^*_{1, \, 1}(H^*_{1, \, 1-\alpha}(A) \cap H^*_{0, \, 0}(J^*_{0, \, 1-\alpha}(A)));$$

$$F_{\alpha, \, \beta}(A)$$
$$= \{\langle x, \; \mu_A(x) + \alpha.\pi_A(x), \; \nu_A(x) + \beta.\pi_A(x)\rangle | x \in E\}$$
$$= \{\langle x, \; 1 - \nu_A(x) - \beta.\pi_A(x), \; \nu_A(x) + \beta.\pi_A(x)\rangle | x \in E\}$$
$$\cap \{\langle x, \; \mu_A(x) + \alpha.(1 - \mu_A(x) - \nu_A(x)), \; \nu_A(x)\rangle | x \in E\}$$
$$= J^*_{1, \, 1}(\{\langle x, \; 0, \; \nu_A(x) + \beta.(1 - \mu_A(x) - \nu_A(x))\rangle | x \in E\} \cap J^*_{\alpha, \, 1}(A)$$

$$= J^*_{1, \, 1}(\{\langle x, \; \mu_A(x), \; \nu_A(x) + \beta.(1 - \mu_A(x) - \nu_A(x))\rangle | x \in E\}$$
$$\cap \{\langle x, \; 0, \; \beta.\nu_A(x)\rangle | x \in E\}) \cap J^*_{\alpha, \, 1}(A)$$
$$= J^*_{1, \, 1}(H^*_{1, \, \beta}(A) \cap H^*_{0, \, 0}(\{\langle x, \; \mu_A(x), \; \beta.\nu_A(x)\rangle | x \in E\})) \cap J^*_{\alpha, \, 1}(A)$$
$$= J^*_{1, \, 1}(H^*_{1, \, \beta}(A) \cap H^*_{0, \, 0}(J^*_{0, \, \beta}(A))) \cap J^*_{\alpha, \, 1}(A);$$

$$G_{\alpha, \, \beta}(A)$$
$$= \{\langle x, \; \alpha.\mu_A(x), \; \beta.\nu_A(x)\rangle | x \in E\})$$
$$= H^*_{\alpha, \, 0}(\{\langle x, \; \mu_A(x), \; \beta.\nu_A(x)\rangle | x \in E\})$$
$$= H^*_{\alpha, \, 0}(J^*_{0, \, \beta}(A));$$

$$H_{\alpha, \, \beta}(A)$$
$$= \{\langle x, \; \alpha.\mu_A(x), \; \nu_A(x) + \beta.(1 - \mu_A(x) - \nu_A(x))\rangle | x \in E\}$$
$$= \{\langle x, \; \mu_A(x), \; \nu_A(x) + \beta.(1 - \mu_A(x) - \nu_A(x))\rangle | x \in E\}$$
$$\cap \{\langle x, \; \alpha.\mu_A(x), \; \beta.\nu_A(x)\rangle | x \subset E\})$$
$$= H^*_{1, \, \beta}(A) \cap H^*_{\alpha, \, 0}(\{\langle x, \; \mu_A(x), \; \beta.\nu_A(x)\rangle | x \in E\})$$
$$= H^*_{1, \, \beta}(A) \cap H^*_{\alpha, \, 0}(J^*_{0, \, \beta}(A));$$

$$J_{\alpha, \, \beta}(A)$$
$$= \{\langle x, \; \mu_A(x) + \alpha.\pi_A(x), \; \beta.\nu_A(x)\rangle | x \in E\}$$
$$= \{\langle x, \; \mu_A(x) + \alpha.(1 - \mu_A(x) - \nu_A(x)), \; \nu_A(x)\rangle | x \in E\}$$
$$\cup \{\langle x, \; \alpha.\mu_A(x), \; \beta.\nu_A(x)\rangle | x \in E\})$$
$$= J_{\alpha, \, 1}^*(A) \cup H_{\alpha, \, 0}^*(\{\langle x, \; \mu_A(x), \; \beta.\nu_A(x)\rangle | x \in E\})$$
$$= J_{\alpha, \, 1}^*(A) \cup H_{\alpha, \, 0}^*(J_{0, \, \beta}^*(A)).$$

Therefore, it is shown that each of the above pairs is a basic couple. Their uniqueness is proved below.

Let an universe E be fixed. Let an element $x \in E$ be fixed, too, and let us construct for the fixed set $A \subseteq E$ the following IFS (below we use only IFSs A, so that no notational collision would arise):

$$A = \{\langle x, \mu_A(x), \nu_A(x)\rangle | x \in E\}$$

for which $\mu_A(x) > 0$ and $\nu_A(x) > 0$. It can be directly seen that the element

$$\langle x, 0, 0 \rangle \in G_{0,0}(A) = \{\langle x, 0, 0 \rangle | x \in E\}$$

cannot be represented by any of the following pairs of operators,

$$(D, F), (D, H), (D, H^*), (D, J), (D, J^*), (F, H), (F, H^*), (F, J), (F, J^*),$$

because none of their combinations applied to the set A yields as a result $\langle x, 0, 0 \rangle$ for $x \in E$. This fact can be seen, e.g., from the geometrical interpretations of the individual operators. Similarly, the element

$$\langle x, 1, 0 \rangle \in J_{1,0}^*(A) = \{\langle x, 1, 0 \rangle | x \in E\}$$

cannot be represented by any of the following pairs of operators: (G, H), (G, H^*), (H, H^*), and the element

$$\langle x, 0, 1 \rangle \in H_{0,1}^*(A) = \{\langle x, 0, 1 \rangle | x \in E\}$$

cannot be represented by any one of the pairs of operators (G, J), (G, J^*), (J, J^*).

Therefore, all the pairs of basic couple operators are unique.

Corollary 5.1: The only possible basic 3-tuples (triples) of operators are,

$$
\begin{array}{lllll}
(D, F, G), & (D, G, H), & (D, G, H^*), & (D, G, J), & (D, G, J^*), \\
(D, H, J), & (D, H, J^*), & (D, H^*, J), & (D, H^*, J^*), & (F, G, H), \\
(F, G, H^*), & (F, G, J), & (F, G, J^*), & (F, H, J), & (F, H, J^*), \\
(F, H^*, J), & (F, H^*, J^*), & (G, H, J), & (G, H, J^*), & (G, H^*, J), \\
(G, H^*, J^*), & (H, H^*, J), & (H, H^*, J^*), & (H, J, J^*), & (H^*, J, J^*).
\end{array}
$$

Proof: These triples are basic triples of operators, since each of them contains a basic couple of operators. For the set A, as in the proof of the Theorem 5.3, it is again seen that the element $\langle x, 0, 0 \rangle \in G_{0,0}(A)$ cannot be represented by any of the triples (D, F, H), (D, F, H^*), (D, F, J), (D, F, J^*), (D, H, H^*), (D, J, J^*), (F, H, H^*), (F, J^*, J^*); the element $\langle x, 1, 0 \rangle \in J_{1,0}(A)$ cannot be represented by (G, H, H^*); and the element $\langle x, 0, 1 \rangle \in H^*_{0,1}(A)$ cannot be represented by (G, J, J^*).

Corollary 5.2: (D, F, H, H^*), (D, F, J, J^*) are the only 4-tuples of operators that are not basic 4-tuples of operators.

Indeed, only these two 4-tuples of operators do not contain basic couples or triples of operators and it can be proved that they cannot represent the elements $\langle x, 1, 0 \rangle \in J^*_{1,0}(A)$ and $\langle x, 0, 1 \rangle \in H^*_{0,1}(A)$, respectively.

Corollary 5.3: All 5-tuples of operators are basic 5-tuples of operators.

It must be noted that from the definitions, it directly follows the fact that the "$\subseteq$" relation is monotonic about every one of the operators, because for every two IFSs A and B, for every $X \in \{D_\alpha, F_{\alpha,\beta}, H_{\alpha,\beta}, H^*_{\alpha,\beta}, J_{\alpha,\beta}, J^*_{\alpha,\beta}\}$, if $A \subseteq B$, then $X(A) \subseteq X(B)$, for every $\alpha, \beta \in [0, 1]$.

An IFS A over the universe E is called *proper*, if there exists at least one $x \in E$ for which $\pi_A(x) > 0$.

The following theorems complete this Section. In [39] it was proved

Theorem 5.4: Let A, B be two proper IFSs for which there exist $y, z \in E$ such that $\mu_A(y) > 0$ and $\nu_B(z) > 0$. If $\mathcal{C}(A) \subseteq \mathcal{I}(B)$, then there are real numbers $\alpha, \beta, \gamma, \delta \in [0, 1]$, such that $J_{\alpha,\beta}(A) \subseteq H_{\gamma,\delta}(B)$.

Here, following [69] we extend it to the following form

Theorem 5.5: Let A, B be two proper IFSs for which there exist $y, z \in E$ such that $\mu_A(y) > 0$ and $\nu_B(z) > 0$. If $\mathcal{C}(A) \subseteq \mathcal{I}(B)$, then there are real numbers $\alpha, \beta, \gamma, \delta \in [0, 1]$, such that $J^*_{\alpha,\beta}(A) \subseteq H^*_{\gamma,\delta}(B)$.

Proof: Let $\mathcal{C}(A) \subset \mathcal{I}(B)$. Therefore,

$$0 < \mu_A(y) \le \sup_{x \in E} \mu_A(x) = K \le k = \inf_{x \in E} \mu_B(x)$$

and

$$\inf_{x \in E} \nu_A(x) = L \ge l = \sup_{x \in E} \nu_B(x) \ge \nu_B(z) > 0.$$

Let

$$a = \sup_{x \in E}(1 - \mu_A(x) - \frac{L+l}{2L} \nu_A(x))$$

and

$$b = \sup_{x \in E}(1 - \frac{k+K}{2k} \mu_B(x) - \nu_B(x)).$$

From the fact that A and B are proper IFSs, i.e., for them there are elements $u, v \in E$ for which $\pi_A(u) > 0$ and $\pi_B(v) > 0$, it follows that

$$\sup_{x \in E}(1 - \mu_A(x) - \frac{L+l}{2L} \nu_A(x)) \ge \sup_{x \in E}(1 - \mu_A(x) - \nu_A(x)) = \sup_{x \in E} \pi_A(x) > 0$$

and

$$\sup_{x \in E}(1 - \frac{k+K}{2k}\mu_B(x) - \nu_B(x)) \geq \sup_{x \in E}(1 - \mu_B(x) - \nu_B(x))$$

$$= \sup_{x \in E} \pi_B(x) > 0.$$

Let

$$\alpha = \min(1, \frac{k-K}{2a}), \quad \beta = \frac{L+l}{2L}, \quad \gamma = \frac{K+k}{2k}, \quad \delta = \min(1, \frac{L-l}{2b}).$$

Then,

$$J^*_{\alpha,\beta} = J^*_{\min(1,\frac{k-K}{2a}),\frac{L+l}{2L}}(A)$$

$$= \{< x, \mu_A(x) + \min(1, \frac{k-K}{2a}).(1 - \mu_A(x) - \frac{L+l}{2L}\nu_A(x)),$$

$$\frac{L+l}{2L}.\nu_A(x) > |x \in E\},$$

$$H^*_{\gamma,\delta} = H^*_{\frac{k+K}{2k},\min(1,\frac{L-l}{2b})}(B)$$

$$= \{\langle x, \frac{k+K}{2k}.\mu_B(x),$$

$$\nu_B(x) + \min(1, \frac{L-l}{2b}).(1 - \frac{k+K}{2k}\mu_B(x) - \nu_B(x))\rangle|x \in E\}.$$

From

$$\mu_A(x) + \min(1, \frac{k-K}{2a}).(1 - \mu_A(x) - \frac{L+l}{2L}\nu_A(x)) \leq K + \frac{k-K}{2a}.a$$

$$= \frac{k+K}{2} \leq \frac{k+K}{2k}.\mu_B(x)$$

and

$$\nu_B(x) + \min(1, \frac{L-l}{2b}).(1 - \frac{k+K}{2k}\mu_B(x) - \nu_B(x)) \leq l + \frac{L-l}{2b}.b$$

$$= \frac{L+l}{2} \leq \frac{L+l}{2L}.\nu_A(x)$$

it follows that $J^*_{\alpha,\beta}(A) \subset H^*_{\gamma,\delta}(B)$.

It can be easily shown that the converse is not always true.

5.7   Operator $X_{a,b,c,d,e,f}$

In 1991, during a lecture given by the author, a question was asked, whether the operators so far constructed can be derived as particular cases of one general operator? In the next lecture, the author gave a positive answer, preparing the following text, which was published in [25].

Here, we construct an operator which is universal for all the above operators. Let

$$X_{a,b,c,d,e,f}(A) = \{\langle x, a.\mu_A(x) + b.(1 - \mu_A(x) - c.\nu_A(x)),$$
$$d.\nu_A(x) + e.(1 - f.\mu_A(x) - \nu_A(x))\rangle | x \in E\} \tag{5.8}$$

where $a, b, c, d, e, f \in [0,1]$ and

$$a + e - e.f \le 1, \tag{5.9}$$

$$b + d - b.c \le 1. \tag{5.10}$$

For every IFS A, and for every $a, b, c, d, e, f \in [0,1]$ satisfying (5.9) and (5.10), $X_{a,b,c,d,e,f}(A)$ is an IFS.

Let $a, b, c, d, e, f \in [0,1]$ satisfying (5.9) and (5.10).

Let A be a fixed IFS. Then,

$$a.\mu_A(x) + b.(1 - \mu_A(x) - c.\nu_A(x)) + d.\nu_A(x)$$
$$+e.(1 - f.\mu_A(x) - \nu_A(x)) \ge a.\mu_A(x) + d.\nu_A(x) \ge 0$$

and from (5.9) and (5.10),

$$a.\mu_A(x) + b.(1 - \mu_A(x) - c.\nu_A(x)) + d.\nu_A(x) + e.(1 - f.\mu_A(x) - \nu_A(x))$$
$$= \mu_A(x).(a - b - e.f) + \nu_A(x).(d - b.c - e) + b + e$$
$$\le \mu_A(x).(1 - b - e) + \nu_A(x).(1 - b - e) + b + e$$
$$= (\mu_A(x) + \nu_A(x)).(1 - b - e) + b + e$$
$$\le 1 - b - e + b + e = 1.$$

All the above operators can be represented by the operator X at suitably chosen values of its parameters. These representations are the following:

$$\square A = X_{1,0,r,1,1,1}(A),$$
$$\lozenge A = X_{1,1,1,1,0,r}(A),$$
$$D_a(A) = X_{1,a,1,1,1-a,1}(A),$$
$$F_{a,b}(A) = X_{1,a,1,1,b,1}(A),$$
$$G_{a,b}(A) = X_{a,0,r,b,0,r}(A),$$
$$H_{a,b}(A) = X_{a,0,r,1,b,1}(A),$$
$$H_{a,b}^*(A) = X_{a,0,r,b,0,a}(A),$$
$$J_{a,b}(A) = X_{1,a,1,b,0,r}(A),$$
$$J_{a,b}^*(A) = X_{1,a,b,b,0,r}(A),$$

where r is an arbitrary real number in $[0,1]$.

Theorem 5.6: Let for $a,b,c,d,e,f,g,h,i,j,k,l \in [0,1]$ such that

$$
\begin{aligned}
u &= a.g \;\;+b \;\;\;\;-b.g \;\;\;-b.c.k+b.c.k.l \geq 0,\\
v &= a.h \;\;+b \;\;\;\;-b.c.k \;-b.h \;\;\;\;\;\;\;\;\;\;\;\;\;\; > 0,\\
w &= a.h.i +b.c.j \;-b.c.k \;-b.h.i \;\;\;\;\;\;\;\;\;\;\;\; \geq 0,\\
x &= d.j \;\;+e \;\;\;\;\;-e.f.h +e.f.h.i-e.j \;\;\; \geq 0,\\
y &= d.k \;\;+e \;\;\;\;\;-e.f.h -e.k \;\;\;\;\;\;\;\;\;\;\;\;\; > 0,\\
z &= d.k.l +e.f.g -e.f.h -e.k.l \;\;\;\;\;\;\;\;\;\;\; \geq 0.
\end{aligned}
$$

Then

$$X_{a,b,c,d,e,f}(X_{g,h,i,j,k,l}(A)) = X_{u,v,w/v,x,y,z/y}.$$

Proof:

$$X_{a,b,c,d,e,f}(X_{g,h,i,j,k,l}(A)) =$$

$$
\begin{aligned}
&= X_{a,b,c,d,e,f}(\{\langle x, g.\mu_A(x) + h.(1-\mu_A(x) - i.\nu_A(x)),\\
&\quad j.\nu_A(x) + k.(1 - l.\mu_A(x) - \nu_A(x))\rangle | x \in E\})\\
&= \{\langle x, a.(g.\mu_A(x) + h.(1-\mu_A(x) - i.\nu_A(x))) + b.(1-g.\mu_A(x)-\\
&\quad h.(1-\mu_A(x) - i.\nu_A(x)) - c.(j.\nu_A(x) + k.(1-l.\mu_A(x)-\\
&\quad \nu_A(x)))), d.(j.\nu_A(x) + k.(1-l.\mu_A(x) - \nu_A(x))) + e.(1-f.\\
&\quad (g.\mu_A(x) + h.(1-\mu_A(x) - i.\nu_A(x))) - (j.\nu_A(x) + k.(1-\\
&\quad l.\mu_A(x) - \nu_A(x))))\rangle | x \in E\}\\
&= \{\langle x, \mu_A(x).(a.g - b.g + b.c.k.l - a.h + b.h) + \nu_A(x).(b.c.k-\\
&\quad b.c.j - a.h.i + b.h.i) + (a.h + b - b.c.k - b.h), \mu_A(x).\\
&\quad (d.k.l + e.f.g - e.f.h - e.k.l) + \nu_A(x).(d.j + e.f.h.i-\\
&\quad e.j - d.k + e.k) + (d.k + e - e.f.h - e.k)\rangle | x \in E\}\\
&= \{\langle x, (a.g + b - b.g - b.c.k + b.c.k.l).\mu_A(x) + (a.h + b - b.c.k\\
&\quad -b.h).(1-\mu_A(x) - (a.h.i + b.c.j - b.c.k - b.h.i)/(a.h + b-\\
&\quad b.c.k - b.h).\nu_A(x)), (d.j + e - e.f.h + e.f.h.i - e.j).\nu_A(x)+\\
&\quad (d.k + e - e.f.h - e.k).(1 - (d.k.l + e.f.g - e.f.h - e.k.l)/\\
&\quad (d.k + e - e.f.h - e.k).\mu_A(x) - \nu_A(x))\rangle | x \in E\}\\
&= X_{u,v,w/v,x,y,z/y}.
\end{aligned}
$$

For example, if $a = c = d = f = g = i = j = l = 1$, $b = \alpha$, $e = \beta$, $h = \gamma$, $k = \delta$ where $\alpha, \beta, \gamma, \delta \in [0,1]$ and $\alpha + \beta \leq 1$, $\gamma + \delta \leq 1$, then

$$
\begin{aligned}
u &= 1,\\
v &= \gamma + \alpha - \alpha.\delta - \alpha.\gamma,\\
w &= \alpha + \gamma - \alpha.\gamma - \alpha.\delta,\\
x &= 1,\\
y &= \beta + \delta - \beta.\gamma - \beta.\delta,\\
z &= \beta + \delta - \beta.\gamma - \beta.\delta.
\end{aligned}
$$

Therefore,

$$F_{\alpha,\beta}(F_{\gamma,\delta}(A)) = X_{1,\alpha,1,1,\beta,1}(X_{1,\gamma,1,1,\delta,1}(A))$$
$$= X_{1,\alpha+\gamma-\alpha.\gamma-\alpha.\delta,1,1,\beta+\delta-\beta.\gamma-\beta.\delta,1}(A)$$
$$= F\alpha + \gamma - \alpha.\gamma - \alpha.\delta, \beta + \delta - \beta.\gamma - \beta.\delta(A)$$

Theorem 5.7: For every two IFSs A and B, and for every $a, b, c, d, e, f \in [0,1]$ satisfying (5.9) and (5.10),

(a) $\overline{X_{a,b,c,d,e,f}(\overline{A})} = X_{d,e,f,a,b,c}(A)$,

(b) $X_{a,b,c,d,e,f}(A \cap B) \subseteq X_{a,b,c,d,e,f}(A) \cap X_{a,b,c,d,e,f}(B)$,

(c) $X_{a,b,c,d,e,f}(A \cup B) \supseteq X_{a,b,c,d,e,f}(A) \cup X_{a,b,c,d,e,f}(B)$,

(d) if $c = f = 1, a \geq b$ and $e \geq d$, then

$$X_{a,b,c,d,e,f}(A + B) \subseteq X_{a,b,c,d,e,f}(A) + X_{a,b,c,d,e,f}(B),$$

(e) if $c = f = 1, a \geq b$ and $e \geq d$, then

$$X_{a,b,c,d,e,f}(A.B) \supseteq X_{a,b,c,d,e,f}(A).X_{a,b,c,d,e,f}(B),$$

(f) $X_{a,b,c,d,e,f}(A@B) = X_{a,b,c,d,e,f}(A)@X_{a,b,c,d,e,f}(B)$.

This operator includes all the rest, but it can be extended even further, as discussed in Section **5.9**.

The interrelation between all the modal type of operators is depicted in Fig. 5.8.

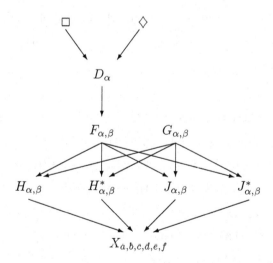

Fig. 5.8

5.8   Extended Modal Operators for Negation

Let A be an IFS. Following [49], we construct seven operators, which are extensions simultaneously of the two standard modal operators and of standard negations. Hence, they are named as *"Extended modal operators for negation"*. They are

$$d_\alpha(A) = \{\langle x, \nu_A(x) + \alpha.\pi_A(x), \mu_A(x) + (1-\alpha).\pi_A(x)\rangle \mid x \in E\}; \qquad (5.11)$$

$$f_{\alpha,\beta}(A) = \{\langle x, \nu_A(x) + \alpha.\pi_A(x), \mu_A(x) + \beta.\pi_A(x)\rangle \mid x \in E\},$$

$$\text{where } \alpha + \beta \le 1; \qquad (5.12)$$

$$g_{\alpha,\beta}(A) = \{\langle x, \alpha.\nu_A(x), \beta.\mu_A(x)\rangle \mid x \in E\}; \qquad (5.13)$$

$$h_{\alpha,\beta}(A) = \{\langle x, \alpha.\nu_A(x), \mu_A(x) + \beta.\pi_A(x)\rangle \mid x \in E\}; \qquad (5.14)$$

$$h^*_{\alpha,\beta}(A) = \{\langle x, \alpha.\nu_A(x), \mu_A(x) + \beta.(1 - \alpha.\nu_A(x) - \mu_A(x))\rangle \mid x \in E\}; \qquad (5.15)$$

$$j_{\alpha,\beta}(A) = \{\langle x, \nu_A(x) + \alpha.\pi_A(x), \beta.\mu_A(x)\rangle \mid x \in E\}; \qquad (5.16)$$

$$j^*_{\alpha,\beta}(A) = \{\langle x, \nu_A(x) + \alpha.(1 - \nu_A(x) - \beta.\mu_A(x)), \beta.\mu_A(x)\rangle \mid x \in E\}. \qquad (5.17)$$

As an immediate consequence of the above operators, for each IFS A,

$$\overline{A} = f_{0,0}(A) = g_{1,1}(A) = h_{1,0}(A) = j_{0,1}(A) = h^*_{1,0}(A) = j^*_{0,1}(A).$$

The following assertions are valid for every IFS A and for every $\alpha, \beta, \gamma \in [0,1]$,

(a) $d_\alpha(A), f_{\alpha,\beta}(A)$ for $\alpha + \beta \le 1, g_{\alpha,\beta}(A), h_{\alpha,\beta}(A), h^*_{\alpha,\beta}(A), j_{\alpha,\beta}(A)$, and $j^*_{\alpha,\beta}(A)$ are IFSs;

(b) $d_\alpha(A) = f_{\alpha,1-\alpha}(A)$;

(c) $\Box \overline{A} = d_0(A) = f_{0,1}(A) = h_{1,1}(A) = h^*_{1,1}(A)$;

(d) $\Diamond \overline{A} = d_1(A) = f_{1,0}(A) = j_{1,1}(A) = j^*_{1,1}(A)$;

(e) if $0 \le \gamma \le \alpha$, then $f_{\gamma,\beta}(A) \subseteq f_{\alpha,\beta}(A)$, where $\alpha + \beta \le 1$;

(f) if $0 \le \gamma \le \beta$, then $f_{\alpha,\beta}(A) \subseteq f_{\alpha,\gamma}(A)$, where $\alpha + \beta \le 1$;

(g) if $\alpha \le \gamma$, then $g_{\alpha,\beta}(A) \subseteq g_{\gamma,\beta}(A)$;

(h) if $\gamma \le \beta$, then $g_{\alpha,\beta}(A) \subseteq g_{\alpha,\gamma}(A)$;

(i) if $\alpha \le \gamma$, then $h_{\alpha,\beta}(A) \subseteq h_{\gamma,\beta}(A)$;

(j) if $\gamma \le \beta$, then $h_{\alpha,\beta}(A) \subseteq h_{\alpha,\gamma}(A)$;

(k) if $\alpha \le \gamma$, then $h^*_{\alpha,\beta}(A) \subseteq h^*_{\gamma,\beta}(A)$;

(l) if $\gamma \le \beta$, then $h^*_{\alpha,\beta}(A) \subseteq h_{\alpha,\gamma}(A)$;

(m) if $\alpha \leq \gamma$, then $j_{\alpha,\beta}$ (A) $\subseteq j_{\gamma,\beta}$ (A);

(n) if $\gamma \leq \beta$, then $j_{\alpha,\beta}$ (A) $\subseteq j_{\alpha,\gamma}$ (A);

(o) if $\alpha \leq \gamma$, then $j^*_{\alpha,\beta}$ (A) $\subseteq j^*_{\gamma,\beta}$ (A);

(p) if $\gamma \leq \beta$, then $j^*_{\alpha,\beta}$ (A) $\subseteq j_{\alpha,\gamma}$ (A);

(q) $\overline{f_{\alpha,\beta}(\overline{A})} = f_{\beta,\alpha}(A)$;

(r) $\overline{g_{\alpha,\beta}(\overline{A})} = g_{\beta,\alpha}(A)$;

(s) $\overline{h_{\alpha,\beta}(\overline{A})} = j_{\beta,\alpha}(A)$;

(t) $\overline{h^*_{\alpha,\beta}(\overline{A})} = j^*_{\beta,\alpha}(A)$;

(u) $\overline{j_{\alpha,\beta}(\overline{A})} = h_{\beta,\alpha}(A)$;

(v) $\overline{j^*_{\alpha,\beta}(\overline{A})} = h^*_{\beta,\alpha}(A)$.

For every IFS A, and for every $\alpha, \beta, \gamma \in [0,1]$ such that $\alpha + \beta \leq 1$,

(a) $\mathcal{C}(f_{\alpha,\beta}(A)) \supseteq f_{\alpha,\beta}\mathcal{C}(A)$,

(b) $\mathcal{I}f_{\alpha,\beta}(A) \subseteq f_{\alpha,\beta}\mathcal{I}(A)$,

(c) $g_{\alpha,\beta}(\mathcal{C}(A)) = \mathcal{C}(g_{\alpha,\beta}(A))$,

(d) $g_{\alpha,\beta}(\mathcal{I}(A)) = \mathcal{I}(g_{\alpha,\beta}(A))$.

For every IFS A, and for every $\alpha, \beta, \gamma, \delta \in [0,1]$,

(a) $f_{\alpha,\beta}(f_{\gamma,\delta}(A)) = f_{\alpha+\gamma-\alpha.\delta-\alpha.\gamma,\beta+\delta-\beta.\gamma-\beta.\delta}(\overline{A})$, where $\alpha + \beta \leq 1$ and $\gamma + \delta \leq 1$;

(b) $d_\alpha(d_\beta(A)) = d_\beta(\overline{A})$,

(c) $d_\alpha(f_{\beta,\gamma}(A)) = d_{\alpha+\beta-\alpha.\beta-\alpha.\gamma}(\overline{A})$, for $\beta + \gamma \leq 1$,

(d) $f_{\alpha,\beta}(d_\gamma(A)) = d_\gamma(\overline{A})$, where $\alpha + \beta \leq 1$,

(e) $g_{\alpha,\beta}(g_{\gamma,\delta}(A)) = g_{\alpha.\gamma,\beta.\delta}(\overline{A}) = g_{\gamma,\delta}(g_{\alpha,\beta}(A))$;

Let $\alpha, \beta, \gamma, \delta \in [0,1]$, such that

$$\alpha + \beta \leq 1 \text{ and } \gamma + \delta \leq 1.$$

Then, for (a) we obtain

$$f_{\alpha,\beta}(f_{\gamma,\delta}(A))$$
$$= f_{\alpha,\beta}(\{\langle x, \nu_A(x) + \gamma.\pi_A(x), \mu_A(x) + \delta.\pi_A(x)\rangle \mid x \in E\})$$

$$= \{\langle x, \mu_A(x) + \delta.\pi_A(x) + \alpha.(1 - \mu_A(x) - \gamma.\pi_A(x) - \nu_A(x) - \delta.\pi_A(x)),$$
$$\nu_A(x) + \gamma.\pi_A(x) + \beta.(1 - \mu_A(x) - \gamma.\pi_A(x) - \nu_A(x) - \delta.\pi_A(x))\rangle$$
$$\mid x \in E\}$$
$$= \{\langle x, \mu_A(x) + (\alpha + \delta - \alpha.\gamma - \alpha.\delta).\pi_A(x),$$
$$\nu_A(x) + (\beta + \gamma - \beta.\gamma - \beta.\delta).\pi_A(x)\rangle \mid x \in E\}$$
$$= f_{\alpha+\gamma-\alpha.\gamma-\alpha.\delta,\beta+\delta-\beta.\gamma-\beta.\delta}(\overline{A}).$$

For every IFS A, and for every $\alpha, \beta, \gamma, \delta, \varepsilon, \eta \in [0, 1]$,

(a) $d_\alpha \square d_\alpha A = \Diamond d_{1-\alpha}\overline{A}$,

(b) $d_\alpha \Diamond d_\alpha A = \square d_{1-\alpha}\overline{A}$,

(c) $f_{\alpha,\beta} \square f_{\alpha,\beta} A = \Diamond f_{\beta,\alpha}\overline{A}$,

(d) $f_{\alpha,\beta} \Diamond f_{\alpha,\beta} A = \square f_{\beta,\alpha}\overline{A}$,

(e) $d_\alpha d_\beta d_\gamma A = d_\gamma A$,

(f) $f_{\alpha,\beta} f_{\gamma,\delta} f_{\varepsilon,\eta} A = f_{a,b}$
where
$$a = \alpha + \gamma + \varepsilon - \alpha\gamma - \alpha\delta + \alpha\varepsilon + \alpha\eta - \gamma\varepsilon - \gamma\eta + \alpha\gamma\varepsilon + \alpha\gamma\eta - \alpha\delta\eta - \alpha\delta\eta,$$
$$b = \beta + \delta + \eta - \beta\gamma - \beta\delta - \beta\varepsilon - \beta\eta - \delta\varepsilon - \delta\eta + \beta\gamma\varepsilon + \beta\gamma\eta + \beta\delta\varepsilon + \beta\delta\eta A,$$

(g) $g_{\alpha,\beta} g_{\gamma,\delta} g_{\varepsilon,\eta} A = g_{\alpha\gamma\varepsilon,\beta\delta\eta} A$.

In [116], Lilija Atanassova formulated and proved the following.

Theorem 5.8: For each IFS A, the following equalities hold:

(a) $F_{\alpha,\beta}(f_{\alpha,\beta}(A)) = f_{\alpha,\beta}(F_{\beta,\alpha}(A))$,

(b) $G_{\alpha,\beta}(g_{\alpha,\beta}(A)) = g_{\alpha,\beta}(G_{\beta,\alpha}(A))$,

(c) $H_{\alpha,\beta}(h_{\alpha,\beta}(A)) = h_{\alpha,\beta}(J_{\beta,\alpha}(A))$,

(d) $J_{\alpha,\beta}(j_{\alpha,\beta}(A)) = j_{\alpha,\beta}(H_{\beta,\alpha}(A))$,

(e) $H^*_{\alpha,\beta}(h^*_{\alpha,\beta}(A)) = h^*_{\alpha,\beta}(J^*_{\beta,\alpha}(A))$,

(f) $J^*_{\alpha,\beta}(j^*_{\alpha,\beta}(A)) = j^*_{\alpha,\beta}(H^*_{\beta,\alpha}(A))$,

(g) $H_{\alpha,\beta}(j_{\alpha,\beta}(A)) = h_{\alpha,\beta}(H_{\beta,\alpha}(A))$,

(h) $J_{\alpha,\beta}(h_{\alpha,\beta}(A)) = j_{\alpha,\beta}(J_{\beta,\alpha}(A))$,

(i) $H^*_{\alpha,\beta}(j^*_{\alpha,\beta}(A)) = h^*_{\alpha,\beta}(H^*_{\beta,\alpha}(A))$,

(j) $J^*_{\alpha,\beta}(h^*_{\alpha,\beta}(A)) = j^*_{\alpha,\beta}(J^*_{\beta,\alpha}(A))$,

(k) $f_{\alpha,\beta}(f_{\gamma,\delta}(A)) = F_{\alpha,\beta}(F_{\delta,\gamma}(A))$,

(l) $g_{\alpha,\beta}(g_{\gamma,\delta}(A)) = G_{\alpha,\beta}(G_{\delta,\gamma}(A))$,

(m) $h_{\alpha,\beta}(j_{\gamma,\delta}(\overline{A})) = \overline{j_{\alpha,\beta}(h_{\delta,\gamma}(A))}$,

(n) $j_{\alpha,\beta}(h_{\gamma,\delta}(\overline{A})) = \overline{h_{\alpha,\beta}(j_{\delta,\gamma}(A))}$,

(o) $h^*_{\alpha,\beta}(j^*_{\gamma,\delta}(\overline{A})) = \overline{j^*_{\alpha,\beta}(h^*_{\delta,\gamma}(A))}$,

(p) $j^*_{\alpha,\beta}(h^*_{\gamma,\delta}(\overline{A})) = \overline{h^*_{\alpha,\beta}(j^*_{\delta,\gamma}(A))}$.

Using definition from Section **5.4**, we prove the following

Theorem 5.9: $(d,g), (f,g), (h,j), (h,j^*), (h^*,j)$ and (h^*,j^*) are the only possible basic couples of operators.

Proof: First, we show that all operators can be represented only by using the operators $f_{\alpha,\beta}$ and $g_{\alpha,\beta}$. Indeed, for $\alpha + \beta \leq 1$ we get,

$$
\begin{aligned}
d_\alpha(A) &= g_{1,\,1}(f_{\alpha,\,1-\alpha}(\overline{A})); \\
h_{\alpha,\,\beta}(A) &= \{\langle x,\ \alpha.\nu_A(x),\ \mu_A(x) + \beta.\pi_A(x)\rangle | x \in E\} \\
&= \{\langle x,\ \min(\nu_A(x),\ \alpha.\nu_A(x)), \\
&\qquad \max(\mu_A(x) + \beta.\pi(x),\ \mu_A(x))\rangle | x \in E\} \\
&= \{\langle x,\ \nu_A(x),\ \mu_A(x) + \beta.\pi_A(x)\rangle | x \in E\} \\
&\quad \cap \{\langle x,\ \alpha.\nu_A(x),\ \mu_A(x)\rangle | x \in E\} \\
&= f_{0,\,\beta}(A) \cap g_{\alpha,\,1}(A); \\
j_{\alpha,\,\beta}(A) &= \{\langle x,\ \nu_A(x) + \alpha.\pi_A(x),\ \beta.\mu_A(x)\rangle | x \in E\} \\
&= \{\langle x,\ \nu_A(x) + \alpha.\pi_A(x),\ \mu_A(x)\rangle | x \in E\} \\
&\quad \cup \{\langle x,\ \nu_A(x),\ \beta.\mu_A(x)\rangle | x \in E\} \\
&= f_{\alpha,\,0}(A) \cup g_{1,\,\beta}(A); \\
h^*_{\alpha,\,\beta}(A) &= \{\langle x,\ \alpha.\nu_A(x),\ \mu_A(x) + \beta.(1 - \alpha.\nu_A(x) - \mu_A(x))\rangle | x \in E\} \\
&= f_{0,\,\beta}(\{\langle x,\ \mu_A(x),\ \alpha.\nu_A(x),\rangle | x \in E\}) \\
&= f_{0,\,\beta}(g_{1,\,\alpha}(\{\langle x,\ \nu_A(x),\ \mu_A(x),\rangle | x \in E\})) \\
&= f_{0,\,\beta}(g_{1,\,\alpha}(\overline{A})); \\
j^*_{\alpha,\,\beta}(A) &= \{\langle x,\ \nu_A(x) + \alpha.(1 - \nu_A(x) - \beta.\mu_A(x)),\ \beta.\mu_A(x)\rangle | x \in E\} \\
&= f_{\alpha,\,0}(\{\langle x,\ \beta.\mu_A(x),\ \nu_A(x)\rangle | x \in E\}) \\
&= f_{\alpha,\,0}(g_{\beta,\,1}(\overline{A})).
\end{aligned}
$$

The proof for the representation by the other pairs is similar to the above one. For example, without proofs we give the following representations.

For the pair $(d,\ g)$, we have (assume $\alpha + \beta \leq 1$),

$$
\begin{aligned}
f_{\alpha,\beta}(A) &= g_{1,0}(d_{1-\alpha}(\overline{A})) \cap d_{1-\beta}(A); \\
h_{\alpha,\,\beta}(A) &= g_{\alpha,\beta}(A) \cap d_{1-\beta}(A); \\
h^*_{\alpha,\beta}(A) &= d_{1-\beta}(g_{1,\alpha}(\overline{A})) \cap g_{\alpha,1}(A); \\
j_{\alpha,\,\beta}(A) &= g_{\alpha,\,\beta}(A) \cup d_\alpha(A); \\
j^*_{\alpha,\,\beta}(A) &= d_\alpha(g_{\beta,1}(\overline{A})) \cup g_{1,\beta}(A).
\end{aligned}
$$

For the pair $(h,\ j)$, we have (let for the second equality below $\alpha + \beta \leq 1$),

$$d_\alpha(A) = j_{0,1}(h_{1,\alpha}(\overline{A}));$$
$$f_{\alpha,\beta}(A) = j_{0,1}(h_{1,\alpha}(\overline{A})) \cup h_{\alpha,\beta}(A);$$
$$g_{\alpha,\beta}(A) = h_{\alpha,0}(h_{\beta,0}(j_{0,1}A));$$
$$h^*_{\alpha,\beta}(A) = h_{1,\beta}(j_{0,1}(h_{\alpha,0}(A)));$$
$$j^*_{\alpha,\ \beta}(A) = j_{\alpha,1}(h_{1,0}(j_{0,\beta}(A))).$$

Their uniqueeness is proved below.

Let us have an one-element universe $E = \{x\}$ and let us construct for the fixed set $A \subseteq E$, the following IFS (below we use only IFS A, i.e., no notational collision will arise),

$$A = \{\langle x, \mu_A(x), \nu_A(x)\rangle | x \in E\}$$

for which $\mu_A(x) > 0$ and $\nu_A(x) > 0$. It can be directly seen that the element

$$\langle x, 0, 0\rangle \in g_{0,0}(A) = \{\langle x, 0, 0\rangle | x \in E\}$$

cannot be represented by any of the following pairs of operators:

$$(d, f), (d, h), (d, h^*), (d, j), (d, j^*), (f, h), (f, h^*), (f, j), (f, j^*),$$

because none of their combinations applied to the set A yields as a result $\langle x, 0, 0\rangle$ for $x \in E$. This fact can be seen, e.g., from the geometrical interpretations of the individual operators. Similarly, the element

$$\langle x, 1, 0\rangle \in j^*_{1,0}(A) = \{\langle x, 1, 0\rangle | x \in E\}$$

cannot be represented by any of the following pairs of operators: (g, h), (g, h^*), (h, h^*), and the element

$$\langle x, 0, 1\rangle \in h^*_{0,1}(A) = \{\langle x, 0, 1\rangle | x \in E\}$$

cannot be represented by any one of the following pairs of operators: (g, j), (g, j^*), (j, j^*).

Therefore, the only pairs of basic couple operators are unique.

Corollary 5.4: The only possible basic triples of operators are:

$$
\begin{array}{lllll}
(d, f, g), & (d, g, h), & (d, g, h^*), & (d, g, j), & (d, g, j^*), \\
(d, h, j), & (d, h, j^*), & (d, h^*, j), & (d, h^*, j^*), & (f, g, h), \\
(f, g, h^*), & (f, g, j), & (f, g, j^*), & (f, h, j), & (f, h, j^*), \\
(f, h^*, j), & (f, h^*, j^*), & (g, h, j), & (g, h, j^*), & (g, h^*, j), \\
(g, h^*, j^*), & (h, h^*, j), & (h, h^*, j^*), & (h, j, j^*), & (h^*, j, j^*).
\end{array}
$$

Corollary 5.5: (d, f, h, h^*), (d, f, j, j^*) are the only 4-tuples of operators that are not basic 4-tuples of operators.

Indeed, only these two 4-tuples of operators do not contain basic couples or triples of operators and it can be proved that they cannot represent the elements $\langle x, 1, 0\rangle \in j^*_{1,0}(A)$ and $\langle x, 0, 1\rangle \in h^*_{0,1}(A)$, respectively.

Corollary 5.6: All 5-tuples of operators are basic 5-tuples of operators.

Open problem 13. *Can all operator representations contain only basic operators, i.e., can they be represented without operations $\cup$ and $\cap$.*

In Section **2.5** the concept of an IFTS is discussed. Now, we introduce two other types of IFSs.

An IFS A is called a *Tautological Set* (TS), iff for every $x \in E, \mu_A(x) = 1$ and $\nu_A(x) = 0$.

For the IFSs A and B, the following set is defined:

$$Insg(A, B) = \{\langle x, \langle 1 - (1 - \mu_B(x)).sg(\mu_A(x) - \mu_B(x)),$$

$$\nu_B(x).sg(\mu_A(x) - \mu_B(x)).sg(\nu_B(x) - \nu_A(x))\rangle \mid x \in E\}. \tag{5.18}$$

It is to be noted that for every two IFSs A and B, $Insg(A, B)$ is an TS iff $A \subseteq B$.

Now, we formulate and prove new properties of the above defined operators. The properties of operators $f_{\alpha,\beta}$, $g_{\alpha,\beta}$, $h_{\alpha,\beta}$, $h^*_{\alpha,\beta}$, $j_{\alpha,\beta}$, $j^*_{\alpha,\beta}$ are analogous to the properties of operators $F_{\alpha,\beta}$, $G_{\alpha,\beta}$, $H_{\alpha,\beta}$, $H^*_{\alpha,\beta}$, $J_{\alpha,\beta}$, $J^*_{\alpha,\beta}$, but there are essential differences, too.

Theorem 5.10: For every IFS A, and for every $\alpha, \beta \in (0, 1]$, and $\alpha + \beta \leq 1$

(a) $Insg(A, f_{\alpha,\beta}f_{\alpha,\beta}(A))$,

(b) $Insg(g_{\alpha,\beta}g_{\alpha,\beta}(A), A)$

(c) $Insg(A, j_{\alpha,1}j_{\alpha,1}(A))$,

(d) $Insg(h_{1,\beta}h_{1,\beta}(A), A)$

(e) $Insg(A, j^*_{\alpha,1}j^*_{\alpha,1}(A))$,

(f) $Insg(h^*_{1,\beta}h^*_{1,\beta}(A), A)$

are TSs.

Proof: (a) Let $\alpha, \beta \in (0, 1]$ be fixed and $\alpha + \beta \leq 1$. Then,

$Insg(A, f_{\alpha,\beta}f_{\alpha,\beta}(A))$

$= Insg(A, f_{\alpha,\beta}(\{\langle x, \nu_A(x) + \alpha.\pi_A(x), \mu_A(x) + \beta.\pi_A(x)\rangle \mid x \in E\})$

$= Insg(A, \{\langle x, \mu_A(x) + \beta.\pi_A(x) + \alpha.(1 - \nu_A(x) - \alpha.\pi_A(x) - \mu_A(x)$

$-\beta.\pi_A(x)), \nu_A(x) + \alpha.\pi_A(x) + \beta.(1 - \nu_A(x) - \alpha.\pi_A(x) - \mu_A(x)$

$-\beta.\pi_A(x))\rangle \mid x \in E\}$

$= Insg(A, \{\langle x, \mu_A(x) + (\alpha + \beta - \alpha\beta - \alpha^2).\pi_A(x),$

$\nu_A(x) + (\alpha + \beta - \alpha\beta - \beta^2).\pi_A(x)\rangle \mid x \in E\}$

$= \{\langle x, 1 - (1 - (\mu_A(x) + (\alpha + \beta - \alpha\beta - \alpha^2).\pi_A(x))).sg(\mu_A(x) \quad (\mu_A(x)$

$+(\alpha + \beta - \alpha\beta - \alpha^2).\pi_A(x))), (\nu_A(x) + (\alpha + \beta - \alpha\beta - \beta^2).\pi_A(x))$

$.sg(\mu_A(x) - (\mu_A(x) + (\alpha + \beta - \alpha\beta - \alpha^2).\pi_A(x)))$

$.sg((\nu_A(x) + (\alpha + \beta - \alpha\beta - \beta^2).\pi_A(x)) - \nu_A(x))\rangle \mid x \in E\},$

$= \{\langle x, 1, 0\rangle \mid x \in E\},$

because
$$\mu_A(x) - (\mu_A(x) + (\alpha + \beta - \alpha\beta - \alpha^2).\pi_A(x))$$
$$= -(\alpha + \beta - \alpha\beta - \alpha^2).\pi_A(x)) \le 0$$

and, therefore,

$$sg(\mu_A(x) - (\mu_A(x) + (\alpha + \beta - \alpha\beta - \alpha^2).\pi_A(x))) = 0.$$

For every IFS A, and for every $\alpha, \beta \in [0, 1)$

(a) $Insg(f_{\alpha,\beta}f_{\alpha,\beta}(A), A)$,

(b) $Insg(A, g_{\alpha,\beta}g_{\alpha,\beta}(A))$

(c) $Insg(j_{\alpha,1}j_{\alpha,1}(A), A)$,

(d) $Insg(A, h_{1,\beta}h_{1,\beta}(A))$

(e) $Insg(j^*_{\alpha,1}j^*_{\alpha,1}(A), A)$,

(f) $Insg(A, h^*_{1,\beta}h^*_{1,\beta}(A))$

  are not TSs.

For every IFS A, and for every $\alpha, \beta \in (0, 1)$

(a) $Insg(\overline{f_{\alpha,\beta}(A)}, A)$,

(b) $Insg(A, \overline{g_{\alpha,\beta}(A)})$

(c) $Insg(\overline{j_{\alpha,1}(A)}, A)$,

(d) $Insg(A, \overline{h_{1,\beta}(A)})$

(e) $Insg(\overline{j^*_{\alpha,1}(A)}, A)$,

(f) $Insg(A, \overline{h^*_{1,\beta}(A)})$

  are not TSs.

For every IFS A, and for every $\alpha, \beta \in (0, 1)$, the following identities

(a) $Insg(\overline{A}, f_{\alpha,\beta}(A))$, where $\alpha + \beta \le 1$,

(b) $Insg(g_{\alpha,\beta}(A), \overline{A})$,

(c) $Insg(h_{\alpha,\beta}(A), \overline{A})$,

(d) $Insg(\overline{A}, j_{\alpha,\beta}(A))$,

(e) $Insg(h^*_{\alpha,\beta}(A), \overline{A})$,

(f) $Insg(\overline{A}, j^*_{\alpha,\beta}(A))$

  are TSs.

For every IFS A, and for every $\alpha, \beta \in (0,1)$, the following identities

(a) $Insg(f_{\alpha,\beta}(A), \overline{A})$, where $\alpha + \beta \leq 1$,

(b) $Insg(\overline{A}, g_{\alpha,\beta}(A))$,

(c) $Insg(\overline{A}, h_{\alpha,\beta}(A))$,

(d) $Insg(j_{\alpha,\beta}(A), \overline{A})$,

(e) $Insg(\overline{A}, h^*_{\alpha,\beta}(A))$,

(f) $Insg(j^*_{\alpha,\beta}(A), \overline{A})$

are not TSs.

As it is shown above, operators d_α and $f_{\alpha,\beta}$ are direct extensions of the operators "necessity" and "possibility", while the other new operators have no analogues in the ordinary modal logic. On the other hand, we have seen that there is a similarity between the behaviour of operators $d_\alpha, f_{\alpha,\beta}, g_{\alpha,\beta}$ and operation "negation" (see (2.9)). In the present form of operation "negation" it satisfies for every IFS A, the equality

$$\overline{\overline{A}} = A.$$

On the other hand, it is seen that the operators $f_{\alpha,\beta}$ and $g_{\alpha,\beta}$ satisfy the latest equality only when, $\alpha = \beta = 0$ for operator $f_{\alpha,\beta}$, $\alpha = \beta = 1$ for operator $g_{\alpha,\beta}$.

Therefore, operators $f_{\alpha,\beta}$ (in particular, d_α) and $g_{\alpha,\beta}$ are extensions of the operation "negation".

Let

$$\mathcal{F} = \{f_{\alpha,\beta} \mid \alpha, \beta, \alpha + \beta \in [0,1]\},$$

$$\mathcal{G} = \{y_{\alpha,\beta} \mid \alpha, \beta \in [0,1]\}.$$

Obviously, set $\mathcal{F} \cap \mathcal{G}$ contains only one element - the classical negation, introduced by (2.9).

Theorem 5.11: For every IFS A, and for every $f_{\alpha,\beta} \in \mathcal{F}, g_{\gamma,\delta} \in \mathcal{G}: g_{\gamma,\delta}(A) \to f_{\alpha,\beta}(A)$ is a TS.

5.9 Operator $x_{a,b,c,d,e,f}$

Similar to with operator $X_{a,b,c,d,e,f}$ (see (5.8)), here we introduce an operator, that is universal for the extended modal operators $f_{\alpha,\beta}, g_{\alpha,\beta}, h_{\alpha,\beta}, h^*_{\alpha,\beta}, j_{\alpha,\beta}, j^*_{\alpha,\beta}$ by

$$x_{a,b,c,d,e,f}(A) = \{\langle x, a.\nu_A(x) + b.(1 - \nu_A(x)) - c.\mu_A(x)),$$
$$d.\mu_A(x) + e.(1 - f.\nu_A(x) - \mu_A(x)))|x \in E\} \tag{5.19}$$

where $a, b, c, d, e, f \in [0,1]$ satisfy (5.9) and (5.10).

For every IFS A and for every $a, b, c, d, e, f \in [0, 1]$ satisfying (5.9) and (5.10), $x_{a,b,c,d,e,f}(A)$ is an IFS.

All operators $f_{\alpha,\beta}, g_{\alpha,\beta}, h_{\alpha,\beta}, h^*_{\alpha,\beta}, j_{\alpha,\beta}, j^*_{\alpha,\beta}$ can be represented by the operator $x_{a,b,c,d,e,f}$ at suitably chosen values of its parameters. These representations are the following:

$$f_{a,b}(A) = x_{1,a,1,1,b,1}(A),$$
$$g_{a,b}(A) = x_{a,0,r,b,0,s}(A),$$
$$h_{a,b}(A) = x_{a,0,r,1,b,1}(A),$$
$$h^*_{a,b}(A) = x_{a,0,r,1,b,a}(A),$$
$$j_{a,b}(A) = x_{1,a,1,b,0,r}(A),$$
$$j^*_{a,b}(A) = x_{1,a,b,b,0,r}(A),$$

where r and s are arbitrary real numbers in interval $[0, 1]$.

Let for $a, b, c, d, e, f, g, h, i, j, k, l \in [0, 1]$ it holds that

$$
\begin{array}{llll}
u = & a.j & +b & -b.j \quad -b.e.h + b.c.h.i \geq 0, \\
v = & a.k & +b & -b.c.h \; -b.k \qquad\qquad\quad > 0, \\
w = & b.k.l & +b.c.h & -a.k.l \; -b.c.g \qquad\quad\; \geq 0, \\
x = & d.g & +e & -e.f.k +e.f.k.l - e.g \; \geq 0, \\
y = & d.h & +e & -e.f.k \; -e.h \qquad\qquad\;\; > 0, \\
z = & e.f.k & +e.f.k & -d.h.i \; -e.f.j \qquad\quad\;\; \geq 0, \\
\end{array}
$$

$$v \geq w,$$
$$y \geq z.$$

Then

$$x_{a,b,c,d,e,f}(x_{g,h,i,j,k,l}(A)) = x_{u,v,w/v,x,y,z/y}.$$

For every two IFSs A and B and for every $a, b, c, d, e, f \in [0, 1]$ satisfying (5.9) and (5.10),

$$\overline{x_{a,b,c,d,e,f}(\overline{A})} = x_{d,e,f,a,b,c}(A),$$
$$x_{a,b,c,d,e,f}(\overline{A}) = X_{a,b,c,d,e,f}(A),$$
$$x_{a,b,c,d,e,f}(A@B) = x_{a,b,c,d,e,f}(A)@x_{a,b,c,d,e,f}(B).$$

For $a, b, c, d, e, f, g, h, i, j, k, l \in [0, 1]$ such that (5.9), (5,10) and

$$g + k - k.l \leq 1,$$
$$h + j - h.i \leq 1$$

are valid, and for a given IFS A:

$$x_{a,b,c,d,e,f}(A) \cap x_{g,h,i,j,k,l}(A)$$

$$\supset x_{\min(a,g),\min(b,h),\max(c,i),\max(d,j),\max(e,k),\min(f,l)}(A),$$

$$x_{a,b,c,d,e,f}(A) \cup x_{g,h,i,j,k,l}(A)$$

$$\subset x_{\max(a,g),\max(b,h),\min(c,i),\min(d,j),\min(e,k),\max(f,l)}(A),$$

$$x_{a,b,c,d,e,f}(A)@x_{g,h,i,j,k,l}(A) = x_{\frac{a+g}{2}+\frac{b+h}{2}+\frac{bc+hi}{b+h}+\frac{d+j}{2}+\frac{e+k}{2}+\frac{ef+kl}{e+k}}(A).$$

In [73], where this operator is introduced for the first time, there were some misprints, which are corrected here.

5.10 Partial Extension of the Extended Modal Operators

Let A and B be two IFSs. Following [48], we generalize the operators from the previous two chapters to the forms,

$$F_B(A) = \{\langle x, \mu_A(x) + \mu_B(x).\pi_A(x), \nu_A(x) + \nu_B(x).\pi_A(x)\rangle \mid x \in E\}; \tag{5.20}$$

$$G_B(A) = \{\langle x, \mu_B(x).\mu_A(x), \nu_B(x).\nu_A(x)\rangle \mid x \in E\}; \tag{5.21}$$

$$H_B(A) = \{\langle x, \mu_B(x).\mu_A(x), \nu_A(x) + \nu_B(x).\pi_A(x)\rangle \mid x \in E\}, \tag{5.22}$$

$$H_B^*(A) = \{\langle x, \mu_B(x).\mu_A(x), \nu_A(x) + \nu_B(x).(1 - \mu_B(x) \\ .\mu_A(x) - \nu_A(x))\rangle \mid x \in E\}, \tag{5.23}$$

$$J_B(A) = \{\langle x, \mu_A(x) + \mu_B(x).\pi_A(x), \nu_B(x).\nu_A(x)\rangle \mid x \in E\}, \tag{5.24}$$

$$J_B^*(A) = \{\langle x, \mu_A(x) + \mu_B(x).(1 - \mu_A(x) - \nu_B(x).\nu_A(x)), \\ \nu_B(x).\nu_A(x)\rangle \mid x \in E\}, \tag{5.25}$$

$$f_B(A) = \{\langle x, \nu_A(x) + \mu_B(x).\pi_A(x), \mu_A(x) + \nu_B(x).\pi_A(x)\rangle \mid x \in E\}; \tag{5.26}$$

$$g_B(A) = \{\langle x, \mu_B(x).\nu_A(x), \nu_B(x).\mu_A(x)\rangle \mid x \in E\}; \tag{5.27}$$

$$h_B(A) = \{\langle x, \mu_B(x).\nu_A(x), \mu_A(x) + \nu_B(x).\pi_A(x)\rangle \mid x \in E\}, \tag{5.28}$$

$$h_B^*(A) = \{\langle x, \mu_B(x).\nu_A(x), \mu_A(x) + \nu_B(x).(1 - \mu_B(x) \\ .\nu_A(x) - \nu_A(x))\rangle \mid x \in E\}, \tag{5.29}$$

$$j_B(A) = \{\langle x, \nu_A(x) + \mu_B(x).\pi_A(x), \nu_B(x).\mu_A(x)\rangle \mid x \in E\}, \tag{5.30}$$

$$j_B^*(A) = \{\langle x, \nu_A(x) + \mu_B(x).(1 - \nu_A(x) - \nu_B(x).\nu_A(x)), \\ \nu_B(x).\mu_A(x)\rangle \mid x \in E\}. \tag{5.31}$$

Each of these sets is an IFS.

Obviously, from some point of view, the present forms of the modal operators are extensions of their previous forms, but these extensions are not total. For example,

$$G_{1,1}(A) = \{\langle x, 1.\mu_A(x), 1.\nu_A(x)\rangle \mid x \in E\} = \{\langle x, \mu_A(x), \nu_A(x)\rangle \mid x \in E\} = A,$$

while

$$G_B(A) = \{\langle x, \mu_B(x).\mu_A(x), \nu_B(x).\nu_A(x)\rangle \mid x \in E\} \neq A,$$

because in this case $\mu_B(x) + \nu_B(x) \leq 1$.

Only operator F_B includes as a particular case operator $F_{\alpha,\beta}$.

Therefore, the new operators are the partial extensions of the extended modal operators and hence the topic of this Section.

Two IFSs A and B are said to be *"like IFSs"*, iff for each $x \in E$

$$\mu_A(x).\nu_B(x) = \nu_A(x).\mu_B(x)). \tag{5.32}$$

For every two IFSs A and B, they are like IFSs iff
(a) $F_A(B) = F_B(A)$,

(b) $H_A(B) = H_B(A)$,

(c) $J_A(B) = J_B(A)$.

For every two IFSs A and B,

$$G_B(A) = G_A(B).$$

For every two IFSs A and B,
(a) $H_B^*(A) = H_A^*(B)$ iff for each $x \in E$ $\nu_A(x) = \nu_B(x)$),

(b) $J_B^*(A) = J_A^*(B)$ iff for each $x \in E$ $\mu_A(x) = \mu_B(x)$).

For every three IFSs A, B and C,
(a) $F_{F_A(B)}(C) = F_A(F_B(C))$,

(b) $G_{G_A(B)}(C) = G_A(G_B(C))$,

(c) $H_{H_A(B)}(C) = H_A(H_B(C))$,

(d) $J_{J_A(B)}(C) = J_A(J_B(C))$.

For every three IFSs A, B, C,

(a) if $B \subseteq C$, then $f_B(A) \subseteq f_C(A)$,

(b) if $B \subseteq C$, then $g_B(A) \subseteq g_C(A)$,

(c) if $B \subseteq C$, then $h_B(A) \subseteq h_C(A)$,

(d) if $B \subseteq C$, then $h_B^*(A) \subseteq h_C^*(A)$,

(e) if $B \subseteq C$, then $j_B(A) \subseteq j_C(A)$,

(f) if $B \subseteq C$, then $j_B^*(A) \subseteq j_C^*(A)$,

(g) $\overline{f_B(\overline{A})} = f_{\overline{B}}(A)$,

(h) $\overline{g_B(\overline{A})} = g_{\overline{B}}(A)$,

(i) $\overline{h_B(\overline{A})} = j_{\overline{B}}(A)$,

(j) $\overline{h_B^*(\overline{A})} = j_{\overline{B}}^*(A)$,

(k) $\overline{j_B(\overline{A})} = h_{\overline{B}}(A)$,

(l) $\overline{j_B^*(\overline{A})} = h_{\overline{B}}^*(A)$.

For every two IFSs A and B,

(a) $C(f_B(A)) \supseteq f_B(C(A))$,

(b) $\mathcal{I}(f_B(A)) \subseteq f_B(\mathcal{I}(A))$,

(c) $g_B(C(A)) = C(g_B(A))$,

(d) $g_B(\mathcal{I}(A)) = \mathcal{I}(g_B(A))$.

For every three IFSs A, B, C,

(a) $f_B(f_C(A)) = f_{f_B(C)}(A)$,

(b) $g_B(g_C(A)) = g_{g_B(C)}(\overline{A}) = g_C(g_B(A))$.

Let B, C be fixed IFSs. Then for (a), we have

$$
\begin{aligned}
&f_B(f_C(A)) \\
&= f_B(\{\langle x, \nu_A(x) + \mu_C(x).\pi_A(x), \mu_A(x) + \nu_C(x).\pi_A(x)\rangle \mid x \in E\}) \\
&= \{\langle x, \mu_A(x) + \nu_C(x).\pi_A(x) \\
&\quad + \mu_B(x).(1 - \mu_A(x) - \mu_C(x).\pi_A(x) - \nu_A(x) - \nu_C(x).\pi_A(x)), \\
&\quad \nu_A(x) + \mu_C(x).\pi_A(x) + \nu_B(x).(1 - \mu_A(x) \\
&\quad - \mu_C(x).\pi_A(x) - \nu_A(x) - \delta.\pi_A(x))\rangle \mid x \in E\} \\
&= \{\langle x, \mu_A(x) + (\nu_C(x) + \mu_B(x).\pi_C(x)).\pi_A(x), \\
&\quad \nu_A(x) + (\mu_C(x) + \nu_B(x).\pi_C(x)).\pi_A(x)\rangle \mid x \in E\} \\
&= f_{f_B(C)}(A).
\end{aligned}
$$

For every three IFSs A, B, C,

(a) $F_B(f_C(A)) = f_{F_B(C)}(A) = F_{F_B(C)}(\overline{A})$,

(b) $f_B(F_C(A)) = f_{f_B(C)}(A) = f_{F_B(\overline{C})}(A) = F_{f_B(C)}(\overline{A}) = F_{F_B(\overline{C})}(\overline{A})$,

(c) $G_B(g_C(A)) = g_{G_B(C)}(A) = G_{G_B(C)}(\overline{A})$,

(d) $g_B(G_C(A)) = g_{g_B(C)}(A) = g_{G_B(\overline{C})}(A) = G_{g_B(C)}(\overline{A}) = F_{G_B(\overline{C})}(\overline{A})$.

6

Other Types of Operators

6.1 IFSs of Certain Level

Following the idea of a fuzzy set of level α (see, e.g. [301]), in [39] the definition of a set of (α, β)-level, generated by an IFS A, has been introduced, where $\alpha, \beta \in [0, 1]$ are fixed numbers for which $\alpha + \beta \leq 1$. Formally, this set has the form,

$$N_{\alpha,\beta}(A) = \{\langle x, \mu_A(x), \nu_A(x) \rangle | x \in E \ \& \ \mu_A(x) \geq \alpha \ \& \ \nu_A(x) \leq \beta\}. \quad (6.1)$$

In [39], two particular cases of the above set are defined:

$$N_\alpha(A) = \{\langle x, \mu_A(x), \nu_A(x) \rangle | x \in E \ \& \ \mu_A(x) \geq \alpha\} \quad (6.2)$$

called *"a set of level of membership α"* generated by A;

$$N^\alpha(A) = \{\langle x, \mu_A(x), \nu_A(x) \rangle | x \in E \ \& \ \nu_A(x) \leq \alpha\} \quad (6.3)$$

called *"a set of level of non-membership α"* generated by A.

From the above definition, for every IFS A, and for every $\alpha, \beta \in [0, 1]$, such that $\alpha + \beta \leq 1$ directly follows the validity of

$$N_{\alpha,\beta} \subseteq \left\{ \begin{array}{c} N^\beta(A) \\ N_\alpha(A) \end{array} \right\} \subseteq A$$

where "$\subset$" is a relation in set-theoretical sense and

$$N_{\alpha,\beta}(A) = N_\alpha(A) \cap N^\beta(A).$$

Now, following [412], we introduce four different extensions of the above set. It is seen directly that N-operator decreases the number of elements of the given IFS A, preserving only these elements of E that have degrees, satisfying the condition $\mu_A(x) \geq \alpha$ and $\nu_A(x) \leq \beta$.

K.T. Atanassov: On Intuitionistic Fuzzy Sets Theory, STUDFUZZ 283, pp. 113–131.
springerlink.com   © Springer-Verlag Berlin Heidelberg 2012

Let A and B be two IFSs defined over the same universe. Then,

$$N_B(A) = \{\langle x, \mu_A(x), \nu_A(x)\rangle | (x \in E) \& (\mu_A(x) \geq \mu_B(x)) \& (\nu_A(x) \leq \nu_B(x))\},$$

$$N_B^*(A) = \{\langle x, \mu_A(x), \nu_A(x)\rangle | (x \in E) \& (\mu_A(x) \leq \mu_B(x)) \& (\nu_A(x) \geq \nu_B(x))\}.$$

Obviously, for each IFS A,

$$A = N_{O^*}(A) = N_{E^*}^*(A),$$

where sets O^* and E^* are defined by (2.13) and (2.14).

For every two IFSs A and B

(a) $A = N_B(A)$ iff $B \subset A$,

(b) $A = N_B^*(A)$ iff $A \subset B$,

(c) $N_B(A) = \emptyset$ iff $(\forall x \in E)(\mu_A(x) < \mu_B(x)) \vee (\nu_A(x) > \nu_B(x))$,

(d) $N_B^*(A) = \emptyset$ iff $(\forall x \in E)(\mu_A(x) > \mu_B(x)) \vee (\nu_A(x) < \nu_B(x))$,

(e) $A \subset N_B(A) \cup N_B^*(A)$.

For every three IFSs A, B, C,

(a) $N_C(N_B(A)) = N_{B \cup C}(A)$.

(b) $N_C^*(N_B^*(A)) = N_{B \cap C}^*(A)$.

Let the three IFSs A, B, C be given. Then for (a) we obtain

$$N_C(N_B(A))$$

$$= N_C(\{\langle x, \mu_A(x), \nu_A(x)\rangle | (x \in E) \& (\mu_A(x) \geq \mu_B(x)) \& (\nu_A(x) \leq \nu_B(x))\})$$

$$= \{\langle x, \mu_A(x), \nu_A(x)\rangle | (x \in E) \& (\mu_A(x) \geq \mu_B(x))$$

$$\& (\mu_A(x) \geq \mu_C(x)) \& (\nu_A(x) \leq \nu_B(x)) \& (\nu_A(x) \leq \nu_C(x))\})$$

$$= \{\langle x, \mu_A(x), \nu_A(x)\rangle | (x \in E) \& (\mu_A(x) \geq \max(\mu_B(x), \mu_C(x))$$

$$\& (\nu_A(x) \leq \min(\nu_B(x), \nu_C(x))\}) = N_{B \cup C}(A).$$

For every three IFSs A, B, C,

(a) $N_C(A \cap B) = N_C(A) \cap^* N_C(B)$.

(b) $N_C^*(A \cup B) = N_C^*(A) \cup^* N_C^*(B)$,

where here and below $\cup^*$ and $\cap^*$ are set theoretic operations *"union"* and *"intersection"*.

For every two IFSs A and B,

$$N_B^*(A) = \overline{N_{\overline{B}}(\overline{A})}.$$

For every two IFSs A and B,

(a) $N_{\mathcal{C}(B)}(A) = \bigcap_{P \subset \mathcal{C}(B)}^* N_P \mathcal{C}(A),$

(b) $N^*_{\mathcal{I}(B)}(A) = \bigcup^*_{\mathcal{I}(B)\subset P} N^*_P(A),$

(c) $N_B(\mathcal{I}(A)) = \bigcup^*_{P\subset\mathcal{I}(A)} N_B(P),$

(d) $N^*_B(\mathcal{C}(A)) = \bigcap^*_{\mathcal{C}(A)\subset P} N^*_B(P).$

where operators $\mathcal{C}$ and $\mathcal{I}$ are defined by (4.7) and (4.10).

(a) $N_{J_C(B)}(A) \subset^* N_B(A),$

(b) $N_{H_C(B)}(A) \supset^* N_B(A),$

(c) $N_{J^*_C(B)}(A) \subset^* N_B(A),$

(d) $N_{H^*_C(B)}(A) \supset^* N_B(A),$

(e) $N^*_{J_C(B)}(A) \supset^* N^*_B(A),$

(f) $N^*_{H_C(B)}(A) \subset^* N^*_B(A),$

(g) $N^*_{J^*_C(B)}(A) \supset^* N^*_B(A),$

(h) $N^*_{H^*_C(B)}(A) \subset^* N^*_B(A),$

where $\subset^*$ and $\supset^*$ are set theoretical relations "*inclusion*".

For every two IFSs A and B,
(a) $N_{\mathcal{C}(A)}(B) = N_{K,L}(B),$

(b) $N^*_{\mathcal{I}(A)}(B) = N_{k,l}(B),$

where numbers K, L, k, l are defined by (4.8), (4.9), (4.11) and (4.12), respectively.

Let

$$f : \{\langle a, b\rangle | a, b \in [0, 1] \ \& \ a + b \leq 1\} \ \to \ \{\langle a, b\rangle | a, b \in [0, 1] \ \& \ a + b \leq 1\}$$

be a function. Define

$$N^f_{\alpha,\beta}(A) = \{\langle x, \mu_A(x), \nu_A(x)\rangle | x \in E \ \& f(\mu_A(x), \nu_A(x)) \geq \langle \alpha, \beta\rangle\}. \quad (6.4)$$

Obviously, when f is the identity, $N^f_{\alpha,\beta}(A)$ coincides with $N_{\alpha,\beta}(A)$.

Third, we extend the first modification, replacing the function f with a predicate p, i.e.,

$$N^p_{\alpha,\beta}(A) = \{\langle x, \mu_A(x), \nu_A(x)\rangle | x \in E \ \& p((\mu_A(x), \nu_A(x)), (\alpha, \beta))\}. \quad (6.5)$$

When predicate p has the form

$$p((a,b),(c,d)) = \text{“}f(a,b) \geq (c,d)\text{”},$$

then $N^p_{\alpha,\beta}(A)$ coincides with $N^f_{\alpha,\beta}(A)$.

In the second and third cases, numbers α, β are given in advance and fixed. Now, we introduce extensions for each of the second and third cases. As in the first case, substitute the two given constants α and β with elements of a given IFS B. The two new N-operators have the respective forms:

$$N^f_B(A) = \{\langle x, \mu_A(x), \nu_A(x)\rangle | x \in E$$

$$\& f(\mu_A(x), \nu_A(x)) \geq \langle \mu_B(x), \nu_B(x)\rangle\}, \tag{6.7}$$

$$N^p_B(A) = \{\langle x, \mu_A(x), \nu_A(x)\rangle | x \in E$$

$$\& p((\mu_A(x), \nu_A(x)), (\mu_B(x), \nu_B(x)))\}. \tag{6.8}$$

For every three IFSs A, B and C,
(a) if $B \subseteq C$, then

$$N_B(A) \supseteq N_C(A),$$

$$N^f_B(A) \supseteq N^f_C(A),$$

(b) if $A \subseteq B$, then

$$N_C(A) \subseteq N_C(B),$$

$$N^f_C(A) \subseteq N^f_C(B).$$

For every two IFSs A and B,

$$A = N^p_B(A) \cup N^{\neg p}_B(A),$$

where $\neg p$ is the negation of predicate p.

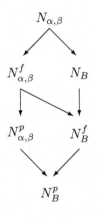

Fig. 6.1

For every two IFSs A and B, and for every n predicates $p_1, p_2, ..., p_n$,

$$N_B^{p_1 \vee p_2 \vee ... \vee p_n}(A) = \bigcup_{1 \leq i \leq n} N_B^{p_i}(A).$$

$$N_B^{p_1 \& p_2 \& ... \& p_n}(A) = \bigcap_{1 \leq i \leq n} N_B^{p_i}(A).$$

The same inclusions are valid for the particular cases, discussed in the beginning of the Section.

Fig. 6.1 describes the interrelation between the operator $N_{\alpha,\beta}$ and its generalizations. Here, the symbol $X \to Y$ means that operator X generalizes operator Y.

6.2 Level Type of Operators

Here, we introduce two new operators:

$$P_{\alpha,\beta}(A) = \{\langle x, \max(\alpha, \mu_A(x)), \min(\beta, \nu_A(x))\rangle | x \in E\}, \tag{6.9}$$

$$Q_{\alpha,\beta}(A) = \{\langle x, \min(\alpha, \mu_A(x)), \max(\beta, \nu_A(x))\rangle | x \in E\}, \tag{6.10}$$

for $\alpha, \beta \in [0,1]$ and $\alpha + \beta \leq 1$.

The degrees of membership and non-membership of the elements of a given universe to its subset can be directly changed by these operators.

Obviously, for every IFS A, and for $\alpha, \beta \in [0,1]$ and $\alpha + \beta \leq 1$,

$$P_{\alpha,\beta}(A) = A \cup \{\langle x, \alpha, \beta\rangle | x \in E\},$$
$$Q_{\alpha,\beta}(A) = A \cap \{\langle x, \alpha, \beta\rangle | x \in E\},$$
$$Q_{\alpha,\beta}(A) \subseteq A \subseteq P_{\alpha,\beta}(A).$$

The geometrical interpretations of both operators are shown in Figs. 6.2 and 6.3.

The new operators satisfy the following

Theorem 6.1: For every IFS A, and for every $\alpha, \beta, \gamma, \delta \in [0,1]$, such that $\alpha + \beta \leq 1, \gamma + \delta \leq 1$:

(a) $\overline{P_{\alpha,\beta}(\overline{A})} = Q_{\beta,\alpha}(A);$

(b) $P_{\alpha,\beta}(P_{\gamma,\delta}(A)) = P_{\max(\alpha,\gamma),\min(\beta,\delta)}(A);$

(c) $P_{\alpha,\beta}(Q_{\gamma,\delta}(A)) = Q_{\max(\alpha,\gamma),\min(\beta,\delta)}(P_{\alpha,\beta}(A));$

(d) $Q_{\alpha,\beta}(P_{\gamma,\delta}(A)) = P_{\min(\alpha,\gamma),\max(\beta,\delta)}(Q_{\alpha,\beta}(A));$

(e) $Q_{\alpha,\beta}(Q_{\gamma,\delta}(A)) = Q_{\min(\alpha,\gamma),\max(\beta,\delta)}(A).$

Proof:

(c) $P_{\alpha,\beta}(Q_{\gamma,\delta}(A))$

$$= P_{\alpha,\beta}(\{\langle x, \min(\gamma, \mu_A(x)), \max(\delta, \nu_A(x))\rangle | x \in E\})$$
$$= \{\langle x, \max(\alpha, \min(\gamma, \mu_A(x))), \min(\beta, \max(\delta, \nu_A(x)))\rangle | x \in E\}$$
$$= \{\langle x, \min(\max(\alpha, \gamma), \max(\alpha, \mu_A(x))), \max(\min(\beta, \delta), \max(\beta, \nu_A(x)))\rangle$$
$$| x \in E\}$$
$$= Q_{\max(\alpha,\gamma),\min(\beta,\delta)}(\{\langle x, \max(\alpha, \mu_A(x)), \max(\beta, \nu_A(x))\rangle | x \in E\})$$
$$= Q_{\max(\alpha,\gamma),\min(\beta,\delta)}(P_{\alpha,\beta}(A)).$$

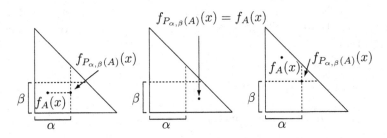

Fig. 6.2

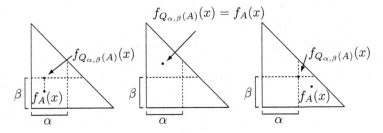

Fig. 6.3

Also, for every two IFSs A and B, and for every $\alpha, \beta \in [0,1]$ such that $\alpha + \beta \leq 1$, the following equalities hold.

(a) $P_{\alpha,\beta}(A \cap B) = P_{\alpha,\beta}(A) \cap P_{\alpha,\beta}(B)$,

(b) $P_{\alpha,\beta}(A \cup B) = P_{\alpha,\beta}(A) \cup P_{\alpha,\beta}(B)$,

(c) $Q_{\alpha,\beta}(A \cap B) = Q_{\alpha,\beta}(A) \cap Q_{\alpha,\beta}(B)$,

(d) $Q_{\alpha,\beta}(A \cup B) = Q_{\alpha,\beta}(A) \cup Q_{\alpha,\beta}(B)$.

(e) $\mathcal{C}(P_{\alpha,\beta}(A)) = P_{\alpha,\beta}(\mathcal{C}(A))$,

(f) $\mathcal{I}(P_{\alpha,\beta}(A)) = P_{\alpha,\beta}(\mathcal{I}(A))$,

(g) $\mathcal{C}(Q_{\alpha,\beta}(A)) = Q_{\alpha,\beta}(\mathcal{C}(A))$,

(h) $\mathcal{I}(Q_{\alpha,\beta}(A)) = Q_{\alpha,\beta}(\mathcal{I}(A))$.

It is worth noting that there are no relations between operators O_1 and O_2, where $O_1 \in \{D_\alpha, F_{\alpha,\beta}, G_{\alpha,\beta}, H_{\alpha,\beta}, H^*_{\alpha,\beta}, J_{\alpha,\beta}, J^*_{\alpha,\beta}\}$ and $O_2 \in \{P_{\alpha,\beta}, Q_{\alpha,\beta}\}$.

For every IFS A, and for every two real numbers $\alpha, \beta \in [0,1]$, such that $\alpha + \beta \leq 1$,

(a) $N_\alpha(P_{\alpha,\beta}(A)) = P_{\alpha,\beta}(A)$,

(b) $N^\beta(P_{\alpha,\beta}(A)) = P_{\alpha,\beta}(A)$,

(c) $N_{\alpha,\beta}(P_{\alpha,\beta}(A)) = P_{\alpha,\beta}(A)$.

Moreover, for every γ, δ, such that $0 \leq \gamma < \alpha$ and $\beta < \delta \leq 1$, we have

(a) $N_\alpha(Q_{\gamma,\beta}(A)) = \emptyset$,

(b) $N^\beta(Q_{\alpha,\delta}(A)) = \emptyset$,

(c) $N_{\alpha,\beta}(Q_{\gamma,\delta}(A)) = \emptyset$.

Theorem 6.2: For every two IFSs A and B, $\mathcal{C}(A) \subseteq \mathcal{I}(B)$, iff there exist two real numbers $\alpha, \beta \in [0,1]$ such that $\alpha + \beta \leq 1$ and $P_{\alpha,\beta}(A) \subseteq Q_{\alpha,\beta}(B)$.

Below, we formulate and prove extended form of this theorem.

Let us define the following two *extended level operators:*

$$P_B(A) = \{\langle x, \max(\mu_B(x), \mu_A(x)), \min(\nu_B(x), \nu_A(x))\rangle \mid x \in E\}, \quad (6.11)$$

$$Q_B(A) = \{\langle x, \min(\mu_B(x), \mu_A(x)), \max(\nu_B(x), \nu_A(x))\rangle \mid x \in E\}. \quad (6.12)$$

Then

$$P_B(A) = B \cup A = A \cup B = P_A(B),$$

$$Q_B(A) = B \cap A = A \cap B = Q_A(B).$$

Therefore, the two IF-operations can be interpreted as IF-operators. For every three IFSs A, B and C,

(a) $\overline{P_B(A)} = Q_{\overline{B}}(\overline{A})$,

(b) $\overline{Q_B(A)} = P_{\overline{B}}(\overline{A})$,

(c) $P_C(P_B(A)) = P_{C \cup B}(A)$,

(d) $Q_C(Q_B(A)) = Q_{C \cap B}(A)$,

(e) $P_C(Q_B(A)) = Q_{C \cap B}(C \cap A)$,

(f) $Q_C(P_B(A)) = P_{C \cup B}(C \cup A)$,

(g) $\mathcal{C}(P_B(A)) = P_{\mathcal{C}(B)}(\mathcal{C}(A))$,

(h) $\mathcal{I}(Q_B(A)) = Q_{\mathcal{I}(B)}(\mathcal{C}(A))$.

Theorem 6.3: For every two IFSs A and B, $A \subset B$, iff there exists an IFS C, such that $P_C(A) \subset Q_C(B)$.

Proof: Let $A \subset B$. Therefore, for all $x \in E$

$$\mu_A(x) \leq \mu_B(x) \ \& \ \nu_A(x) \geq \nu_B(x)).$$

Let
$$C = \{\langle x, \mu_C(x), \nu_C(x)\rangle \mid x \in E\},$$

be defined so that
$$\mu_C(x) = \frac{\mu_A(x) + \mu_B(x)}{2},$$

$$\nu_C(x) = \frac{\nu_A(x) + \nu_B(x)}{2}.$$

Then,
$$P_C(A) = C \cup A = C = C \cap B = Q_C(B),$$

i.e.,
$$P_C(A) \subset Q_C(B).$$

In the opposite case, let there exists an IFS C, such that
$$P_C(A) \subset Q_C(B).$$

Then, for all $x \in E$:
$$\mu_A(x) \le \max(\mu_A(x), \mu_C(x)) \le \min(\mu_B(x), \mu_C(x)) \le \mu_B(x)),$$

and
$$\nu_A(x) \ge min(\nu_A(x), \nu_C(x)) \ge \max(\nu_B(x), \nu_C(x)) \ge \nu_B(x)),$$

i.e.,
$$A \subset B.$$

In [539], a planar geometric representation of two operators $P_{\alpha,\beta}$ and $Q_{\alpha,\beta}$ is given (see Figs. 6.4 and 6.5).

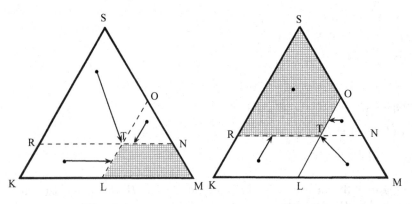

Fig. 6.4   Fig. 6.5

Here the area RTOS corresponds to all x for which $\mu_A(x) \leq \alpha$ & $\nu_A(x) \geq \beta$ (OL is the line defined by $\mu_A(x) = \alpha$ and RN is the line defined by $\nu_A(x) = \beta$). The area KRTL corresponds to all x for which $\mu_A(x) \leq \alpha$ & $\nu_A(x) \leq \beta$. The area TON corresponds to all x for which $\mu_A(x) \geq \alpha$ & $\nu_A(x) \geq \beta$. The area MNTL corresponds to all x for which $\mu_A(x) \geq \alpha$ & $\nu_A(x) \leq \beta$. The shaded areas represent the set of points which are unchanged by the application of the respective operator.

Let A and B be two like IFSs, i.e., for each $x \in E$ they satisfy equality (5.32). Then,

(a) $F_A(B) = F_B(A)$,

(b) $H_A(B) = H_B(A)$,

(c) $J_A(B) = J_B(A)$.

For every two IFSs A and B,

$$G_B(A) = G_A(B).$$

For every two IFSs A and B,

(a) $H_B^*(A) = H_A^*(B)$ iff for each $x \in E$ $\nu_A(x) = \nu_B(x)$,

(b) $J_B^*(A) = J_A^*(B)$ iff for each $x \in E$ $\mu_A(x) = \mu_B(x)$.

In [539], three-dimensional interpretation of operators $P_{\alpha,\beta}$ and $Q_{\alpha,\beta}$ is given (see Figs. 6.6 and 6.7).

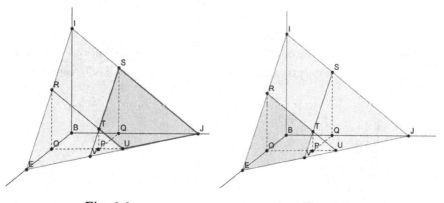

Fig. 6.6   Fig. 6.7

6.3   Other Types of Modal Operators

In this Section five types of other *modal-like operators,* following [35, 39, 53, 61, 66, 70, 197] are defined and the consequences of their generalizations are discussed. We formulate the properties of these operators which hold for them but do not hold for their extensions.

The following are the two operators of modal type, which are similar to the operators in Section **4.1** (A is an IFS).

$$\boxplus A = \{\langle x, \frac{\mu_A(x)}{2}, \frac{\nu_A(x)+1}{2}\rangle | x \in E\}, \qquad (6.13)$$

$$\boxtimes A = \{\langle x, \frac{\mu_A(x)+1}{2}, \frac{\nu_A(x)}{2}\rangle | x \in E\}. \qquad (6.14)$$

All of their properties are valid for their first extensions. For a given real number $\alpha \in [0,1]$ and IFS A,

$$\boxplus_\alpha A = \{\langle x, \alpha.\mu_A(x), \alpha.\nu_A(x)+1-\alpha\rangle | x \in E\}, \qquad (6.15)$$

$$\boxtimes_\alpha A = \{\langle x, \alpha.\mu_A(x)+1-\alpha, \alpha.\nu_A(x)\rangle | x \in E\}. \qquad (6.16)$$

Obviously,

$$0 \le \alpha.\mu_A(x) + \alpha.\nu_A(x) + 1 - \alpha = 1 - \alpha.(1 - \mu_A(x) - \alpha.\nu_A(x)) \le 1.$$

For every IFS A,

$$\boxplus_{0.5}A = \boxplus A,$$

$$\boxtimes_{0.5}A = \boxtimes A.$$

Therefore, the new operators "$\boxplus_\alpha$" and "$\boxtimes_\alpha$" are generalizations of $\boxplus$ and $\boxtimes$. Their graphical interpretations are given in Figs. 6.8 and 6.9, respectively.

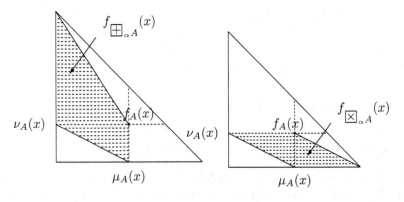

Fig. 6.8   Fig. 6.9

For every IFS A, and for every $\alpha \in [0,1]$,

(a) $\boxplus_\alpha A \subseteq A \subseteq \boxtimes_\alpha A$,

(b) $\overline{\boxplus_\alpha \overline{A}} = \boxtimes_\alpha A$,

(c) $\boxplus_\alpha \boxplus_\alpha A \subseteq \boxplus_\alpha A$,

(d) $\boxtimes_\alpha \boxtimes_\alpha A \supseteq \boxtimes_\alpha A$.

For every two IFSs A and B,

(a) $\boxplus_\alpha(A+B) \subseteq \boxplus_\alpha A + \boxplus_\alpha B$,

(b) $\boxtimes_\alpha(A+B) \supseteq \boxtimes_\alpha A + \boxtimes_\alpha B$,

(c) $\boxplus_\alpha(A.B) \supseteq \boxplus_\alpha A.\boxplus_\alpha B$,

(d) $\boxtimes_\alpha(A.B) \subseteq \boxtimes_\alpha A.\boxtimes_\alpha B$.

For every IFS A, and for every two real numbers $\alpha, \beta \in [0,1]$,

(a) $\boxplus_\alpha \boxplus_\beta A = \boxplus_\beta \boxplus_\alpha A$,

(b) $\boxtimes_\alpha \boxtimes_\beta A = \boxtimes_\beta \boxtimes_\alpha A$,

(c) $\boxtimes_\alpha \boxplus_\beta A \supseteq \boxplus_\beta \boxtimes_\alpha A$.

For every IFS A, and for every three real numbers $\alpha, \beta, \gamma \in [0,1]$,

(a) $\boxplus_\alpha D_\beta(A) = D_\beta(\boxplus_\alpha A)$,

(b) $\boxplus_\alpha F_{\beta,\gamma}(A) = F_{\beta,\gamma}(\boxplus_\alpha A)$, where $\beta + \gamma \leq 1$,

(c) $\boxplus_\alpha G_{\beta,\gamma}(A) \subseteq G_{\beta,\gamma}(\boxplus_\alpha A)$,

(d) $\boxplus_\alpha H_{\beta,\gamma}(A) = H_{\beta,\gamma}(\boxplus_\alpha A)$,

(e) $\boxplus_\alpha H^*_{\beta,\gamma}(A) = H^*_{\beta,\gamma}(\boxplus_\alpha A)$,

(f) $\boxplus_\alpha J_{\beta,\gamma}(A) - J_{\beta,\gamma}(\boxplus_\alpha A)$,

(g) $\boxplus_\alpha J^*_{\beta,\gamma}(A) = J^*_{\beta,\gamma}(\boxplus_\alpha A)$,

(h) $\boxtimes_\alpha D_\beta(A) = D_\beta(\boxtimes_\alpha A)$,

(i) $\boxtimes_\alpha F_{\beta,\gamma}(A) = F_{\beta,\gamma}(\boxtimes_\alpha A)$, where $\beta + \gamma \leq 1$,

(j) $\boxtimes_\alpha G_{\beta,\gamma}(A) \subseteq G_{\beta,\gamma}(\boxtimes_\alpha A)$,

(k) $\boxtimes_\alpha H_{\beta,\gamma}(A) = H_{\beta,\gamma}(\boxtimes_\alpha A)$,

(l) $\boxtimes_\alpha H^*_{\beta,\gamma}(A) = H^*_{\beta,\gamma}(\boxtimes_\alpha A)$,

(m) $\boxtimes_\alpha J_{\beta,\gamma}(A) = J_{\beta,\gamma}(\boxtimes_\alpha A)$,

(n) $\boxtimes_\alpha J^*_{\beta,\gamma}(A) = J^*_{\beta,\gamma}(\boxtimes_\alpha A)$.

The second extension was introduced in [197] by Katerina Dencheva. She extended the last two operators to the forms:

$$\boxplus_{\alpha,\beta} A = \{\langle x, \alpha.\mu_A(x), \alpha.\nu_A(x) + \beta \rangle | x \in E\}, \tag{6.17}$$

$$\boxtimes_{\alpha,\beta} A = \{\langle x, \alpha.\mu_A(x) + \beta, \alpha.\nu_A(x)\rangle | x \in E\}, \tag{6.18}$$

where $\alpha, \beta, \alpha + \beta \in [0, 1]$.

Obviously, for every IFS A,

$$\boxplus A = \boxplus A_{0.5,0.5},$$

$$\boxtimes A = \boxtimes A_{0.5,0.5},$$

$$\boxplus A_\alpha = \boxplus A_{\alpha,1-\alpha},$$

$$\boxtimes A_\alpha = \boxtimes A_{\alpha,1-\alpha}.$$

For every IFS A, and for every $\alpha, \beta, \alpha + \beta \in [0, 1]$,

(a) $\overline{\boxplus_{\alpha,\beta} \overline{A}} = \boxtimes_{\alpha,\beta} A$,

(b) $\overline{\boxtimes_{\alpha,\beta} \overline{A}} = \boxplus_{\alpha,\beta} A$.

For every IFS A, and for every $\alpha, \beta \in [0, 1]$, each of the inclusions

(a) $\boxplus_{\alpha,\beta} \boxplus_{\alpha,\beta} A \subseteq \boxplus_{\alpha,\beta} A$,

(b) $\boxtimes_{\alpha,\beta} \boxtimes_{\alpha,\beta} A \supseteq \boxtimes_{\alpha,\beta} A$,

is valid if and only if $\alpha + \beta = 1$.

For every IFS A, and for every $\alpha \in [0, 1]$,

$$\boxplus_{\alpha,\beta} \boxtimes_{\alpha,\beta} A = \boxtimes_{\alpha,\beta} \boxplus_{\alpha,\beta} A \quad \text{iff} \quad \beta = 0.$$

For every IFS A, and for every $\alpha, \beta, \gamma, \delta \in [0, 1]$ such that $\alpha + \beta, \gamma + \delta \in [0, 1]$,

$$\boxplus_{\alpha,\beta} \boxtimes_{\gamma,\delta} A \subseteq \boxtimes_{\gamma,\delta} \boxplus_{\alpha,\beta} A.$$

For every two IFSs A and B,

(a) $\boxplus_{\alpha,\beta}(A + B) \subseteq \boxplus_{\alpha,\beta} A + \boxplus_{\alpha,\beta} B$, iff $2\alpha + \beta \le 1$,

(b) $\boxtimes_{\alpha,\beta}(A + B) \subseteq \boxtimes_{\alpha,\beta} A + \boxtimes_{\alpha,\beta} B$, iff $2\alpha + \beta \le 1$,

(c) $\boxplus_{\alpha,\beta}(A.B) \supseteq \boxplus_{\alpha,\beta} A.\boxplus_{\alpha,\beta} B$, iff $2\alpha + \beta \le 1$,

(d) $\boxtimes_{\alpha,\beta}(A.B) \supseteq \boxtimes_{\alpha,\beta} A.\boxtimes_{\alpha,\beta} B$, iff $2\alpha + \beta \le 1$,

Now, the third extension of the above operators are as follows:

$$\boxplus_{\alpha,\beta,\gamma} A = \{\langle x, \alpha.\mu_A(x), \beta.\nu_A(x) + \gamma\rangle | x \in E\}, \tag{6.19}$$

$$\boxtimes_{\alpha,\beta,\gamma} A = \{\langle x, \alpha.\mu_A(x) + \gamma, \beta.\nu_A(x)\rangle | x \in E\}, \tag{6.20}$$

where $\alpha, \beta, \gamma \in [0, 1]$ and $\max(\alpha, \beta) + \gamma \le 1$.

Obviously, for every IFS A,

$$\boxplus A = \boxplus A_{0.5,0.5,0.5},$$

$$\boxtimes A = \boxtimes A_{0.5,0.5,0.5},$$

$$\boxplus A_\alpha = \boxplus A_{\alpha,\alpha,1-\alpha},$$

$$\boxtimes A_\alpha = \boxtimes A_{\alpha,1-\alpha},$$

$$\boxplus A_{\alpha,\beta} = \boxplus A_{\alpha,\alpha,\beta},$$

$$\boxtimes A_{\alpha,\beta} = \boxtimes A_{\alpha,\alpha,\beta}.$$

For every IFS A, and for every $\alpha,\beta,\gamma \in [0,1]$ for which $\max(\alpha,\beta) + \gamma \le 1$,

(a) $\overline{\boxplus_{\alpha,\beta,\gamma}\overline{A}} = \boxtimes_{\beta,\alpha,\gamma}A,$

(b) $\overline{\boxtimes_{\alpha,\beta,\gamma}\overline{A}} = \boxplus_{\beta,\alpha,\gamma}A.$

For every IFS A, and for every $\alpha,\beta,\gamma \in [0,1]$ for which $\max(\alpha,\beta) + \gamma \le 1$,

(a) $\boxplus_{\alpha,\beta,\gamma}\boxplus_{\alpha,\beta,\gamma}A \subseteq \boxplus_{\alpha,\beta,\gamma}A$ is valid if and only if $\beta + \gamma = 1$,

(b) $\boxtimes_{\alpha,\beta,\gamma}\boxtimes_{\alpha,\beta,\gamma}A \supseteq \boxtimes_{\alpha,\beta,\gamma}A$ is valid if and only if $\alpha + \gamma = 1$.

For every IFS A, and for every $\alpha,\beta,\alpha+\beta \in [0,1]$,

$$\boxplus_{\alpha,\beta,\gamma}\boxtimes_{\alpha,\beta,\gamma}A = \boxtimes_{\alpha,\beta,\gamma}\boxplus_{\alpha,\beta,\gamma}A \ \text{ iff } \ \gamma = 0.$$

For every IFS A, and for every $\alpha,\beta,\gamma \in [0,1]$ for which $\max(\alpha,\beta) + \gamma \le 1$, the four properties

(a) $\boxplus_{\alpha,\beta,\gamma}\,\square\, A = \square\,\boxplus_{\alpha,\beta,\gamma}A,$

(b) $\boxtimes_{\alpha,\beta,\gamma}\,\square\, A = \square\,\boxtimes_{\alpha,\beta,\gamma}A,$

(c) $\boxplus_{\alpha,\beta,\gamma}\Diamond A = \Diamond\boxplus_{\alpha,\beta,\gamma}A,$

(d) $\boxtimes_{\alpha,\beta,\gamma}\Diamond A = \Diamond\boxtimes_{\alpha,\beta,\gamma}A,$

are valid if and only if $\alpha = \beta$ and $\alpha + \gamma = 1$.

For every two IFSs A and B, and for every $\alpha,\beta,\gamma \in [0,1]$ for which $\max(\alpha,\beta) + \gamma \le 1$,

(a) $\boxplus_{\alpha,\beta,\gamma}(A \cap B) = \boxplus_{\alpha,\beta,\gamma}A \cap \boxplus_{\alpha,\beta,\gamma}B,$

(b) $\boxtimes_{\alpha,\beta,\gamma}(A \cap B) = \boxtimes_{\alpha,\beta,\gamma}A \cap \boxtimes_{\alpha,\beta,\gamma}B,$

(c) $\boxplus_{\alpha,\beta,\gamma}(A \cup B) = \boxplus_{\alpha,\beta,\gamma}A \cup \boxplus_{\alpha,\beta,\gamma}B,$

(d) $\boxtimes_{\alpha,\beta,\gamma}(A \cup B) = \boxtimes_{\alpha,\beta,\gamma}A \cup \boxtimes_{\alpha,\beta,\gamma}B,$

(e) $\boxplus_{\alpha,\beta,\gamma}(A@B) = \boxplus_{\alpha,\beta,\gamma}A@\boxplus_{\alpha,\beta,\gamma}B,$

(f) $\boxtimes_{\alpha,\beta,\gamma}(A@B) = \boxtimes_{\alpha,\beta,\gamma}A@\boxtimes_{\alpha,\beta,\gamma}B,$

For every IFS A, and for every $\alpha, \beta, \gamma \in [0,1]$ for which $\max(\alpha, \beta) + \gamma \leq 1$,

(a) $\boxplus_{\alpha,\beta,\gamma} \mathcal{C}(A) = \mathcal{C}(\boxplus_{\alpha,\beta,\gamma} A)$,

(b) $\boxtimes_{\alpha,\beta,\gamma} \mathcal{C}(A) = \mathcal{C}(\boxtimes_{\alpha,\beta,\gamma} A)$,

(c) $\boxplus_{\alpha,\beta,\gamma} \mathcal{I}(A) = \mathcal{I}(\boxplus_{\alpha,\beta,\gamma} A)$,

(d) $\boxtimes_{\alpha,\beta,\gamma} \mathcal{I}(A) = \mathcal{I}(\boxtimes_{\alpha,\beta,\gamma} A)$.

A natural extension of the last two operators is the operator

$$\boxdot_{\alpha,\beta,\gamma,\delta} A = \{\langle x, \alpha.\mu_A(x) + \gamma, \beta.\nu_A(x) + \delta\rangle | x \in E\}, \tag{6.21}$$

where $\alpha, \beta, \gamma, \delta \in [0,1]$ and $\max(\alpha, \beta) + \gamma + \delta \leq 1$.

It is the fourth type of operator from the current group.

Obviously, for every IFS A,

$$\boxplus A = \boxdot A_{0.5,0.5,0,0.5},$$

$$\boxtimes A = \boxdot A_{0.5,0.5,0.5,0},$$

$$\boxplus A_\alpha = \boxdot A_{\alpha,\alpha,0,1-\alpha},$$

$$\boxtimes A_\alpha = \boxdot A_{\alpha,\alpha,1-\alpha,0},$$

$$\boxplus A_{\alpha,\beta} = \boxdot A_{\alpha,\alpha,0,\beta},$$

$$\boxtimes A_{\alpha,\beta} = \boxdot A_{\alpha,\alpha,\beta,0}.$$

$$\boxplus A_{\alpha,\beta,\gamma} = \boxdot A_{\alpha,\beta,0,\gamma},$$

$$\boxtimes A_{\alpha,\beta,\gamma} = \boxdot A_{\alpha,\beta,\gamma,0}.$$

For every IFS A, and for every $\alpha, \beta, \gamma, \delta \in [0,1]$ for which $\max(\alpha, \beta) + \gamma + \delta \leq 1$:

(a) $\overline{\boxdot_{\alpha,\beta,\gamma,\delta} \overline{A}} = \boxdot_{\beta,\alpha,\delta,\gamma} A$,

(b) $\boxdot_{\alpha,\beta,\gamma,\delta}(\boxdot_{\varepsilon,\zeta,\eta,\theta} A) = \boxdot_{\alpha\varepsilon,\beta\zeta,\alpha\eta+\gamma,\beta\theta+\delta} A$,

(c) $\boxdot_{\alpha,\beta,\gamma,\delta} \square A \supseteq \square \boxdot_{\alpha,\beta,\gamma,\delta} A$,

(d) $\boxdot_{\alpha,\beta,\gamma,\delta} \Diamond A \subseteq \Diamond \boxdot_{\alpha,\beta,\gamma,\delta} A$.

For every two IFSs A and B, and for every $\alpha, \beta, \gamma, \delta \in [0,1]$ for which $\max(\alpha, \beta) + \gamma + \delta \leq 1$,

(a) $\boxdot_{\alpha,\beta,\gamma,\delta}(A \cap B) = \boxdot_{\alpha,\beta,\gamma,\delta} A \cap \boxdot_{\alpha,\beta,\gamma,\delta} B$,

(b) $\boxdot_{\alpha,\beta,\gamma,\delta}(A \cup B) = \boxdot_{\alpha,\beta,\gamma,\delta} A \cup \boxdot_{\alpha,\beta,\gamma,\delta} B$,

(c) $\boxdot_{\alpha,\beta,\gamma,\delta}(A@B) = \boxdot_{\alpha,\beta,\gamma,\delta} A@\boxdot_{\alpha,\beta,\gamma,\delta} B$.

In [187], Gökhan Çuvalcioğlu introduced operator $E_{\alpha,\beta}$ by

$$E_{\alpha,\beta}(A) = \{\langle x, \beta(\alpha.\mu_A(x) + 1 - \alpha), \alpha(\beta.\nu_A(x) + 1 - \beta)\rangle | x \in E\}, \quad (6.22)$$

where $\alpha, \beta \in [0, 1]$, and he studied some of its properties. Obviously,

$$E_{\alpha,\beta}(A) = \boxed{\bullet}_{\alpha\beta,\alpha\beta,(1-\alpha)\beta,(1-\beta)\alpha} A.$$

For every IFS A, and for every $\alpha, \beta, \gamma, \delta \in [0, 1]$ for which $\max(\alpha, \beta) + \gamma + \delta \leq 1$,

(a) $\boxed{\bullet}_{\alpha,\beta,\gamma,\delta} \mathcal{C}(A) = \mathcal{C}(\boxed{\bullet}_{\alpha,\beta,\gamma,\delta} A)$,

(b) $\boxed{\bullet}_{\alpha,\beta,\gamma,\delta} \mathcal{I}(A) = \mathcal{I}(\boxed{\bullet}_{\alpha,\beta,\gamma,\delta} A)$,

(c) $\boxed{\bullet}_{\alpha,\beta,\gamma,\delta} P_{\varepsilon,\zeta}((A)) = P_{\alpha\varepsilon+\gamma,\beta\zeta+\delta}(\boxed{\bullet}_{\alpha,\beta,\gamma,\delta} A)$,

(d) $\boxed{\bullet}_{\alpha,\beta,\gamma,\delta} P_{\varepsilon,\zeta}((A)) = Q_{\alpha\varepsilon+\gamma,\beta\zeta+\delta}(\boxed{\bullet}_{\alpha,\beta,\gamma,\delta} A)$.

A new (final?) extension of the above operators is the operator

$$\boxed{\circ}_{\alpha,\beta,\gamma,\delta,\varepsilon,\zeta} A = \{\langle x, \alpha.\mu_A(x) - \varepsilon.\nu_A(x) + \gamma,$$

$$\beta.\nu_A(x) - \zeta.\mu_A(x) + \delta\rangle | x \in E\}, \quad (6.23)$$

where $\alpha, \beta, \gamma, \delta, \varepsilon, \zeta \in [0, 1]$ and

$$\max(\alpha - \zeta, \beta - \varepsilon) + \gamma + \delta \leq 1, \quad (6.24)$$

$$\min(\alpha - \zeta, \beta - \varepsilon) + \gamma + \delta \geq 0. \quad (6.25)$$

Assume that in the particular cases, $\alpha - \zeta > -\varepsilon, \beta = \delta = 0$ and $\delta - \zeta < \beta, \gamma = \varepsilon = 0$, the inequalities

$$\gamma \geq \varepsilon \quad \text{and} \quad \beta + \delta \leq 1$$

hold. Obviously, for every IFS A,

$$\boxplus A = \boxed{\circ}_{0.5,0.5,0,0.5,0,0} A,$$

$$\boxtimes A = \boxed{\circ}_{0.5,0.5,0.5,0,0,0} A,$$

$$\boxplus_\alpha A = \boxed{\circ}_{\alpha,\alpha,0,1-\alpha,0,0} A,$$

$$\boxtimes_\alpha A = \boxed{\circ}_{\alpha,\alpha,1-\alpha,0,0,0} A,$$

$$\boxplus_{\alpha,\beta} A = \boxed{\circ}_{\alpha,\alpha,0,\beta,0,0} A,$$

$$\boxtimes_{\alpha,\beta} A = \boxed{\circ}_{\alpha,\alpha,\beta,0,0,0} A.$$

$$\boxplus_{\alpha,\beta,\gamma} A = \boxed{\circ}_{\alpha,\beta,0,\gamma,0,0} A,$$

$$\boxtimes_{\alpha,\beta,\gamma} A = \boxed{\circ}_{\alpha,\beta,\gamma,0,0,0} A.$$

$$\boxed{\bullet}_{\alpha,\beta,\gamma,\delta} A = \boxed{\circ}_{\alpha,\beta,\gamma,\delta,0,0} A.$$

$$E_{\alpha,\beta} A = \boxed{\circ}_{\alpha\beta,\alpha\beta,\beta(1-\alpha),\alpha(1-\beta)} A.$$

For every IFS A, and for every $\alpha, \beta, \gamma, \delta, \varepsilon, \zeta \in [0,1]$ for which (6.24) and (6.25) are valid, the equality

$$\boxed{\circ}_{\alpha,\beta,\gamma,\delta,\varepsilon,\zeta} \overline{A} = \boxed{\circ}_{\beta,\alpha,\delta,\gamma,\zeta,\varepsilon} A$$

holds.

For every IFS A, and for every $\alpha_1, \beta_1, \gamma_1, \delta_1, \varepsilon_1, \zeta_1, \alpha_2, \beta_2, \gamma_2, \delta_2, \varepsilon_2, \zeta_2 \in [0,1]$ for which conditions that are similar to (6.24) and (6.25) are valid, the equality

$$\boxed{\circ}_{\alpha_1,\beta_1,\gamma_1,\delta_1,\varepsilon_1,\zeta_1} \left(\boxed{\circ}_{\alpha_2,\beta_2,\gamma_2,\delta_2,\varepsilon_2,\zeta_2} A \right)$$

$$= \boxed{\circ}_{\alpha_1.\alpha_2+\varepsilon_1.\zeta_2,\beta_1.\beta_2+\zeta_1.\varepsilon_2,\alpha_1.\gamma_2-\varepsilon_1.\delta_2+\gamma_1,\beta_1.\delta_2-\zeta_1.\gamma_2+\delta_1,\alpha_1.\varepsilon_2+\varepsilon_1.\beta_2,\beta_1.\zeta_2+\zeta_1.\alpha_2} A.$$

holds.

For every two IFSs A and B, and for every $\alpha, \beta, \gamma, \delta, \varepsilon, \zeta \in [0,1]$ for which (6.24) and (6.25) are valid, the equality

$$\boxed{\circ}_{\alpha,\beta,\gamma,\delta,\varepsilon,\zeta} (A@B) = \boxed{\circ}_{\alpha,\beta,\gamma,\delta,\varepsilon,\zeta} A @ \boxed{\circ}_{\alpha,\beta,\gamma,\delta,\varepsilon,\zeta} B$$

holds.

It must be noted that the equalities

$$\boxed{\circ}_{\alpha,\beta,\gamma,\delta,\varepsilon,\zeta} (A \cap B) = \boxed{\circ}_{\alpha,\beta,\gamma,\delta,\varepsilon,\zeta} A \cap \boxed{\circ}_{\alpha,\beta,\gamma,\delta,\varepsilon,\zeta} B$$

and

$$\boxed{\circ}_{\alpha,\beta,\gamma,\delta,\varepsilon,\zeta} (A \cup B) = \boxed{\circ}_{\alpha,\beta,\gamma,\delta,\varepsilon,\zeta} A \cup \boxed{\circ}_{\alpha,\beta,\gamma,\delta,\varepsilon,\zeta} B$$

which are valid for operator $\boxed{\bullet}_{\alpha,\beta,\gamma,\delta}$, are not always valid.

6.4   Relations between the Most General Intuitionistic Fuzzy Modal Operators

Theorem. 6.4 Operators $X_{a,b,c,d,e,f}$ and $\boxed{\circ}_{\alpha,\beta,\gamma,\delta,\varepsilon,\zeta}$ are equivalent.
Proof Let $a, b, c, d, e, f \in [0,1]$ and satisfy (6.24) and (6.25). Let

$$\alpha = a - b, \quad \beta = d - e, \quad \gamma = b, \quad \delta = e, \quad \varepsilon = bc, \quad \zeta = ef.$$

Also, let

$$X \equiv \alpha.\mu_A(x) - \varepsilon.\nu_A(x) + \gamma = (a-b).\mu_A(x) - b.c.\nu_A(x) + b,$$

$$Y \equiv \beta.\nu_A(x) - \zeta.\mu_A(x) + \delta = (d-e).\nu_A(x) - e.f.\mu_A(x) + e.$$

Then,

$$X \geq (a - b).0 - b.c.1 + b = b.(1 - c) \geq 0,$$
$$X \leq (a - b).1 - b.c.0 + b = a \leq 1,$$
$$Y \geq (d - e).0 - e.f.1 + e = e.(1 - f) \geq 0,$$
$$Y \leq (d - e).1 - e.f.0 + e = d \leq 1$$

and

$$X + Y = (a - b).\mu_A(x) - b.c.\nu_A(x) + b + (d - e).\nu_A(x) - e.f.\mu_A(x) + e$$

$$= (a - b - e.f).\mu_A(x) + (d - e - b.c).\nu_A(x) + b + e$$
$$\leq (a - b - e.f).\mu_A(x) + (d - e - b.c).(1 - \mu_A(x)) + b + e$$
$$= d - e - b.c + b + e + (a - b - e.f - d + e + b.c).\mu_A(x)$$
$$\leq d - b.c + b + a - b - e.f - d + e + b.c$$
$$= a - e.f + e = \alpha + \gamma - \zeta + \delta \leq 1$$

(from (6.24)).

Thus, we obtain

$$\boxed{\bigcirc}_{\alpha,\beta,\gamma,\delta,\varepsilon,\zeta} A = \{\langle x, \alpha.\mu_A(x) - \varepsilon.\nu_A(x) + \gamma, \beta.\nu_A(x) - \zeta.\mu_A(x) + \delta \rangle | x \in E\}$$

$$= \{\langle x, (a - b).\mu_A(x) - b.c.\nu_A(x) + b, (d - e).\nu_A(x) - e.f.\mu_A(x) + e \rangle | x \in E\}$$

$$= \{\langle x, a.\mu_A(x) + b.(1 - \mu_A(x) - c.\nu_A(x)),$$
$$d.\nu_A(x) + e.(1 - f.\mu_A(x) - \nu_A(x)) \rangle | x \in E\}$$

$$= X_{a,b,c,d,e,f}(A).$$

Conversely, let $\alpha, \beta, \gamma, \delta, \varepsilon, \zeta \in [0, 1]$ and satisfy (6.24) and (6.25). From (6.25) it follows that for $\alpha - \beta - \delta - \zeta - 0 : \varepsilon \leq \gamma$, while for $\alpha - \beta - \gamma - \varepsilon - 0 : \zeta \leq \delta$; from (6.24) it follows that for $\beta = \delta = \varepsilon = \zeta = 0 : \alpha + \gamma \leq 1$, while for $\alpha = \gamma = \varepsilon = \zeta = 0 : \beta + \delta \leq 1$. Then, let

$$a = \alpha + \gamma \ (\leq 1),$$

$$b = \gamma,$$

$$c = \frac{\varepsilon}{\gamma} \ (\leq 1),$$

$$d = \beta + \delta \ (\leq 1),$$

$$e = \delta,$$

$$f = \frac{\zeta}{\delta} \ (\leq 1).$$

Let

$$X \equiv a.\mu_A(x) + b.(1 - \mu_A(x) - c.\nu_A(x))$$

$$= (\alpha + \gamma).\mu_A(x) + \gamma.(1 - \mu_A(x) - \frac{\varepsilon}{\gamma}.\nu_A(x)),$$

$$Y \equiv d.\nu_A(x) + e.(1 - f.\mu_A(x) - \nu_A(x))$$

$$= (\beta + \delta).\nu_A(x) + \delta.(1 - \frac{\zeta}{\delta}.\mu_A(x) - \nu_A(x)).$$

Then, we obtain,

$$0 \le \gamma - \varepsilon \le X = \alpha.\mu_A(x) + \gamma - \varepsilon.\nu_A(x) \le \alpha + \gamma \le 1,$$

$$0 \le \delta - \zeta \le Y = \beta.\nu_A(x) + \delta - \zeta.\mu_A(x) \le \beta + \delta \le 1,$$

$$X + Y = \alpha.\mu_A(x) + \gamma - \varepsilon.\nu_A(x) + \beta.\nu_A(x) + \delta - \zeta.\mu_A(x)$$

$$= (\alpha - \zeta).\mu_A(x) - (\beta - \varepsilon).\nu_A(x) + \gamma + \delta$$

$$\le (\alpha - \zeta).\mu_A(x) - (\beta - \varepsilon).(1 - \mu_A(x)) + \gamma + \delta$$

$$= (\alpha - \zeta + \beta - \varepsilon).\mu_A(x) - \beta + \gamma + \delta + \varepsilon$$

$$\le \alpha - \zeta + \beta - \varepsilon - \beta + \gamma + \delta + \varepsilon$$

$$= \alpha - \zeta + \gamma + \delta$$

$$\le \max(\alpha - \zeta, \beta - \varepsilon) + \gamma + \delta \le 1$$

(from (6.24)).

Then, we obtain

$$X_{a,b,c,d,e,f}(A)$$

$$= \{\langle x, a.\mu_A(x) + b.(1 - \mu_A(x) - c.\nu_A(x)),$$

$$d.\nu_A(x) + e.(1 - f.\mu_A(x) - \nu_A(x))\rangle | x \in E\}$$

$$= \{\langle x, (\alpha + \gamma).\mu_A(x) + \gamma.(1 - \mu_A(x) - \frac{\varepsilon}{\gamma}.\nu_A(x)),$$

$$(\beta + \delta).\nu_A(x) + \delta.(1 - \frac{\zeta}{\delta}.\mu_A(x) - \nu_A(x))\rangle | x \in E\}$$

$$= \{\langle x, (\alpha + \gamma).\mu_A(x) + \gamma - \gamma.\mu_A(x) - \varepsilon.\nu_A(x),$$

$$(\beta + \delta).\nu_A(x) + \delta - \zeta.\mu_A(x) - \delta.\nu_A(x)\rangle | x \in E\}$$

$$= \{\langle x, \alpha.\mu_A(x) - \varepsilon.\nu_A(x) + \gamma, \beta.\nu_A(x) - \zeta.\mu_A(x) + \delta\rangle | x \in E\}$$

$$= \boxed{\bigcirc}_{\alpha,\beta,\gamma,\delta,\varepsilon,\zeta} A.$$

Therefore, the two operators are equivalent.

Finally, we construct the Fig. 6.10 in which "$X \to Y$" denotes that operator X represents operator Y, while the converse is not valid.

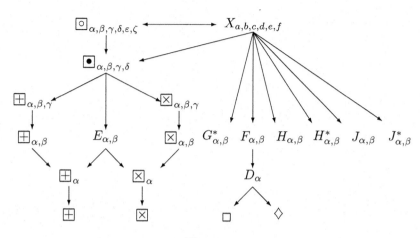

Fig. 6.10

Norms and Metrics over IFSs or Their Elements

7.1 Standard IF-Norms and Metrics

First, let us emphasize that here we do not study the usual set-theoretic properties of the IFSs (i.e. properties which follow directly from the fact that IFSs are sets in the sense of the set theory – see Section **2.1**). For example, given a metric space E, one can study the metric properties of the IFSs over E. This can be done directly by topological methods (see e.g. [421]) and the essential properties of the IFSs are not used. On the other hand, all IFSs (and hence, all fuzzy sets) over a fixed universe E generate a metric space (in the sense of [421]), but with a special metric (cf., e.g., [301]): one that is not related to the elements of E and to the values of the functions μ_A and ν_A defined for these elements.

This peculiarity is based on the fact that the "norm" of a given IFS' element is not actually a norm in the sense of [421]. Rather, it is in some sense a "pseudo-norm" which assigns a number to every element $x \in E$. This number depends on the values of the functions μ_A and ν_A (which are calculated for this element).

Thus, the important conditions,

$$\|x\| = 0 \ \text{ iff } \ x = 0,$$

and

$$\|x\| = \|y\| \ \text{ iff } \ x = y,$$

do not hold here.

Instead, the following is valid

$$\|x\| = \|y\| \ \text{ iff } \ \mu_A(x) = \mu_A(y) \text{ and } \nu_A(x) = \nu_A(y).$$

Actually, the value of $\mu_A(x)$ plays the role of a norm (more precisely, a pseudo-norm) for the element $x \in E$ in every fuzzy set over E. In the intuitionistic fuzzy case, the existence of the second functional component – the function ν_A – gives rise to different options for the definition of the

K.T. Atanassov: On Intuitionistic Fuzzy Sets Theory, STUDFUZZ 283, pp. 133–146.
springerlink.com

concept of norm (in the sense of pseudo-norm) over the subsets and the elements of a given universe E.

The *first intuitionistic fuzzy norm* for every $x \in E$ with respect to a fixed set $A \subseteq E$ is

$$\sigma_{1,A}(x) = \mu_A(x) + \nu_A(x). \tag{7.1}$$

It represents the *degree of definiteness* of the element x. From

$$\pi_A(x) = 1 - \mu_A(x) - \nu_A(x)$$

it follows that

$$\sigma_{1,A}(x) = 1 - \pi_A(x).$$

For every two IFSs A and B, and for every $x \in E$,

(1.1) $\sigma_{1,\overline{A}}(x) = \sigma_{1,A}(x)$;

(1.2) $\sigma_{1,A\cap B}(x) \geq \min(\sigma_{1,A}(x), \sigma_{1,B}(x))$;

(1.3) $\sigma_{1,A\cup B}(x) \leq \max(\sigma_{1,A}(x), \sigma_{1,B}(x))$;

(1.4) $\sigma_{1,A+B}(x) \geq \sigma_{1,A}(x).\sigma_{1,B}(x)$;

(1.5) $\sigma_{1,A.B}(x) \geq \sigma_{1,A}(x).\sigma_{1,B}(x)$;

(1.6) $\sigma_{1,A@B}(x) = \dfrac{(\sigma_{1,A}(x) + \sigma_{1,B}(x))}{2}$;

(1.7) $\sigma_{1,\square A}(x) = 1$;

(1.8) $\sigma_{1,\diamond A}(x) = 1$;

(1.9) $\sigma_{1,C(A)}(x) \geq \max\limits_{x\in E} \sigma_{1,A}(x)$;

(1.10) $\sigma_{1,\mathcal{I}(A)}(x) \leq \min\limits_{x\in E} \sigma_{1,A}(x)$;

(1.11) $\sigma_{1,D_\alpha(A)}(x) = 1$, for every $\alpha \in [0,1]$;

(1.12) $\sigma_{1,F_{\alpha,\beta}(A)}(x) = \alpha + \beta + (1 - \alpha - \beta).\sigma_{1,A}(x)$ for every $\alpha, \beta \in [0,1]$ and $\alpha + \beta \leq 1$;

(1.13) $\sigma_{1,G_{\alpha,\beta}(A)}(x) \leq \sigma_{1,A}(x)$, for every $\alpha, \beta \in [0,1]$;

(1.14) $\sigma_{1,H_{\alpha,\beta}(A)}(x) \leq \beta + (\alpha + \beta).\sigma_{1,A}(x)$, for every $\alpha, \beta \in [0,1]$;

(1.15) $\sigma_{1,H^*_{\alpha,\beta}(A)}(x) \leq \beta + (1 - \beta).\sigma_{1,A}(x)$, for every $\alpha, \beta \in [0,1]$;

(1.16) $\sigma_{1,J_{\alpha,\beta}(A)}(x) \leq \alpha + (\alpha + \beta).\sigma_{1,A}(x)$, for every $\alpha, \beta \in [0,1]$;

(1.17) $\sigma_{1,J^*_{\alpha,\beta}(A)}(x) \leq \alpha + (1 - \alpha).\sigma_{1,A}(x)$, for every $\alpha, \beta \in [0,1]$;

(1.18) $\sigma_{1,X_{a,b,c,d,e,f}A}(x) \begin{cases} \leq \max(a - b - ef, d - e - bc) + b + e \\ \geq \min(a - b - ef, d - e - bc) + b + e \end{cases}$,

(1.19) $\sigma_{1,P_{\alpha,\beta}(A)}(x) \geq \alpha$, for every $\alpha, \beta \in [0,1]$ and $\alpha + \beta \leq 1$;

(1.20) $\sigma_{1,Q_{\alpha,\beta}(A)}(x) \geq \beta$, for every $\alpha, \beta \in [0,1]$ and $\alpha + \beta \leq 1$;

(1.21) $\sigma_{1,\boxplus A}(x) = \frac{1}{2}\sigma_{1,A}(x) + \frac{1}{2}$;

(1.22) $\sigma_{1,\boxtimes A}(x) = \frac{1}{2}\sigma_{1,A}(x) + \frac{1}{2}$;

(1.23) $\sigma_{1,\boxplus_\alpha A}(x) = \alpha.\sigma_{1,A}(x) + 1 - \alpha$;

(1.24) $\sigma_{1,\boxtimes_\alpha A}(x) = \alpha.\sigma_{1,A}(x) + 1 - \alpha$;

(1.25) $\sigma_{1,\boxplus_{\alpha,\beta} A}(x) = \alpha.\sigma_{1,A}(x) + \beta$;

(1.26) $\sigma_{1,\boxtimes_{\alpha,\beta} A}(x) = \alpha.\sigma_{1,A}(x) + \beta$;

(1.27) $\sigma_{1,\boxplus_{\alpha,\beta,\gamma} A}(x) \begin{cases} \leq \max(\alpha,\beta) + \gamma \\ \geq \min(\alpha,\beta) + \gamma \end{cases}$;

(1.28) $\sigma_{1,\boxtimes_{\alpha,\beta,\gamma} A}(x) \begin{cases} \leq \max(\alpha,\beta) + \gamma \\ \geq \min(\alpha,\beta) + \gamma \end{cases}$;

(1.29) $\sigma_{1,E_{\alpha,\beta}(A)}(x) \leq \alpha + \beta - \alpha.\beta$;

(1.30) $\sigma_{1,\bullet_{\alpha,\beta,\gamma,\delta} A}(x) \begin{cases} \leq \max(\alpha,\beta) + \gamma + \delta \\ \geq \min(\alpha,\beta) + \gamma + \delta \end{cases}$;

(1.31) $\sigma_{1,\square_{\alpha,\beta,\gamma,\delta,\varepsilon,\zeta} A}(x) \begin{cases} \leq \max(\alpha - \zeta, \beta - \varepsilon) + \gamma + \delta \\ \geq \min(\alpha - \zeta, \beta - \varepsilon) + \gamma + \delta \end{cases}$.

For example, we prove (1.5)

$$\sigma_{1,A+B}(x) = \mu_A(x) + \mu_B(x) - \mu_A(x).\mu_B(x) + \nu_A(x).\nu_B(x)$$

$$= (\mu_A(x) + \nu_A(x)).(\mu_B(x) + \nu_B(x)) + \mu_A(x) + \mu_B(x) -$$
$$2.\mu_A(x).\mu_B(x) - \mu_A(x).\nu_B(x) - \nu_A(x).\mu_B(x)$$
$$= \sigma_A(x).\sigma_B(x) + \mu_A(x).(1 - \mu_B(x) - \nu_B(x)) + \nu_B(x).(1 - \mu_A(x) - \nu_A(x))$$
$$\geq \sigma_{1,A}(x).\sigma_{1,B}(x).$$

It must be noted that for every two IFSs A and B, and for every $x \in E$, if $A = B$, then $\sigma_{1,A}(x) = \sigma_{1,B}(x)$.

Obviously, the converse is not true.

The *second intuitionistic fuzzy norm* which is defined for every $x \in E$ with respect to a fixed $A \subseteq E$ is,

$$\sigma_{2,A}(x) = \sqrt{\mu_A(x)^2 + \nu_A(x)^2} \tag{7.2}$$

The two norms are analogous to both basic classical types of norms. For every two IFSs A and B, and for every $x \in E$,

(2.1) $\sigma_{2,\overline{A}}(x) = \sigma_{2,A}(x)$;

(2.2) $\sigma_{2,A \cap B}(x) \geq \min(\sigma_{2,A}(x), \sigma_{2,B}(x))$;

(2.3) $\sigma_{2,A \cup B}(x) \leq \max(\sigma_{2,A}(x), \sigma_{2,B}(x))$;

(2.4) $\sigma_{2,A+B}(x) \geq \sigma_{2,A}(x).\sigma_{2,B}(x)$;

(2.5) $\sigma_{2,A.B}(x) \geq \sigma_{2,A}(x).\sigma_{2,B}(x)$;

(2.6) $\sigma_{2,A@B}(x) \leq \frac{1}{\sqrt{2}}.(\sigma_{2,A}(x) + \sigma_{2,B}(x))$;

(2.7) $\sigma_{2,\square A}(x) \leq 1$;

(2.8) $\sigma_{2,\diamond A}(x) \leq 1$;

(2.9) $\sigma_{2,C(A)}(x) \leq \max\limits_{x \in E} \sigma_{2,A}(x)$;

(2.10) $\sigma_{2,\mathcal{I}(A)}(x) \geq \min\limits_{x \in E} \sigma_{2,A}(x)$;

(2.11) $\sigma_{2,D_\alpha}(x) \geq \sigma_{2,A}(x)$, for every $\alpha \in [0,1]$;

(2.12) $\sigma_{2,F_{\alpha,\beta}}(x) \geq \sigma_{2,A}(x)$, for every $\alpha, \beta \in [0,1]$ such that $\alpha + \beta \leq 1$;

(2.13) $\sigma_{2,G_{\alpha,\beta}}(x) \leq \sigma_{2,A}(x)$, for every $\alpha, \beta \in [0,1]$;

(2.14) $\sigma_{2,H_{\alpha,\beta}}(x) \geq \alpha.\sigma_{2,A}(x)$, for every $\alpha, \beta \in [0,1]$;

(2.15) $\sigma_{2,H^*_{\alpha,\beta}(A)}(x) \geq \alpha.\sigma_{2,A}(x)$, for every $\alpha, \beta \in [0,1]$;

(2.16) $\sigma_{2,J_{\alpha,\beta}}(x) \geq \beta.\sigma_{2,A}(x)$, for every $\alpha, \beta \in [0,1]$;

(2.17) $\sigma_{2,J^*_{\alpha,\beta}(A)}(x) \geq \beta.\sigma_{2,A}(x)$, for every $\alpha, \beta \in [0,1]$;

(2.18) $\sigma_{2,X_{a,b,c,d,e,f}A}(x) \geq \min(a,d).\sigma_{2,A}(x)$,

(2.19) $\sigma_{2,P_{\alpha,\beta}(A)}(x) \geq \alpha$, for every $\alpha, \beta \in [0,1]$ and $\alpha + \beta \leq 1$;

(2.20) $\sigma_{2,Q_{\alpha,\beta}(A)}(x) \geq \beta$, for every $\alpha, \beta \in [0,1]$ and $\alpha + \beta \leq 1$;

(2.21) $\sigma_{2,\boxplus A}(x) \geq \frac{1}{2}\sigma_{2,A}(x)$;

(2.22) $\sigma_{2,\boxtimes A}(x) = \frac{1}{2}\sigma_{2,A}(x)$;

(2.23) $\sigma_{2,\boxplus_\alpha A}(x) = \alpha.\sigma_{2,A}(x)$;

(2.24) $\sigma_{2,\boxtimes_\alpha A}(x) = \alpha.\sigma_{2,A}(x)$;

(2.25) $\sigma_{2,\boxplus_{\alpha,\beta}A}(x) = \alpha.\sigma_{2,A}(x)$;

(2.26) $\sigma_{2,\boxtimes_{\alpha,\beta}A}(x) = \alpha.\sigma_{2,A}(x)$;

(2.27) $\sigma_{2,\boxplus_{\alpha,\beta,\gamma}A}(x) \geq \min(\alpha,\beta).\sigma_{2,A}(x)$;

(2.28) $\sigma_{2,\boxtimes_{\alpha,\beta,\gamma}A}(x) \geq \min(\alpha,\beta).\sigma_{2,A}(x)$;

(2.29) $\sigma_{2,E_{\alpha,\beta}(A)}(x) \leq \max(\alpha.\sqrt{2\beta^2 - 2\beta + 1}, \beta.\sqrt{2\alpha^2 - 2\alpha + 1})$;

(2.30) $\sigma_{2,\boxed{\bullet}_{\alpha,\beta,\gamma,\delta}A}(x) \geq \min(\alpha,\beta).\sigma_{2,A}(x)$;

(2.31) $\sigma_{2,\boxed{\circ}_{\alpha,\beta,\gamma,\delta,\varepsilon,\zeta}A}(x) \geq \min(\alpha,\beta).\sigma_{2,A}(x)$.

It must also be noted that for every two IFSs A and B, and for every $x \in E$, if $A = B$, then $\sigma_{2,A}(x) = \sigma_{2,B}(x)$, and also, that from $A \subseteq B$ it does not follow that $\sigma_{2,A}(x) \leq \sigma_{2,B}(x)$.

For example, if

$$E = \{x\}, A = \{\langle x, 0.3, 0.4\rangle\}, B = \{\langle x, 0.4, 0.09\rangle\},$$

then $A \subseteq B$, but

$$\sigma_{2,A}(x) = 0.5 > 0.41 = \sigma_{2,B}(x).$$

The *third intuitionistic fuzzy norm* norm over the elements of a given IFSs was introduced by D. Tanev in [516] and has the form,

$$\sigma_{3,A}(x) = \frac{\mu_A(x) + 1 - \nu_A(x)}{2}. \tag{7.3}$$

Its properties are similar to the above.

For a given finite universe E, and for a given IFS A the following discrete norms can be defined,

$$n_\mu(A) = \sum_{x \in E} \mu_A(x),$$

$$n_\nu(A) = \sum_{x \in E} \nu_A(x),$$

$$n_\pi(A) = \sum_{x \in E} \pi_A(x),$$

which can be extended to continuous norms. In this case the sum $\sum$ is replaced by an integral over E.

The above norms can be normalized on the interval $[0, 1]$:

$$n_\mu^*(A) = \frac{1}{card(E)} \sum_{x \in E} \mu_A(x),$$

$$n_\nu^*(A) = \frac{1}{card(E)} \sum_{x \in E} \nu_A(x),$$

$$n_\pi^*(A) = \frac{1}{card(E)} \sum_{x \in E} \pi_A(x),$$

where $card(E)$ is the cardinality of the set E.

These norms have similar properties.

In the theory of fuzzy sets (see e.g. [301]), two different types of distances are defined, generated from the following metric

$$m_A(x, y) = |\mu_A(x) - \mu_A(y)|.$$

Therefore, in this case the Hamming and Euclidean metrics coincide.

In the case of the intuitionistic fuzziness, these metrics, in the author's form (see, e.g., [39]), are

- the *intuitionistic fuzzy Hamming metric*

$$h_A(x, y) = \frac{1}{2}.(|\mu_A(x) - \mu_A(y) + \nu_A(x) - \nu_A(y)|) \tag{7.4}$$

- the *intuitionistic fuzzy Euclidean metric*

$$e_A(x, y) = \sqrt{\frac{1}{2}.((\mu_A(x) - \mu_A(y))^2 + (\nu_A(x) - \nu_A(y))^2)} \qquad (7.5)$$

When the equality

$$\nu_A(x) = 1 - \mu_A(x)$$

holds, both metrics are reduced to the metric $m_A(x, y)$. To prove that h_A and e_A are pseudo-metrics over E in the sense of [302, 421], we must prove that for every three elements $x, y, z \in E$:

$$h_A(x, y) + h_A(y, z) \geq h_A(x, z),$$
$$h_A(x, y) = h_A(y, x),$$
$$e_A(x, y) + e_A(y, z) \geq e_A(x, z),$$
$$e_A(x, y) = e_A(y, x).$$

The third equality, which characterizes the metrics (as above) does not hold. Therefore, h_A and e_A are pseudo-mertrics. The proofs of the above four equalities and inequalities are trivial.

Eulalia Szmidt and Janusz Kacprzyk introduced other forms of both metrics, as follows:

$$h_A^{S-K}(x, y)$$

$$= \frac{1}{3}.(|\mu_A(x) - \mu_A(y)| + |\nu_A(x) - \nu_A(y)| + |\pi_A(x) - \pi_A(y)|) \qquad (7.6)$$

(*Szmidt and Kacprzyk's form of intuitionistic fuzzy Hamming metric*)

$$e_A^{S-K}(x, y)$$

$$= \sqrt{\frac{1}{3}.((\mu_A(x) - \mu_A(y))^2 + (\nu_A(x) - \nu_A(y))^2 + (\pi_A(x) - \pi_A(y))^2)}. \qquad (7.7)$$

(*Szmidt and Kacprzyk's form of intuitionistic fuzzy Euclidean metric*)

The two types of distances defined for the fuzzy sets A and B are:

- the Hamming distance

$$H(A, B) = \sum_{x \in E} |\mu_A(x) - \mu_B(x)|$$

- the Euclidean distance

$$E(A, B) = \sqrt{\sum_{x \in E} (\mu_A(x) - \mu_B(x))^2}$$

These distances, transformed over the IFSs, have the respective forms:

$$H(A, B) = \frac{1}{2} \cdot \sum_{x \in E} (|\, \mu_A(x) - \mu_B(x)\, | + |\, \nu_A(x) - \nu_B(x)\, |), \qquad (7.8)$$

(*intuitionistic fuzzy Hamming distance*)

$$H(A, B)^{S-K}$$

$$= \frac{1}{3} \cdot \sum_{x \in E} (|\mu_A(x) - \mu_B(x)| + |\nu_A(x) - \nu_B(x)| + |\pi_A(x) - \pi_B(x)|), \quad (7.9)$$

(*Szmidt and Kacprzyk's form of intuitionistic fuzzy Hamming distance*)

$$E(A, B) = \sqrt{\frac{1}{2} \cdot \sum_{x \in E} ((\mu_A(x) - \mu_B(x))^2 + (\nu_A(x) - \nu_B(x))^2)}, \qquad (7.10)$$

(*intuitionistic fuzzy Euclidean distance*)

$$E(A, B)^{S-K}$$

$$= \sqrt{\frac{1}{3} \cdot \sum_{x \in E} ((\mu_A(x) - \mu_B(x))^2 + (\nu_A(x) - \nu_B(x))^2) + (\pi_A(x) - \pi_B(x))^2)}.$$

$$(7.11)$$

(*Szmidt and Kacprzyk's form of intuitionistic fuzzy Euclidean distance*)

7.2 Cantor's IF-Norms

Here, following [80], we introduce new norms. They are related to the two mentioned IFS-interpretations. These two norms are defined on the basis of one of the most important Georg Cantor's ideas in set theory and by this reason below we call them *"Cantor's intuitionistic fuzzy norms"*. They are essentially different from the Euclidean and Hamming norms, existing in fuzzy set theory.

Let $x \in E$ be fixed universe and let

$$\mu_A(x) = 0.a_1 a_2 ...$$

$$\nu_A(x) = 0.b_1 b_2 ...$$

Then, we bijectively construct the numbers

$$||x||_{\mu,\nu} = 0, a_1 b_1 a_2 b_2 ...$$

and

$$||x||_{\nu,\mu} = 0, b_1 a_1 b_2 a_2 ...$$

for which we see that
1. $||x||_{\mu,\nu}, ||x||_{\nu,\mu} \in [0,1]$
2. having both of them, we can directly reconstruct numbers $\mu_A(x)$ and $\nu_A(x)$.

Let us call numbers $||x||_{\mu,\nu}$ and $||x||_{\nu,\mu}$ Cantor norm of element $x \in E$. In some cases, these norms are denoted by $||x||_{2,\mu,\nu}$ and $||x||_{2,\nu,\mu}$ with aim to emphasize that they are related to the two-dimensional IFS-interpretation.

In the case of Szmidt-Kacprzyk interpretation, i.e., when we use a three-dimensional IFS-interpretation, for point x we have

$$\mu_A(x) = 0.a_1a_2...$$

$$\nu_A(x) = 0.b_1b_2...$$

$$\pi_A(x) = 0.c_1c_2...$$

where, of course, $\mu_A(x) + \nu_A(x) + \pi_A(x) = 1$. Now, in principle, we introduce six different Cantor norms:

$$||x||_{3,\mu,\nu,\pi} = 0.a_1b_1c_1a_2b_2c_2...,$$

$$||x||_{3,\mu,\pi,\nu} = 0.a_1c_1b_1a_2c_2b_2...,$$

$$||x||_{3,\nu,\mu,\pi} = 0.b_1a_1c_1b_2a_2c_2...,$$

$$||x||_{3,\nu,\pi,\mu} = 0.b_1c_1a_1b_2c_2a_2...,$$

$$||x||_{3,\pi,\mu,\nu} = 0.c_1a_1b_1c_2a_2b_2...,$$

$$||x||_{3,\pi,\nu,\mu} = 0.c_1b_1a_1c_2b_2a_2....$$

Therefore, for a given three-dimensional Cantor norm we can again reconstruct bijectively the three degrees of element $x \in E$.

In each of these cases, the norm has a standard form from mathematical point of view.

In Section **9.4**, other norms and distances are discussed.

7.3   Intuitionistic Fuzzy Histograms

Following [120], we introduce the concept of Intuitionistic Fuzzy Histogram (IFH) and the three derivatives of its, by discussing two illustrative examples.

Example 1. Let us take a sudoku puzzle that was being solved, no matter correctly or not, and let some of its cells be still vacant. For instance, the sudoku puzzle in Fig. 7.1 contains a lot of mistakes and is not complete, but it serves us well as illustration.

Let us separate the 9×9 grid into nine 3×3 sub-grids, and let us arrange vertically these sub-grids one over another (see Fig. 7.2). Let the rows and columns of each sub-grid be denoted by "i" and "j", and let each of the 9 cells in a sub-grid be denoted by the pair "(i,j)", where $i, j = 1, 2, 3$. Let us design the following table from Fig. 7.3, in which over the "(i, j)" indices, that correspond to a cell, nine fields be placed, coloured respectively in white if the digit in the cell is even; black – if the digit is odd; and half-white half-black if no digit has been entered in the cell.

Now let us rearrange the table fields in a way that the black ones are shifted to the bottom positions, the black-and-white cells are placed in the middle and the white fields float to top. Thus, we obtain Fig. 7.4.

This new table has the appearance of a histogram and we can juxtapose to its columns the pair of real numbers $\langle \frac{p}{9}, \frac{q}{9} \rangle$, where p and q are respectively the numbers of the white and the black sudoku cells, while $9 - p - q$ is the number of the empty cells. Let us call this object an intuitionistic fuzzy histogram. It gives us an idea of the kinds of numbers in the sudoku placeholders, and is clearer than the one we would have if we used a standard histogram. Back to the example from Figure 4, the values of the individual columns will be, respectively: $\langle \frac{3}{9}, \frac{3}{9} \rangle$, $\langle \frac{3}{9}, \frac{4}{9} \rangle$, $\langle \frac{4}{9}, \frac{3}{9} \rangle$, $\langle \frac{5}{9}, \frac{4}{9} \rangle$, $\langle 1, 0 \rangle$, $\langle \frac{4}{9}, \frac{2}{9} \rangle$, $\langle \frac{4}{9}, \frac{4}{9} \rangle$, $\langle \frac{5}{9}, \frac{4}{9} \rangle$, $\langle \frac{1}{9}, \frac{4}{9} \rangle$.

In the case of IFH, several situations may possibly rise:

1. The black-and-white cells count as white cells. Then we obtain the histogram on Fig. 7.5. We call it "N-histogram" by analogy with the modal operator "necessity" from Section 4.1.

2. The black-and-white cells will be counted as half cells each, so that two mixed cells yield one black and one white cell. Then we obtain the histogram from Fig. 7.6. We call this histogram "A-histogram", meaning that its values are average with respect to Fig. 7.4.

3. The black-and-white cells count for black cells. Then we obtain the histogram from Fig. 7.7. This histogram will be called "P-histogram" by analogy with the modal operator "possibility" from Section 4.1.

Example 2. Now let us consider the chess board, part of which is illustrated in Fig. 7.8. Each couple of squares on the board is divided by a stripe of non-zero width. The board squares are denoted in the standard way by "a_1", "a_2", ..., "h_8" and they have side length of 2.

Let us place a coin of surface 1 and let us toss it n times, each time having it falling on the chess board. After each tossing, we assign the pairs $\langle a, b \rangle$ to the squares on which the coin has fallen, where a denotes the surface of the coin that belongs to the respective chess square, while b is the surface of the coin that lies on one or more neighbouring squares. Obviously, $a + b < 1$.

In this example, it is possible to have two specific cases:

- If the tossed coin falls on one square only, then it will be assigned the pair of values $\langle 1, 0 \rangle$. This case is possible, because the radius of the coin is $\sqrt{\frac{1}{\pi}}$ while its diameter is $2\sqrt{\frac{1}{\pi}} < 2$.
- If the tossed coin falls on a square and its neighbouring zone, without crossing another board square, then it will be assigned the pair of values $\langle a, 0 \rangle$; where $0 < a < 1$ is the surface of this part of the coin that lies on the respective square.

6	1		8	3		5	4	2
4	7	9	2	1		6	3	
3	1	5		9			2	7
1	9		6	3			8	6
7	5		3	7	9	2	1	5
2	6	3	1		5	4		9
9	8	7	4	2	8	3	9	1
5	3		7	9	3	1	5	
1	2	4	3	5	6		7	

Fig. 7.1

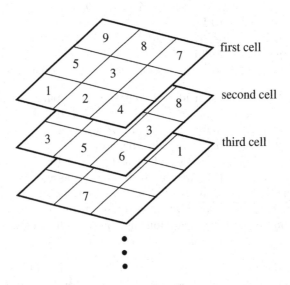

Fig. 7.2

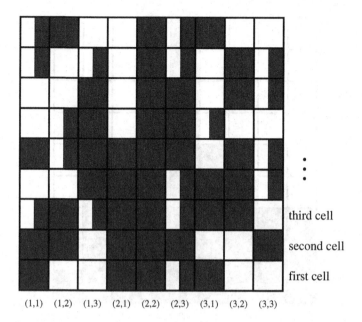

third cell

second cell

first cell

(1,1) (1,2) (1,3) (2,1) (2,2) (2,3) (3,1) (3,2) (3,3)

Fig. 7.3

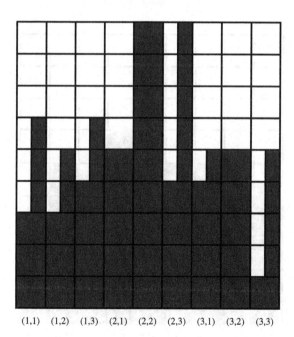

(1,1) (1,2) (1,3) (2,1) (2,2) (2,3) (3,1) (3,2) (3,3)

Fig. 7.4

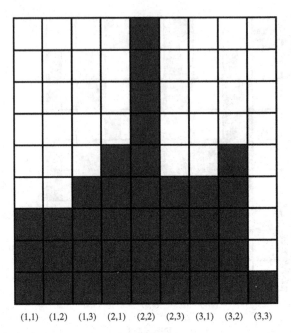

(1,1) (1,2) (1,3) (2,1) (2,2) (2,3) (3,1) (3,2) (3,3)

Fig. 7.5

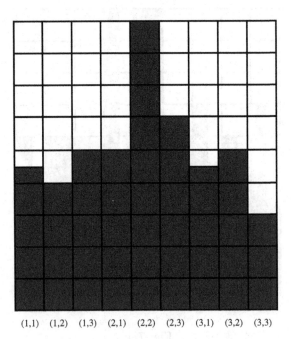

(1,1) (1,2) (1,3) (2,1) (2,2) (2,3) (3,1) (3,2) (3,3)

Fig. 7.6

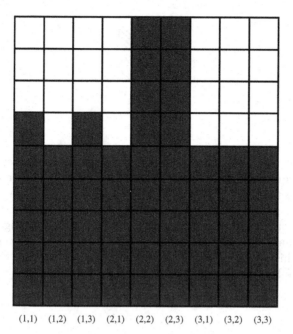

(1,1)   (1,2)   (1,3)   (2,1)   (2,2)   (2,3)   (3,1)   (3,2)   (3,3)

Fig. 7.7

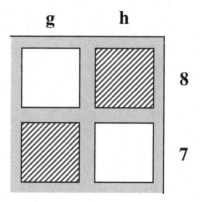

Fig. 7.8

Obviously, the tossed coin cannot fall on more than 4 squares at a time. Let us draw a table, having 64 columns, that will represent the number of the squares on the chess board, and n rows, that will stand for the number of tossings. On every toss, enter pairs of values in no more than 4 columns at a time. However, despite assigning pairs of numbers to each chess square, we may proceed by colouring the square in black (rectangle with width a), white (rectangle of width b) and leave the rest of the square white, as shown in Fig. 7.9.

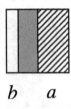

b a

Fig. 7.9

After n tossings of the coin, we rearrange the squares by placing the black and/or grey squares to the bottom of the table. Thus we obtain a histogram that is analogous to the one from the first example. We can also build a histogram of necessity, a histogram of possiblity and an average histogram.

The so constructed histograms can be easily further investigated so that their modes and medians be calculated. Let us note, for instance, that the mode of the histogram in Fig. 7.5 is the column indexed by "(2,2)", whilst in Fig. 7.7, we have two modes of the histograms represented by the columns indexed by "(2,2)" and "(2,3)".

Intuitionistic Fuzzy Relations (IFRs)

8.1 Cartesian Products over IFSs

First, we define six versions of another operation over IFSs – namely, Cartesian products of two IFSs. To introduce the concept of intuitionistic fuzzy relation, we use these operations.

Let E_1 and E_2 be two universes and let

$$A = \{\langle x, \mu_A(x), \nu_A(x)\rangle | x \in E_1\},$$
$$B = \{\langle y, \mu_B(y), \nu_B(y)\rangle | y \in E_2\},$$

be two IFSs over E_1 and over E_2, respectively.

Now, following [39], define,

$$A \times_1 B = \{\langle \langle x, y\rangle, \mu_A(x).\mu_B(y), \nu_A(x).\nu_B(y)\rangle | x \in E_1 \& y \in E_2\}, \tag{8.1}$$

$$A \times_2 B = \{\langle \langle x, y\rangle, \mu_A(x) + \mu_B(y) - \mu_A(x).\mu_B(y), \nu_A(x).\nu_B(y)\rangle \\ | x \in E_1 \& y \in E_2\}, \tag{8.2}$$

$$A \times_3 B = \{\langle \langle x, y\rangle, \mu_A(x).\mu_B(y), \nu_A(x) + \nu_B(y) - \nu_A(x).\nu_B(y)\rangle \\ | x \in E_1 \& y \in E_2\}, \tag{8.3}$$

$$A \times_4 B = \{\langle \langle x, y\rangle, \min(\mu_A(x), \mu_B(y)), \max(\nu_A(x), \nu_B(y))\rangle \\ | x \in E_1 \& y \in E_2\}, \tag{8.4}$$

$$A \times_5 B = \{\langle \langle x, y\rangle, \max(\mu_A(x), \mu_B(y)), \min(\nu_A(x), \nu_B(y))\rangle \\ | x \in E_1 \& y \in E_2\}, \tag{8.5}$$

$$A \times_6 B = \{\langle \langle x, y\rangle, \tfrac{\mu_A(x)+\mu_B(y)}{2}, \tfrac{\nu_A(x)+\nu_B(y)}{2}\rangle) | x \in E_1 \& y \in E_2\}. \tag{8.6}$$

Operation $\times_6$ was introduced by my student Velin Andonov in [8], who studied its properties.

From,

$$0 \le \mu_A(x).\mu_B(y) + \nu_A(x).\nu_B(y) \le \mu_A(x) + \nu_A(x) \le 1,$$

K.T. Atanassov: On Intuitionistic Fuzzy Sets Theory, STUDFUZZ 283, pp. 147–193.
springerlink.com   © Springer-Verlag Berlin Heidelberg 2012

it follows that $A \times_1 B$ is an IFS over the universe $E_1 \times E_2$, where "$\times$" is the classical Cartesian product on ordinary sets (E_1 and E_2). For the five other products the computations are analogous.

For every three universes E_1, E_2 and E_3 and four IFSs A, B (over E_1), C (over E_2) and D (over E_3):

(a) $(A \times C) \times D = A \times (C \times D)$, where $\times \in \{\times_1, \times_2, \times_3, \times_4, \times_5\}$.

(b) $(A \cup B) \times C = (A \times C) \cup (B \times C)$,

(c) $(A \cap B) \times C = (A \times C) \cap (B \times C)$,

(d) $C \times (A \cup B) = (C \times A) \cup (C \times B)$,

(e) $C \times (A \cap B) = (C \times A) \cap (C \times B)$,

where $\times \in \{\times_1, \times_2, \times_3, \times_4, \times_5, \times_6\}$.

For every two universes E_1 and E_2 and three IFSs A, B (over E_1) and C (over E_2):

(a) $(A + B) \times C \subseteq (A \times C) + (B \times C)$,

(b) $(A.B) \times C \supseteq (A \times C).(B \times C)$,

(c) $(A@B) \times C = (A \times C)@(B \times C)$,

(d) $C \times (A + B) \subseteq (C \times A) + (C \times B)$,

(e) $C \times (A.B) \supseteq (C \times A).(C \times B)$,

(f) $C \times (A@B) = (C \times A)@(C \times B)$,

where $\times \in \{\times_1, \times_2, \times_3, \times_6\}$.

If A is an IFS over E_1 and B is an IFS over E_2, then,

(a) $\square (A \times_1 B) \subseteq \square A \times_1 \square B$,

(b) $\lozenge(A \times_1 B) \supseteq \lozenge A \times_1 \lozenge B$,

(c) $\square (A \times_2 B) = \square A \times_2 \square B$,

(d) $\lozenge(A \times_2 B) = \lozenge A \times_2 \lozenge B$,

(e) $\square (A \times_3 B) = \square A \times_3 \square B$,

(f) $\lozenge(A \times_3 B) = \lozenge A \times_3 \lozenge B$,

(g) $\square (A \times_4 B) = \square A \times_4 \square B$,

(h) $\lozenge(A \times_4 B) = \lozenge A \times_4 \lozenge B$,

(i) $\square (A \times_5 B) = \square A \times_5 \square B$,

(j) $\lozenge(A \times_5 B) = \lozenge A \times_5 \lozenge B$.

(k) $\square (A \times_6 B) = \square A \times_6 \square B$,

(l) $\lozenge(A \times_6 B) = \lozenge A \times_6 \lozenge B$.

Formulae similar to De Morgan's laws hold for the above Cartesian products. If A is an IFS over E_1 and B is an IFS over E_2, then,

(a) $\overline{\overline{A} \times_1 \overline{B}} = A \times_1 B,$

(b) $\overline{\overline{A} \times_2 \overline{B}} = A \times_3 B,$

(c) $\overline{\overline{A} \times_3 \overline{B}} = A \times_2 B,$

(d) $\overline{\overline{A} \times_4 \overline{B}} = A \times_5 B,$

(e) $\overline{\overline{A} \times_5 \overline{B}} = A \times_4 B,$

(f) $\overline{\overline{A} \times_6 \overline{B}} = A \times_6 B.$

Therefore, operations $\times_2$ and $\times_3$; $\times_4$ and $\times_5$ are dual and the operations $\times_1$ and $\times_6$ are autodual.

The following inclusions hold for every two universes E_1 and E_2, and two IFSs A (over E_1) and B (over E_2):

(a) $A \times_3 B \subseteq A \times_1 B \subseteq A \times_2 B,$

(b) $A \times_3 B \subseteq A \times_4 B \subseteq A \times_6 B \subseteq A \times_5 B \subseteq A \times_2 B.$

For every two universes E_1 and E_2, and two IFSs A (over E_1) and B (over E_2) and for every $\alpha, \beta \in [0,1]$,

$$G_{\alpha,\beta}(A \times_1 B) = G_{\alpha/\gamma,\beta/\delta}(A) \times_1 G_{\gamma,\delta}(B),$$

for every $0 < \gamma \le 1, 0 < \delta \le 1$ for which $\frac{\alpha}{\gamma}, \frac{\beta}{\delta} \in [0,1]$.

For every two IFSs A and B, and for every $\alpha, \beta \in [0,1]$, the following relations hold:

(a) $D_\alpha(A \times_4 B) \subseteq D_\alpha(A) \times_4 D_\alpha(B),$

(b) $F_{\alpha,\beta}(A \times_4 B) \subseteq F_{\alpha,\beta}(A) \times_4 F_{\alpha,\beta}(B),$ where $\alpha + \beta \le 1,$

(c) $G_{\alpha,\beta}(A \times_4 B) = G_{\alpha,\beta}(A) \times_4 G_{\alpha,\beta}(B),$

(d) $H_{\alpha,\beta}(A \times_4 B) \subseteq H_{\alpha,\beta}(A) \times_4 H_{\alpha,\beta}(B),$

(e) $H^*_{\alpha,\beta}(A \times_4 B) \subseteq H^*_{\alpha,\beta}(A) \times_4 H^*_{\alpha,\beta}(B),$

(f) $J_{\alpha,\beta}(A \times_4 B) \subseteq J_{\alpha,\beta}(A) \times_4 J_{\alpha,\beta}(B),$

(g) $J^*_{\alpha,\beta}(A \times_4 B) \subseteq J^*_{\alpha,\beta}(A) \times_4 J^*_{\alpha,\beta}(B),$

(h) $P_{\alpha,\beta}(A \times_4 B) = P_{\alpha,\beta}(A) \times_4 P_{\alpha,\beta}(B),$

(i) $Q_{\alpha,\beta}(A \times_4 B) = Q_{\alpha,\beta}(A) \times_4 Q_{\alpha,\beta}(B),$

(j) $\blacksquare_{\alpha,\beta,\gamma,\delta}(A \times_4 B) = \blacksquare_{\alpha,\beta,\gamma,\delta}(A) \times_4 \blacksquare_{\alpha,\beta,\gamma,\delta}(B),$

(k) $D_\alpha(A \times_5 B) \supseteq D_\alpha(A) \times_5 D_\alpha(B),$

(l) $F_{\alpha,\beta}(A \times_5 B) \supseteq F_{\alpha,\beta}(A) \times_5 F_{\alpha,\beta}(B),$ where $\alpha + \beta \le 1,$

(m) $G_{\alpha,\beta}(A \times_5 B) = G_{\alpha,\beta}(A) \times_5 G_{\alpha,\beta}(B),$

(n) $H_{\alpha,\beta}(A \times_5 B) \supseteq H_{\alpha,\beta}(A) \times_5 H_{\alpha,\beta}(B),$

(o) $H^*_{\alpha,\beta}(A \times_5 B) \supseteq H^*_{\alpha,\beta}(A) \times_5 H^*_{\alpha,\beta}(B),$

(p) $J_{\alpha,\beta}(A \times_5 B) \supseteq J_{\alpha,\beta}(A) \times_5 J_{\alpha,\beta}(B),$

(q) $J_{\alpha,\beta}^*(A \times_5 B) \supseteq J_{\alpha,\beta}^*(A) \times_5 J_{\alpha,\beta}^*(B)$.

(r) $P_{\alpha,\beta}(A \times_5 B) = P_{\alpha,\beta}(A) \times_5 P_{\alpha,\beta}(B)$,

(s) $Q_{\alpha,\beta}(A \times_5 B) = Q_{\alpha,\beta}(A) \times_5 Q_{\alpha,\beta}(B)$,

(t) $\boxed{\bullet}_{\alpha,\beta,\gamma,\delta}(A \times_5 B) = \boxed{\bullet}_{\alpha,\beta,\gamma,\delta}(A) \times_5 \boxed{\bullet}_{\alpha,\beta,\gamma,\delta}(B)$,

(u) $D_{\alpha}(A \times_6 B) = D_{\alpha}(A) \times_6 D_{\alpha}(B)$,

(v) $F_{\alpha,\beta}(A \times_6 B) = F_{\alpha,\beta}(A) \times_6 F_{\alpha,\beta}(B)$, where $\alpha + \beta \leq 1$,

(w) $G_{\alpha,\beta}(A \times_6 B) = G_{\alpha,\beta}(A) \times_6 G_{\alpha,\beta}(B)$,

(x) $H_{\alpha,\beta}(A \times_6 B) = H_{\alpha,\beta}(A) \times_6 H_{\alpha,\beta}(B)$,

(y) $H_{\alpha,\beta}^*(A \times_6 B) = H_{\alpha,\beta}^*(A) \times_6 H_{\alpha,\beta}^*(B)$,

(z) $J_{\alpha,\beta}(A \times_6 B) = J_{\alpha,\beta}(A) \times_6 J_{\alpha,\beta}(B)$,

(α) $J_{\alpha,\beta}^*(A \times_6 B) = J_{\alpha,\beta}^*(A) \times_6 J_{\alpha,\beta}^*(B)$,

(β) $P_{\alpha,\beta}(A \times_6 B) \subseteq P_{\alpha,\beta}(A) \times_6 P_{\alpha,\beta}(B)$,

(γ) $Q_{\alpha,\beta}(A \times_6 B) \supseteq Q_{\alpha,\beta}(A) \times_5 Q_{\alpha,\beta}(B)$,

(δ) $\boxed{\circ}_{\alpha,\beta,\gamma,\delta,\varepsilon,\zeta}(A \times_6 B) = \boxed{\circ}_{\alpha,\beta,\gamma,\delta,\varepsilon,\zeta}(A) \times_6 \boxed{\circ}_{\alpha,\beta,\gamma,\delta,\varepsilon,\zeta}(B)$.

8.2   Index Matrix

The concept of Index Matrix (IM) was introduced in 1984 in [12, 15]. During the last 25 years, some of its properties were studied (see [75, 76]), but in general it was only used as an auxiliary tool for describing the transitions of the generalized nets (see [23, 65]), the intuitionistic fuzzy relations with finite universes, the intuitionistic fuzzy graphs with finite set of vertrices, as well as in some decision making algorithms based on intuitionistic fuzzy estimations. Some authors found an application of the IMs in the area of number theory, [337].

Following [75, 76], in Subsection **8.2.1**, we give the basic definition of an IM and the operations over two IMs, as well as some properties of IMs. In Subsection **8.2.2**, operations extending those from Subsection **8.2.1** and essentially new ones, are introduced for the first time and their properties has been discussed. Since the proofs of the formulated theorems are based on the respective definitions, only one proof is given as an illustration.

8.2.1   Basic Definitions and Properties

Following [15], the basic definitions and properties related to IMs are given.

Let I be a fixed set of indices and $\mathcal{R}$ be the set of all real numbers. By IM with index sets K and L $(K, L \subset I)$, we mean the object,

$$[K, L, \{a_{k_i, l_j}\}] \equiv \begin{array}{c|cccc} & l_1 & l_2 & \dots & l_n \\ \hline k_1 & a_{k_1, l_1} & a_{k_1, l_2} & \dots & a_{k_1, l_n} \\ k_2 & a_{k_2, l_1} & a_{k_2, l_2} & \dots & a_{k_2, l_n} \\ \vdots & & & & \\ k_m & a_{k_m, l_1} & a_{k_m, l_2} & \dots & a_{k_m, l_n} \end{array},$$

where $K = \{k_1, k_2, ..., k_m\}$, $L = \{l_1, l_2, ..., l_n\}$, for $1 \le i \le m$, and $1 \le j \le n : a_{k_i, l_j} \in \mathcal{R}$.

For the IMs $A = [K, L, \{a_{k_i, l_j}\}]$, $B = [P, Q, \{b_{p_r, q_s}\}]$, operations that are analogous of the usual matrix operations of addition and multiplication are defined, as well as other, specific ones.

(a) **addition** $A \oplus B = [K \cup P, L \cup Q, \{c_{t_u, v_w}\}]$, where

$$c_{t_u, v_w} = \begin{cases} a_{k_i, l_j}, & \text{if } t_u = k_i \in K \text{ and } v_w = l_j \in L - Q \\ & \text{or } t_u = k_i \in K - P \text{ and } v_w = l_j \in L; \\[2mm] b_{p_r, q_s}, & \text{if } t_u = p_r \in P \text{ and } v_w = q_s \in Q - L \\ & \text{or } t_u = p_r \in P - K \text{ and } v_w = q_s \in Q; \\[2mm] a_{k_i, l_j} + b_{p_r, q_s}, & \text{if } t_u = k_i = p_r \in K \cap P \\ & \text{and } v_w = l_j = q_s \in L \cap Q \\[2mm] 0, & \text{otherwise} \end{cases}$$

(b) **termwise multiplication** $A \otimes B = [K \cap P, L \cap Q, \{c_{t_u, v_w}\}]$, where

$$c_{t_u, v_w} = a_{k_i, l_j}.b_{p_r, q_s}, \text{ for } t_u = k_i = p_r \in K \cap P \text{ and } \\ v_w = l_j = q_s \in L \cap Q;$$

(c) **multiplication** $A \odot B = [K \cup (P - L), Q \cup (L - P), \{c_{t_u, v_w}\}]$, where

$$c_{t_u, v_w} = \begin{cases} a_{k_i, l_j}, & \text{if } t_u = k_i \in K \text{ and } v_w = l_j \in L - P \\[2mm] b_{p_r, q_s}, & \text{if } t_u = p_r \in P - L \text{ and } v_w = q_s \in Q \\[2mm] \displaystyle\sum_{l_j = p_r \in L \cap P} a_{k_i, l_j}.b_{p_r, q_s}, & \text{if } t_u = k_i \in K \text{ and } v_w = q_s \in Q \\[2mm] 0, & \text{otherwise} \end{cases}$$

(d) **structural subtraction** $A \ominus B = [K - P, L - Q, \{c_{t_u, v_w}\}]$, where "$-$" is the set–theoretic difference operation and

$$c_{t_u, v_w} = a_{k_i, l_j}, \text{ for } t_u = k_i \in K - P \text{ and } v_w = l_j \in L - Q.$$

(e) **multiplication with a constant** $\alpha.A = [K, L, \{\alpha.a_{k_i, l_j}\}]$, where α is a constant.

(f)   termwise subtraction $A - B = A \oplus (-1).B$.
For example, if we have the IMs X and Y

$$X = \frac{\begin{array}{c|ccc} & c & d & e \\ \hline a & 1 & 2 & 3 \\ b & 4 & 5 & 6 \end{array}}{}, \quad Y = \frac{\begin{array}{c|cc} & c & r \\ \hline a & 10 & 11 \\ p & 12 & 13 \\ q & 14 & 15 \end{array}}{},$$

then

$$X \oplus Y = \frac{\begin{array}{c|cccc} & c & d & e & r \\ \hline a & 11 & 2 & 3 & 11 \\ b & 4 & 5 & 6 & 0 \\ p & 12 & 0 & 0 & 13 \\ q & 14 & 0 & 0 & 15 \end{array}}{}$$

and

$$X \otimes Y = \frac{\begin{array}{c|c} & c \\ \hline a & 10 \end{array}}{}$$

If IM Z has the form

$$Z = \frac{\begin{array}{c|c} & u \\ \hline c & 10 \\ d & 11 \\ s & 12 \\ t & 13 \end{array}}{},$$

then

$$X \odot Z = \frac{\begin{array}{c|cc} & e & u \\ \hline a & 3 & 1 \times 10 + 2 \times 11 \\ b & 6 & 4 \times 10 + 5 \times 11 \\ s & 0 & 12 \\ t & 0 & 13 \end{array}}{} = \frac{\begin{array}{c|cc} & e & u \\ \hline a & 3 & 32 \\ b & 6 & 95 \\ s & 0 & 12 \\ t & 0 & 13 \end{array}}{}.$$

Now, it is seen that when

$$K = P = \{1, 2, ..., m\},$$

$$L = Q = \{1, 2, ..., n\},$$

we obtain the definitions for standard matrix operations. In IMs, we use different symbols as indices of the rows and columns and they, as we have seen above, give us additional information and possibilities for description.

Let $\mathcal{IM}_{\mathcal{R}}$ be the set of all IMs with their elements being real numbers, $\mathcal{IM}_{\{0,1\}}$ be the set of all $(0, 1)$-IMs. i.e., IMs with elements only 0 or 1, and $\mathcal{IM}_{\mathcal{P}}$ be the class of all IMs with elements – predicates[1]. The problem with the "zero"-IM is more complex than in the standard matrix case. We introduce "zero"-IM for $\mathcal{IM}_{\mathcal{R}}$ as

---

[1] All IMs over the class of the predicates also generate a class in Neuman-Bernaus-Gödel set theoretical sense.

$$I_0 = [K, L, \{0\}],$$

an IM whose elements are equal to 0 and $K, L \subset I$ are arbitrary index sets, as well as the IM

$$I_\emptyset = [\emptyset, \emptyset, \{a_{k_i, l_j}\}].$$

In the second case there are no matrix cells where the elements a_{k_i, l_j} may be inserted. In both cases, for each IM $A = [K, L, \{b_{k_i, l_j}\}]$ and for $I_0 = [K, L, \{0\}]$ with the same index sets, we have,

$$A \oplus I_0 = A = I_0 \oplus A.$$

The situation with $\mathcal{IM}_{\{0,1\}}$ is similar, while in the case of $\mathcal{IM}_\mathcal{P}$ the "zero"-IM can be either IM

$$I_f = [K, L, \{\text{``}false\text{''}\}],$$

or the IM

$$I_\emptyset = [\emptyset, \emptyset, \{a_{k_i, l_j}\}]$$

where the elements a_{k_i, l_j} are arbitrary predicates.

Let

$$I_1 = [K, L, \{1\}]$$

denote the IM, whose elements are equal to 1, and where $K, L \subset I$ are arbitrary index sets.

The operations defined above are oriented to IMs, whose elements are real or complex numbers. Let us denote these operations, respectively, by $\oplus_+$, $\otimes_\times$, $\odot_{+,.}$, $\ominus_-$.

The following properties of the IM are discussed in [15].

Theorem 8.1: (a) $\langle \mathcal{IM}_\mathcal{R}, \oplus_+ \rangle$ is a commutative semigroup,
(b) $\langle \mathcal{IM}_\mathcal{R}, \otimes_\times \rangle$ is a commutative semigroup,
(c) $\langle \mathcal{IM}_\mathcal{R}, \odot_{+,\times} \rangle$ is a semigroup,
(d) $\langle \mathcal{IM}_\mathcal{R}, \oplus_+, I_\emptyset \rangle$ is a commutative monoid.

8.2.2 Other Definitions and Properties

Now, a series of new operations and relations over IMs are introduced. They have been collected by the author during the last 10-15 years, but now they are published for the first time.

8.2.2.1. Modifications of the IM-operations

It is well-known that the $(0, 1)$-matrices have applications in the areas of discrete mathematics and combinatorial analysis. When we choose to work with this kind of matrices, the above operations have the following forms.

(a') $A \oplus_{\max} B = [K \cup P, L \cup Q, \{c_{t_u, v_w}\}]$, where

$$
c_{t_u,v_w} =
\begin{cases}
a_{k_i,l_j}, & \text{if } t_u = k_i \in K \text{ and } v_w = l_j \in L - Q \\
& \text{or } t_u = k_i \in K - P \text{ and } v_w = l_j \in L; \\[2ex]
b_{p_r,q_s}, & \text{if } t_u = p_r \in P \text{ and } v_w = q_s \in Q - L \\
& \text{or } t_u = p_r \in P - K \text{and } v_w = q_s \in Q; \\[2ex]
\max(a_{k_i,l_j}, b_{p_r,q_s}), & \text{if } t_u = k_i = p_r \in K \cap P \\
& \text{and } v_w = l_j = q_s \in L \cap Q \\[2ex]
0, & \text{otherwise}
\end{cases}
$$

(b') $A \otimes_{\min} B = [K \cap P, L \cap Q, \{c_{t_u,v_w}\}]$, where

$$
c_{t_u,v_w} = \min(a_{k_i,l_j}, b_{p_r,q_s}), \text{ for } t_u = k_i = p_r \in K \cap P \text{ and } v_w = l_j = q_s \in L \cap Q;
$$

(c') $A \odot_{\max,\min} B = [K \cup (P - L), Q \cup (L - P), \{c_{t_u,v_w}\}]$, where

$$
c_{t_u,v_w} =
\begin{cases}
a_{k_i,l_j}, & \text{if } t_u = k_i \in K \\
& \text{and } v_w = l_j \in L - P \\[2ex]
b_{p_r,q_s}, & \text{if } t_u = p_r \in P - L \\
& \text{and } v_w = q_s \in Q \\[2ex]
\max\limits_{l_j = p_r \in L \cap P} \min(a_{k_i,l_j}, b_{p_r,q_s}), & \text{if } t_u = k_i = p_r \in K \\
& \text{and } v_w = q_s \in Q \\[2ex]
0, & \text{otherwise}
\end{cases}
$$

Operation (d) from Subsection **8.2.1** preserves its form, operation (e) is possible only in the case when $\alpha \in \{0, 1\}$, while operation (f) is impossible.

The three operations are applicable also to the IMs, whose elements are real numbers.

Theorem 8.2: (a) $\langle \mathcal{IM}_{\{0,1\}}, \oplus_{\max} \rangle$ is a commutative semigroup,

(b) $\langle \mathcal{IM}_{\{0,1\}}, \otimes_{\min} \rangle$ is a commutative semigroup,

(c) $\langle \mathcal{IM}_{\{0,1\}}, \odot_{\max,\min} \rangle$ is a semigroup,

(d) $\langle \mathcal{IM}_{\{0,1\}}, \oplus_{\max}, I_\emptyset \rangle$ is a commutative monoid.

When working with matrices, whose elements are sentences or predicates, the forms of the above operations become

(a'') $A \oplus_\vee B = [K \cup P, L \cup Q, \{c_{t_u,v_w}\}]$, where

$$c_{t_u,v_w} = \begin{cases} a_{k_i,l_j}, & \text{if } t_u = k_i \in K \text{ and } v_w = l_j \in L - Q \\ & \text{or } t_u = k_i \in K - P \text{ and } v_w = l_j \in L; \\[2ex] b_{p_r,q_s}, & \text{if } t_u = p_r \in P \text{ and } v_w = q_s \in Q - L \\ & \text{or } t_u = p_r \in P - K \text{ and } v_w = q_s \in Q; \\[2ex] a_{k_i,l_j} \vee b_{p_r,q_s}, & \text{if } t_u = k_i = p_r \in K \cap P \\ & \text{and } v_w = l_j = q_s \in L \cap Q \\[2ex] false, & \text{otherwise} \end{cases}$$

(b'') $A \otimes_\wedge B = [K \cap P, L \cap Q, \{c_{t_u,v_w}\}]$, where

$$c_{t_u,v_w} = a_{k_i,l_j} \wedge b_{p_r,q_s}, \text{ for } t_u = k_i = p_r \in K \cap P \text{ and} \\ v_w = l_j = q_s \in L \cap Q;$$

(c'') $A \odot_{\vee,\wedge} B = [K \cup (P - L), Q \cup (L - P), \{c_{t_u,v_w}\}]$, where

$$c_{t_u,v_w} = \begin{cases} a_{k_i,l_j}, & \text{if } t_u = k_i \in K \text{ and } v_w = l_j \in L - P \\[2ex] b_{p_r,q_s}, & \text{if } t_u = p_r \in P - L \text{ and } v_w = q_s \in Q \\[2ex] \displaystyle\bigvee_{l_j=p_r \in L \cap P}(a_{k_i,l_j} \wedge b_{p_r,q_s}), & \text{if } t_u = k_i = p_r \in K \text{ and } v_w = q_s \in Q \\[2ex] false, & \text{otherwise} \end{cases}$$

Operation (d) from Subsection **8.2.1** preserves its form, while operations (e) and (f) are impossible.

Theorem 8.3: (a) $\langle \mathcal{IM}_\mathcal{P}, \oplus_\vee \rangle$ is a commutative semigroup,

(b) $\langle \mathcal{IM}_\mathcal{P}, \otimes_\wedge \rangle$ is a commutative semigroup,

(c) $\langle \mathcal{IM}_\mathcal{P}, \odot_{\vee,\wedge} \rangle$ is a semigroup,

(d) $\langle \mathcal{IM}_\mathcal{P}, \oplus_\vee, I_\emptyset \rangle$ is a commutative monoid.

8.2.2.2. Relations over IMs

Let two IMs $A = [K, L, \{a_{k,l}\}]$ and $B = [P, Q, \{b_{p,q}\}]$ be given. We introduce the following (new) definitions where $\subset$ and $\subseteq$ denote the relations *"strong inclusion"* and *"weak inclusion"*.

The strict relation *"inclusion about dimension"* is

$$A \sqsubset_d B \text{ iff } ((K \subset P) \& (L \subset Q)) \vee (K \subseteq P) \& (L \subset Q) \vee (K \subset P) \& (L \subseteq Q))$$

$$\& (\forall k \in K)(\forall l \in L)(a_{k,l} = b_{k,l}).$$

The non-strict relation *"inclusion about dimension"* is

$$A \subseteq_d B \text{ iff } (K \subseteq P) \& (L \subseteq Q) \& (\forall k \in K)(\forall l \in L)(a_{k,l} = b_{k,l}).$$

The strict relation *"inclusion about value"* is

$$A \subset_v B \text{ iff } (K = P)\&(L = Q)\&(\forall k \in K)(\forall l \in L)(a_{k,l} < b_{k,l}).$$

The non-strict relation *"inclusion about value"* is

$$A \subseteq_v B \text{ iff } (K = P)\&(L = Q)\&(\forall k \in K)(\forall l \in L)(a_{k,l} \leq b_{k,l}).$$

The strict relation *"inclusion"* is

$$A \subset B \text{ iff } (((K \subset P)\&(L \subset Q)) \vee ((K \subseteq P)\&(L \subset Q)) \vee ((K \subset P)$$

$$\&(L \subseteq Q))) \& (\forall k \in K)(\forall l \in L)(a_{k,l} < b_{k,l}).$$

The non-strict relation *"inclusion"* is

$$A \subseteq B \text{ iff } (K \subseteq P)\&(L \subseteq Q)\&(\forall k \in K)(\forall l \in L)(a_{k,l} \leq b_{k,l}).$$

It is obvious that for every two IMs A and B,

- if $A \subset_d B$, then $A \subseteq_d B$;
- if $A \subset_v B$, then $A \subseteq_v B$;
- if $A \subset B$, $A \subseteq_d B$, or $A \subseteq_v B$, then $A \subseteq B$;
- if $A \subset_d B$ or $A \subset_v B$, then $A \subseteq B$.

8.2.2.3. Operations "Reduction" over an IM

First, we introduce operations $(k,*)$-reduction and $(*,l)$-reduction of a given IM $A = [K, L, \{a_{k_i,l_j}\}]$,

$$A_{(k,*)} = [K - \{k\}, L, \{c_{t_u,v_w}\}]$$

where

$$c_{t_u,v_w} = a_{k_i,l_j} \text{ for } t_u = k_i \in K - \{k\} \text{ and } v_w = l_j \in L$$

and

$$A_{(*,l)} = [K, L - \{l\}, \{c_{t_u,v_w}\}],$$

where

$$c_{t_u,v_w} = a_{k_i,l_j} \text{ for } t_u = k_i \in K \text{ and } v_w = l_j \in L - \{l\}.$$

Second, we define (k, l)-reduction

$$A_{(k,l)} = (A_{(k,*)})_{(*,l)} = (A_{(*,l)})_{(k,*)},$$

i.e.,

$$A_{(k,l)} = [K - \{k\}, L - \{l\}, \{c_{t_u,v_w}\}],$$

where

$$c_{t_u,v_w} = a_{k_i,l_j} \text{ for } t_u = k_i \in K - \{k\} \text{ and } v_w = l_j \in L - \{l\}.$$

For every IM A and for every $k_1, k_2 \in K$, $l_1, l_2 \in L$,

$$(A_{(k_1,l_1)})_{(k_2,l_2)} = (A_{(k_2,l_2)})_{(k_1,l_1)}.$$

Third, let $P = \{k_1, k_2, ..., k_s\} \subseteq K$ and $Q = \{q_1, q_2, ..., q_t\} \subseteq L$. Now, we define the following three operations:

$$A_{(P,*)} = (...((A_{(k_1,*)})_{(k_2,*)})...)_{(k_s,*)},$$

$$A_{(*,Q)} = (...((A_{(*,l_1)})_{(*,l_2)})...)_{(*,l_t)},$$

$$A_{(P,Q)} = (A_{(P,*)})_{(*,Q)} = (A_{(*,Q)})_{(P,*)}.$$

Obviously,

$$A_{(K,L)} = I_\emptyset,$$

$$A_{(\emptyset,\emptyset)} = A.$$

For every two IMs $A = [K, L, \{a_{k_i,l_j}\}]$ and $B = [P, Q, \{b_{p_r,q_s}\}]$:

$$A \subseteq_d B \quad \text{iff} \quad A = B_{(P-K,Q-L)}.$$

Let $A \subseteq_d B$. Therefore, $K \subseteq P$ and $L \subseteq Q$, and for every $k \in K, l \in L$: $a_{k,l} = b_{k,l}$. From the definition,

$$B_{(P-K,Q-L)} = (...((B_{(p_1,q_1)})_{(p_1,q_2)})...)_{(p_r,q_s)},$$

where $p_1, p_2, ..., p_r \in P - K$, i.e., $p_1, p_2, ..., p_r \in P$, and $p_1, p_2, ..., p_r \notin K$, and $q_1, q_2, ..., q_s \in Q - L$, i.e., $q_1, q_2, ..., q_s \in Q$, and $q_1, q_2, ..., q_s \notin L$. Therefore,

$$B_{(P-K,Q-L)} = [P - (P - K), Q - (Q - L), \{b_{k,l}\}]$$

$$= [K, L, \{b_{k,l}\}] = [K, L, \{a_{k,l}\}] = A,$$

because, by definition the elements of the two IMs which are indexed by equal symbols coincide.

Conversely, if $A = B_{(P-K,Q-L)}$, then

$$A = B_{(P-K,Q-L)} \subseteq_d B_{\emptyset,\emptyset} = B.$$

8.2.2.4.   Operation "Projection" over an IM

Let $M \subseteq K$ and $N \subseteq L$. Then,

$$pr_{M,N} A = [M, N, \{b_{k_i,l_j}\}],$$

where

$$(\forall k_i \in M)(\forall l_j \in N)(b_{k_i,l_j} = a_{k_i,l_j}).$$

For every IM A, and sets $M_1 \subseteq M_2 \subseteq K$ and $N_1 \subseteq N_2 \subseteq L$, the equality

$$pr_{M_1,N_1}(pr_{M_2,N_2} A) = pr_{M_1,N_1} A$$

holds.

8.2.2.5. Hierarhical Operations over IMs

Let A be an ordinary IM, and let its element a_{k_f,e_g} be an IM by itself,

$$a_{k_f,l_g} = [P, Q, \{b_{p_r,q_s}\}],$$

where

$$K \cap P = L \cap Q = \emptyset.$$

Here, introduce the following hierarchical operation

$$A|(a_{k_f,l_g}) = [(K - \{k_f\}) \cup P, (L - \{l_g\}) \cup Q, \{c_{t_u,v_w}\}],$$

where

$$
c_{t_u,v_w} = \begin{cases}
a_{k_i,l_j}, & \text{if } t_u = k_i \in K - \{k_f\} \text{ and } v_w = l_j \in L - \{l_g\} \\
b_{p_r,q_s}, & \text{if } t_u = p_r \in P \text{ and } v_w = q_s \in Q \\
0, & \text{otherwise}
\end{cases}
$$

Assume that, if a_{k_f,l_g} is not an element of IM A, then

$$A|(a_{k_f,l_g}) = A.$$

Therefore,

$$A|(a_{k_f,l_g})$$

$$=$$

	l_1	$\cdots$	l_{g-1}	q_1	$\cdots$	q_u	l_{g+1}	$\cdots$	l_n
k_1	a_{k_1,l_1}	$\cdots$	$a_{k_1,l_{g-1}}$	0	$\cdots$	0	$a_{k_1,l_{g+1}}$	$\cdots$	a_{k_1,l_n}
$\vdots$	$\vdots$	$\vdots$	$\vdots$	$\vdots$	$\vdots$	$\vdots$	$\vdots$	$\vdots$	$\vdots$
k_{f-1}	a_{k_{f-1},l_1}	$\cdots$	$a_{k_{f-1},l_{g-1}}$	0	$\cdots$	0	$a_{k_{f-1},l_{g+1}}$	$\cdots$	a_{k_{f-1},l_n}
p_1	0	$\cdots$	0	b_{p_1,q_1}	$\cdots$	b_{p_1,q_v}	0	$\cdots$	0
$\vdots$	$\vdots$	$\vdots$	$\vdots$	$\vdots$	$\vdots$	$\vdots$	$\vdots$	$\vdots$	$\vdots$
p_u	0	$\cdots$	0	b_{p_u,q_1}	$\cdots$	b_{p_u,q_v}	0	$\cdots$	0
k_{f+1}	a_{k_{f+1},l_1}	$\cdots$	$a_{k_{f+1},l_{g-1}}$	0	$\cdots$	0	$a_{k_{f+1},l_{g+1}}$	$\cdots$	a_{k_{f+1},l_n}
$\vdots$	$\vdots$	$\vdots$	$\vdots$	$\vdots$	$\vdots$	$\vdots$	$\vdots$	$\vdots$	$\vdots$
k_m	a_{k_m,l_1}	$\cdots$	$a_{k_m,l_{g-1}}$	0	$\cdots$	0	$a_{k_m,l_{g+1}}$	$\cdots$	a_{k_m,l_n}

From this form of the IM $A|(a_{k_f,l_g})$ we see that for the hierarchical operation the following equality holds.

$$A|(a_{k_f,l_g}) = (A \ominus [\{k_f\}, \{l_g\}, \{0\}]) \oplus a_{k_f,l_g}.$$

We see that the elements $a_{k_f,l_1}, a_{k_f,l_2}, ..., a_{k_f,l_{g-1}}, a_{k_f,l_{g+1}}, ..., a_{k_f,l_n}$ in the IM A now are changed with 0. Therefore, in a result of this operation information is lost.

Below, we modify the hierarchical operation, so, all information from the IMs, participating in it, to be kept. The new form of this operation, for the above defined IM A and its fixed element a_{k_f,l_g}, is

$$A|^*(a_{k_f,l_g})$$

	l_1	$\cdots$	l_{g-1}	q_1	$\cdots$	q_u	l_{g+1}	$\cdots$	l_n
k_1	a_{k_1,l_1}	$\cdots$	$a_{k_1,l_{g-1}}$	a_{k_1,l_g}	$\cdots$	a_{k_1,l_g}	$a_{k_1,l_{g+1}}$	$\cdots$	a_{k_1,l_n}
$\vdots$	$\vdots$	$\vdots$	$\vdots$	$\vdots$	$\vdots$	$\vdots$	$\vdots$	$\vdots$	$\vdots$
k_{f-1}	a_{k_{f-1},l_1}	$\cdots$	$a_{k_{f-1},l_{g-1}}$	a_{k_{f-1},l_g}	$\cdots$	a_{k_{f-1},l_g}	$a_{k_{f-1},l_{g+1}}$	$\cdots$	a_{k_{f-1},l_n}
p_1	a_{k_f,l_1}	$\cdots$	$a_{k_f,l_{g-1}}$	b_{p_1,q_1}	$\cdots$	b_{p_1,q_v}	$a_{k_f,l_{g+1}}$	$\cdots$	a_{k_f,l_n}
$\vdots$	$\vdots$	$\vdots$	$\vdots$	$\vdots$	$\vdots$	$\vdots$	$\vdots$	$\vdots$	$\vdots$
p_u	a_{k_f,l_1}	$\cdots$	$a_{k_f,l_{g-1}}$	b_{p_u,q_1}	$\cdots$	b_{p_u,q_v}	$a_{k_f,l_{g+1}}$	$\cdots$	a_{k_f,l_n}
k_{f+1}	a_{k_{f+1},l_1}	$\cdots$	$a_{k_{f+1},l_{g-1}}$	a_{k_{f+1},l_g}	$\cdots$	a_{k_{f+1},l_g}	$a_{k_{f+1},l_{g+1}}$	$\cdots$	a_{k_{f+1},l_n}
$\vdots$	$\vdots$	$\vdots$	$\vdots$	$\vdots$	$\vdots$	$\vdots$	$\vdots$	$\vdots$	$\vdots$
k_m	a_{k_m,l_1}	$\cdots$	$a_{k_m,l_{g-1}}$	a_{k_m,l_g}	$\cdots$	a_{k_m,l_g}	$a_{k_m,l_{g+1}}$	$\cdots$	a_{k_m,l_n}

Now, the following equality is valid.

$$A|^*(a_{k_f,l_g}) = (A \ominus [\{k_f\}, \{l_g\}, \{0\}]) \oplus a_{k_f,l_g} \oplus [P, L - \{l_g\}, \{c_{x,l_j}\}]$$

$$\oplus [K - \{k_f\}Q, \{d_{k_i,y}\}],$$

where for each $t \in P$ and for each $l_j \in L - \{l_g\}$,

$$c_{x,l_j} = a_{k_f,l_j}$$

and for each $k_i \in K - \{k_f\}$ and for each $y \in Q$,

$$d_{k_i,y} = a_{k_i,l_g},$$

Let for $i = 1, 2, ..., s$,

$$a^i_{k_{i,f},l_{i,g}} = [P_i, Q_i, \{b^i_{p_{i,r},q_{i,s}}\}],$$

where for every i, j $(1 \le i < j \le s)$,

$$P_i \cap P_j = Q_i \cap Q_j = \emptyset,$$

$$P_i \cap K = Q_i \cap L = \emptyset.$$

Then, for $k_{1,f}, k_{2,f}, ..., k_{s,f} \in K$ and $l_{1,g}, l_{2,g}, ..., l_{s,g} \in L$,

$$A|(a^1_{k_{1,f},l_{1,g}}, a^2_{k_{2,f},l_{2,g}}, ..., a^s_{k_{s,f},l_{s,g}})$$

$$= (...((A|(a^1_{k_{1,f},l_{1,g}}))|(a^2_{k_{2,f},l_{2,g}}))...)|(a^s_{k_{s,f},l_{s,g}})$$

and

$$A|^*(a^1_{k_1,f,l_1,g}, a^2_{k_2,f,l_2,g}, \ldots, a^s_{k_s,f,l_s,g})$$
$$= (\ldots((A|^*(a^1_{k_1,f,l_1,g}))|^*(a^2_{k_2,f,l_2,g}))\ldots)|^*(a^s_{k_s,f,l_s,g}).$$

Let the IM A be given and let for $i = 1, 2$: $k_{1,f} \neq k_{2,f}$ and $l_{1,g} \neq l_{2,g}$ and

$$a^i_{k_{i,f},l_{i,g}} = [P_i, Q_i, \{b^i_{p_{i,r},q_{i,s}}\}],$$

where

$$P_1 \cap P_2 = Q_1 \cap Q_2 = \emptyset,$$
$$P_i \cap K = Q_i \cap L = \emptyset.$$

Then,

$$A|(a^1_{k_1,f,l_1,g}, a^2_{k_2,f,l_2,g}) = A|(a^2_{k_2,f,l_2,g}, a^1_{k_1,f,l_1,g})$$

and

$$A|^*(a^1_{k_1,f,l_1,g}, a^2_{k_2,f,l_2,g}) = A|^*(a^2_{k_2,f,l_2,g}, a^1_{k_1,f,l_1,g}).$$

Let A and a_{k_f,l_g} be as above, let b_{m_d,n_e} be the element of the IM a_{k_f,l_g}, and let

$$b_{m_d,n_e} = [R, S, \{c_{t_u,v_w}\}],$$

where

$$K \cap R = L \cap S = P \cap R = Q \cap S = K \cap P = L \cap Q = \emptyset.$$

Then,

$$(A|(a_{k_f,l_g}))|(b_{m_d,n_e})$$
$$= [(K - \{k_f\}) \cup (P - \{m_d\}) \cup R, (L - \{l_g\}) \cup (Q - \{n_e\} \cup S \{\alpha_{\beta_\gamma,\delta_\varepsilon}\}],$$

where

$$\alpha_{\beta_\gamma,\delta_\varepsilon} = \begin{cases} a_{k_i,l_j}, & \text{if } \beta_\gamma = k_i \in K - \{k_f\} \text{ and } \delta_\varepsilon = l_j \in L - \{l_g\} \\ \\ b_{p_r,q_s}, & \text{if } \beta_\gamma = p_r \in P - \{m_d\} \text{ and } \delta_\varepsilon = q_s \in Q - \{n_e\} \\ \\ c_{t_u,v_w}, & \text{if } \beta_\gamma = t_u \in R \text{ and } \delta_\varepsilon = v_w \in S \\ \\ 0, & \text{otherwise} \end{cases}$$

For the above IMs A, a_{k_f,l_g} and b_{m_d,n_e}

$$(A|(a_{k_f,l_g}))|(b_{m_d,n_e}) = A|((a_{k_f,l_g})|(b_{m_d,n_e})).$$

8.2.2.6.   Operation "Substitution" over an IM

Let IM $A = [K, L, \{a_{k,l}\}]$ be given.

First, local substitution over the IM is defined for the couples of indices (p, k) and/or (q, l), respectively, by

$$[\frac{p}{k}]A = [(K - \{k\}) \cup \{p\}, L, \{a_{k,l}\}],$$

$$[\frac{q}{l}]A = [K, (L - \{l\}) \cup \{q\}, \{a_{k,l}\}],$$

Secondly,

$$[\frac{p\ q}{k\ l}]A = [\frac{p}{k}][\frac{q}{l}]A,$$

i.e.

$$[\frac{p\ q}{k\ l}]A = [(K - \{k\}) \cup \{p\}, (L - \{l\}) \cup \{q\}, \{a_{k,l}\}].$$

Obviously, for the above indices k, l, p, q,

$$[\frac{k}{p}]([\frac{p}{k}]A) = [\frac{l}{q}]([\frac{q}{l}]A) = [\frac{k\ l}{p\ q}]([\frac{p\ q}{k\ l}]A) = A,$$

Let the sets of indices $P = \{p_1, p_2, ..., p_m\}$, $Q = \{q_1, q_2, ..., q_n\}$ be given. Third, for them define sequentially,

$$[\frac{P}{K}]A = [\frac{p_1\ p_2}{k_1\ k_2}...\frac{p_n}{k_n}]A,$$

$$[\frac{Q}{L}]A = ([\frac{q_1\ q_2}{l_1\ l_2}...\frac{q_n}{l_n}]A),$$

$$[\frac{P\ Q}{K\ L}]A) = [\frac{P}{K}][\frac{Q}{L}]A,$$

i.e.,

$$[\frac{P\ Q}{K\ L}]A = [\frac{p_1\ p_2}{k_1\ k_2}...\frac{p_m}{k_m}\frac{q_1\ q_2}{l_1\ l_2}...\frac{q_n}{l_n}]A = [P, Q, \{a_{k,l}\}]$$

Obviously, for the sets K, L, P, Q:

$$[\frac{K}{P}]([\frac{P}{K}]A) = [\frac{L}{Q}]([\frac{Q}{L}]A) = [\frac{K\ L}{P\ Q}]([\frac{P\ Q}{K\ L}]A) = A.$$

For every four sets of indices P_1, P_2, Q_1, Q_2

$$[\frac{P_2\ Q_2}{P_1\ Q_1}][\frac{P_1\ Q_1}{K\ L}]A = [\frac{P_2\ Q_2}{K\ L}]A.$$

8.2.3   Intuitionistic Fuzzy IMs (IFIMs)

In this Section, basic definitions and properties related to IFIMs are given, by extending the results from the previous Section.

8.2.3.1.   Basic Definitions and Properties

Now, the new object – the IFIM – has the form

$$[K, L, \{\langle \mu_{k_i,l_j}, \nu_{k_i,l_j} \rangle\}]$$

$$\equiv \begin{array}{c|cccc} & l_1 & l_2 & \ldots l_n \\ \hline k_1 & \langle \mu_{k_1,l_1}, \nu_{k_1,l_1} \rangle & \langle \mu_{k_1,l_2}, \nu_{k_1,l_2} \rangle & \cdots & \langle \mu_{k_1,l_n}, \nu_{k_1,l_n} \rangle \\ k_2 & \langle \mu_{k_2,l_1}, \nu_{k_2,l_1} \rangle & \langle \mu_{k_2,l_2}, \nu_{k_2,l_2} \rangle & \cdots & \langle \mu_{k_2,l_n}, \nu_{k_2,l_n} \rangle \\ \vdots & & & \\ k_m & \langle \mu_{k_m,l_1}, \nu_{k_m,l_1} \rangle & \langle \mu_{k_m,l_2}, \nu_{k_m,l_2} \rangle & \cdots & \langle \mu_{k_m,l_n}, \nu_{k_m,l_n} \rangle \end{array},$$

where for every $1 \leq i \leq m, 1 \leq j \leq n$: $0 \leq \mu_{k_i,l_j}, \nu_{k_i,l_j}, \mu_{k_i,l_j} + \nu_{k_i,l_j} \leq 1$.

For the IFIMs $A = [K, L, \{\langle \mu_{k_i,l_j}, \nu_{k_i,l_j} \rangle\}], B = [P, Q, \{\langle \rho_{p_r,q_s}, \sigma_{p_r,q_s} \rangle\}]$, operations that are analogous of the usual matrix operations of addition and multiplication are defined, as well as other specific ones.

(a)   **addition** $A \oplus B = [K \cup P, L \cup Q, \{\langle \varphi_{t_u,v_w}, \psi_{t_u,v_w} \rangle\}]$, where

$$\langle \varphi_{t_u,v_w}, \psi_{t_u,v_w} \rangle =$$

$$= \begin{cases} \langle \mu_{k_i,l_j}, \nu_{k_i,l_j} \rangle, & \text{if } t_u = k_i \in K \text{ and } v_w = l_j \in L - Q \\ & \text{or } t_u = k_i \in K - P \text{ and } v_w = l_j \in L; \\[2ex] \langle \rho_{p_r,q_s}, \sigma_{p_r,q_s} \rangle, & \text{if } t_u = p_r \in P \text{ and } v_w = q_s \in Q - L \\ & \text{or } t_u = p_r \in P - K \text{ and } v_w = q_s \in Q; \\[2ex] \langle \max(\mu_{k_i,l_j}, \rho_{p_r,q_s}), & \text{if } t_u = k_i = p_r \in K \cap P \\ \min(\nu_{k_i,l_j}, \sigma_{p_r,q_s}) \rangle, & \text{and } v_w = l_j = q_s \in L \cap Q \\[2ex] \langle 0, 1 \rangle, & \text{otherwise} \end{cases}$$

(b)   **termwise multiplication** $A \otimes B = [K \cap P, L \cap Q, \langle \varphi_{t_u,v_w}, \psi_{t_u,v_w} \rangle\}]$, where

$$\langle \varphi_{t_u,v_w}, \psi_{t_u,v_w} \rangle = \langle \min(\mu_{k_i,l_j}, \rho_{p_r,q_s}), \max(\nu_{k_i,l_j}, \sigma_{p_r,q_s}) \rangle,$$

if $t_u = k_i = p_r \in K \cap P$ and $v_w = l_j = q_s \in L \cap Q$.

(c)   **multiplication** $A \odot B = [K \cup (P - L), Q \cup (L - P), \langle \varphi_{t_u,v_w}, \psi_{t_u,v_w} \rangle\}]$, where

$$\langle \varphi_{t_u,v_w}, \psi_{t_u,v_w} \rangle =$$

$$= \begin{cases} \langle \mu_{k_i,l_j}, \nu_{k_i,l_j} \rangle, & \text{if } t_u = k_i \in K \text{ and } v_w = l_j \in L - P \\[2mm] \langle \rho_{p_r,q_s}, \sigma_{p_r,q_s} \rangle, & \text{if } t_u = p_r \in P - L \text{ and } v_w = q_s \in Q \\[2mm] \langle \displaystyle\max_{l_j = p_r \in L \cap P} (\min(\mu_{k_i,l_j}, \rho_{p_r,q_s})), & \text{if } t_u = k_i \in K \text{ and } v_w = q_s \in Q \\[2mm] \quad \displaystyle\min_{l_j = p_r \in L \cap P} (\max(\nu_{k_i,l_j}, \sigma_{p_r,q_s})) \rangle, & \\[2mm] \langle 0, 1 \rangle, & \text{otherwise} \end{cases}$$

(d) **structural subtraction** $A \ominus B = [K - P, L - Q, \{\langle \varphi_{t_u,v_w}, \psi_{t_u,v_w} \rangle\}]$, where "$-$" is the set–theoretic difference operation and

$$\langle \varphi_{t_u,v_w}, \psi_{t_u,v_w} \rangle = \langle \mu_{k_i,l_j}, \nu_{k_i,l_j} \rangle, \text{ for } t_u = k_i \in K - P \text{ and } v_w = l_j \in L - Q.$$

(e) **negation of an IFIM** $\neg A = [K, L, \{\neg \langle \mu_{k_i,l_j}, \nu_{k_i,l_j} \rangle\}]$, where $\neg$ is one of the negations, defined in Subsection **9.2.1**.

(f) **termwise subtraction** $A - B = A \oplus \neg B$.

For example, consider two IFIMs X and Y

$$X = \begin{array}{c|cc} & c & d \\ \hline a & \langle 0.5, 0.3 \rangle & \langle 0.4, 0.2 \rangle \\ b & \langle 0.1, 0.8 \rangle & \langle 0.7, 0.1 \rangle \end{array}, \quad Y = \begin{array}{c|cc} & c & g \\ \hline a & \langle 0.3, 0.1 \rangle & \langle 0.6, 0.2 \rangle \\ e & \langle 0.3, 0.6 \rangle & \langle 0.3, 0.6 \rangle \\ f & \langle 0.5, 0.2 \rangle & \langle 0.6, 0.1 \rangle \end{array},$$

then

$$X \oplus Y = \begin{array}{c|ccc} & c & d & g \\ \hline a & \langle 0.5, 0.1 \rangle & \langle 0.4, 0.2 \rangle & \langle 0.6, 0.2 \rangle \\ b & \langle 0.1, 0.8 \rangle & \langle 0.7, 0.1 \rangle & \langle 0.0, 1.0 \rangle \\ e & \langle 0.3, 0.6 \rangle & \langle 0.0, 1.0 \rangle & \langle 0.3, 0.6 \rangle \\ f & \langle 0.5, 0.2 \rangle & \langle 0.0, 1.0 \rangle & \langle 0.6, 0.1 \rangle \end{array}.$$

Obviously when

$$K = P = \{1, 2, ..., m\},$$
$$L = Q = \{1, 2, ..., n\}$$

we obtain the definitions for standard matrix operations with intuitionistic fuzzy pairs. In the IFIM case, we use different symbols as indices of the rows and columns and they, as we have seen above, give us additional information and possibilities for description.

Let $\mathcal{IM}_{IF}$ be the set of all IFIMs with their elements being intuitionistic fuzzy pairs. The problem with the "zero"-IFIM is more complex than in the standard matrix case. We introduce "zero"-IFIM for $\mathcal{IM}_{IF}$ as the IFIM

$$I_0 = [K, L, \{\langle 0.0, 1.0 \rangle\}]$$

whose elements are equal to $\langle 0.0, 1.0 \rangle$ and $K, L \subset I$ are arbitrary index sets, as well as the IFIM

$$I_\emptyset = [\emptyset, \emptyset, \{a_{k_i,l_j}\}].$$

In the second case, there are no matrix cells where the elements a_{k_i,l_j} may be inserted. In both cases, for each IFIM $A = [K, L, \{b_{k_i,l_j}\}]$ and for I_0 with the same index sets, we obtain

$$A \oplus I_0 = A = I_0 \oplus A.$$

Let $I_1 = [K, L, \{\langle 1.0, 0.0\rangle\}]$ denote the IFIM, whose elements are equal to $\langle 1.0, 0.0\rangle$, and where $K, L \subset I$ are arbitrary index sets.

The following properties of the IFIM are valid, similar from Section **8.2.1**

Theorem 8.4: (a) $\langle \mathcal{IM_{IF}}, \oplus\rangle$ is a commutative semigroup,
(b) $\langle \mathcal{IM_{IF}}, \otimes\rangle$ is a commutative semigroup,
(c) $\langle \mathcal{IM_{IF}}, \odot\rangle$ is a semigroup,
(d) $\langle \mathcal{IM_{IF}}, \oplus, I_\emptyset\rangle$ is a commutative monoid.

8.2.3.2.   Relations over IFIMs

Let the two IFIMs $A = [K, L, \{\langle a_{k,l}, b_{k,l}\rangle\}]$ and $B = [P, Q, \{\langle c_{p,q}, d_{p,q}\rangle\}]$ be given. We introduce the following (new) definitions where $\subset$ and $\subseteq$ denote the relations *"strong inclusion"* and *"weak inclusion"*, respectively.

The strict relation *"inclusion about dimension"* is

$$A \subset_d B \text{ iff } ((K \subset P)\&(L \subset Q)) \vee (K \subseteq P)\&(L \subset Q) \vee (K \subset P)\&(L \subseteq Q))$$

$$\&(\forall k \in K)(\forall l \in L)(\langle a_{k,l}, b_{k,l}\rangle = \langle c_{k,l}, d_{k,l}\rangle).$$

The non-strict relation *"inclusion about dimension"* is

$$A \subseteq_d B \text{ iff } (K \subseteq P)\&(L \subseteq Q)\&(\forall k \in K)(\forall l \in L)(\langle a_{k,l}, b_{k,l}\rangle = \langle c_{k,l}, d_{k,l}\rangle).$$

The strict relation *"inclusion about value"* is

$$A \subset_v B \text{ iff } (K = P)\&(L = Q)\&(\forall k \in K)(\forall l \in L)(\langle a_{k,l}, b_{k,l}\rangle < \langle c_{k,l}, d_{k,l}\rangle).$$

The non-strict relation *"inclusion about value"* is

$$A \subseteq_v B \text{ iff } (K = P)\&(L = Q)\&(\forall k \in K)(\forall l \in L)(\langle a_{k,l}, b_{k,l}\rangle \leq \langle c_{k,l}, d_{k,l}\rangle).$$

The strict relation *"inclusion"* is

$$A \subset B \text{ iff } ((K \subset P)\&(L \subset Q)) \vee (K \subseteq P)\&(L \subset Q) \vee (K \subset P)\&(L \subseteq Q))$$

$$\&(\forall k \in K)(\forall l \in L)(\langle a_{k,l}, b_{k,l}\rangle < \langle c_{k,l}, d_{k,l}\rangle).$$

The non-strict relation *"inclusion"* is

$$A \subseteq B \text{ iff } (K \subseteq P)\&(L \subseteq Q)\&(\forall k \in K)(\forall l \in L)(\langle a_{k,l}, b_{k,l}\rangle \leq \langle c_{k,l}, d_{k,l}\rangle).$$

Obviously, for every two IFIMs A and B,

- if $A \subset_d B$, then $A \subseteq_d B$;
- if $A \subset_v B$, then $A \subseteq_v B$;
- if $A \subset B$, $A \subseteq_d B$, or $A \subseteq_v B$, then $A \subseteq B$;
- if $A \subset_d B$ or $A \subset_v B$, then $A \subseteq B$.

Operations "reduction", "projection" and "substitution" coincide with the respective operations defined over IMs, while hierarhical operations over IMs are not applied here.

8.3 Intuitionistic Fuzzy Relations (IFRs)

The concept of Intuitionistic Fuzzy Relation (IFR) is based on the definition of the IFSs. It was introduced in different forms and approached from different starting points, and independently, in 1984 and 1989 by the author (in two partial cases; see [13, 39]), in 1989 in [148] by Buhaescu and in 1992-1995 in [156, 154, 155, 157] by Bustince and Burillo. We must note that the approaches in the various IFR definitions differ in the different authors' researches. On the other hand, the author's results were not widely known; first he got acquainted with Stoyanova's results and after this he learned about Buhaescu's (obtained earlier); then he sent parts of the above works to Burillo and Bustince after they had obtained their own results.

Thus the idea of IFR was generated independently in four different places (Sofia and Varna in Bulgaria, Romania and Spain). The Spanish authors' approach is in some sense the most general. In the present form it includes Buhaescu's results.

First, we introduce Burillo and Bustince's definition of the concept of IFR, following [154, 155].

Let X and Y be arbitrary finite non-empty sets.

An $\circ$-IFR (or briefly, IFR, for a fixed operation $\circ \in \{\times_1, \times_2, \ldots, \times_6\}$) will mean an IFS $R \subseteq X \times Y$ of the form:

$$R = \{\langle \langle x, y \rangle, \mu_R(x, y), \nu_R(x, y) \rangle | x \in X \& y \in Y\},$$

where $\mu_R : X \times Y \to [0, 1], \nu_R : X \times Y \to [0, 1]$ are degrees of membership and non-membership as in the ordinary IFSs (or degrees of truth and falsity) of the relation R, and for all $\langle x, y \rangle \in X \times Y$,

$$0 \leq \mu_R(x, y) + \nu_R(x, y) \leq 1,$$

where the "$\times$" operation is the standard Cartesian product and the form of μ_R and ν_R is related to the form of the Cartesian product $\circ$.

Now, we introduce an index matrix approach of IFR.

Let $IFR_\circ(X, Y)$ be the set of all IFRs over the set $X \times Y$, where $X = \{x_1, x_2, \ldots, x_m\}$ and $Y = \{y_1, y_2, \ldots, y_n\}$ are fixed finite sets (universes), the $\times$ operation between them is the standard Cartesian product, and $\circ \in \{\times_1, \times_2, \ldots, \times_6\}$. Therefore, the set $R \in IFR_\circ(X, Y)$ can be represented in the form [32],

	y_1	$\cdots$	y_n
x_1	$\langle \mu_R(x_1, y_1), \nu_R(x_1, y_1) \rangle$	$\cdots$	$\langle \mu_R(x_1, y_n), \nu_R(x_1, y_n) \rangle$
x_2	$\langle \mu_R(x_2, y_1), \nu_R(x_2, y_1) \rangle$	$\cdots$	$\langle \mu_R(x_2, y_n), \nu_R(x_2, y_n) \rangle$
$\vdots$	$\cdots$		
x_m	$\langle \mu_R(x_m, y_1), \nu_R(x_m, y_1) \rangle$	$\cdots$	$\langle \mu_R(x_m, y_n), \nu_R(x_m, y_n) \rangle$

This IM-representation allows for a more pictorial description of the elements of R and their degrees of membership and non-membership. Let $R \in IFR_\circ(X_1, Y_1)$ and $S \in IFR_\circ(X_2, Y_2)$, where X_1, Y_1, X_2 and Y_2 are fixed finite sets and $X_1 \cap X_2 \cap Y_1 = X_1 \cap X_2 \cap Y_2 = X_1 \cap Y_1 \cap Y_2 = X_2 \cap Y_1 \cap Y_2 = \emptyset$.

Using the definitions of the operations over IMs, we shall define three operations over IFRs:

$$1.\ R \cup S \in IFR_\circ(X_1 \cup X_2, Y_1 \cup Y_2)$$

and has the form

	y_1	$\cdots$	y_N
x_1	$\langle \mu_{R \cup S}(x_1, y_1), \nu_{R \cup S}(x_1, y_1) \rangle$	$\cdots$	$\langle \mu_{R \cup S}(x_1, y_N), \nu_{R \cup S}(x_1, y_N) \rangle$
x_2	$\langle \mu_{R \cup S}(x_2, y_1), \nu_{R \cup S}(x_2, y_1) \rangle$	$\cdots$	$\langle \mu_{R \cup S}(x_2, y_N), \nu_{R \cup S}(x_2, y_N) \rangle$
$\vdots$			
x_M	$\langle \mu_{R \cup S}(x_M, y_1), \nu_{R \cup S}(x_M, y_1) \rangle$	$\cdots$	$\langle \mu_{R \cup S}(x_M, y_N), \nu_{R \cup S}(x_M, y_N) \rangle$

where

$$X_1 \cup X_2 = \{x_1, x_2, \ldots, x_M\} \text{ and } Y_1 \cup Y_2 = \{y_1, y_2, \ldots, y_N\}, \text{ and}$$

$$\langle \mu_{R \cup S}(x_i, y_j), \nu_{R \cup S}(x_i, y_j) \rangle = \begin{cases} \langle \mu_R(x'_a, y'_b), \nu_R(x'_a, y'_b) \rangle, \\ \quad \text{if } x_i = x'_a \in X_1 \text{ and } y_j = y'_b \in Y_1 - Y_2 \\ \quad \text{or } x_i = x'_a \in X_1 - X_2 \text{ and } y_j = y'_b \in Y_1 \\ \langle \mu_S(x''_c, y''_d), \nu_S(x''_c, y''_d) \rangle, \\ \quad \text{if } x_i = x''_c \in X_2 \text{ and } y_j = y''_d \in Y_2 - Y_1 \\ \quad \text{or } x_i = x''_c \in X_2 - X_1 \text{ and } y_j = y''_d \in Y_2 \\ \langle \max(\mu_R(x', y'), \mu_S(x'', y'')), \\ \quad \min(\nu_R(x', y'), \nu_S(x'', y'')) \rangle, \\ \quad \text{if } x_i = x'_a = x''_c \in X_1 \cap X_2 \text{ and} \\ \quad y_j = y'_b = y''_d \in Y_1 \cap Y_2 x \\ \langle 0, 1 \rangle, \text{ otherwise} \end{cases}$$

$$2.\ R \cap S \in IFR_\circ(X_1 \cap X_2, Y_1 \cap Y_2)$$

and has the form of the above IM, but with elements

$$\langle \mu_{R \cap S}(x_i, y_j), \nu_{R \cap S}(x_i, y_j) \rangle$$

$$= \langle \min(\mu_R(x', y'), \mu_S(x'', y'')), \max(\nu_R(x', y'), \nu_S(x'', y'')) \rangle,$$

where $x_i = x'_a = x''_c \in X_1 \cap X_2$ and $y_j = y'_b = y''_d \in Y_1 \cap Y_2$ (therefore $X_1 \cap X_2 = \{x_1, x_2,..., x_M\}$ and $Y_1 \cap Y_2 = \{y_1, y_2,..., y_N\}$).

3. $R \bullet S \in IFR_o(X_1 \cup (X_2 - Y_1), Y_2 \cup (Y_1 - X_2))$

and has the form of the above IM, but with elements

$$\langle \mu_{R \bullet S}(x_i, y_j), \nu_{R \bullet S}(x_i, y_j) \rangle = \begin{cases} \langle \mu_R(x'_a, y'_b), \nu_R(x'_a, y'_b) \rangle, \\ \quad \text{if } x_i = x'_a \in X_1 \text{ and } y_j = y'_b \in Y_1 - X_2 \\[2mm] \langle \mu_S(x''_c, y''_d), \nu_S(x''_c, y''_d) \rangle, \\ \quad \text{or } x_i = x''_c \in X_2 - Y_1 \text{ and } y_j = y''_d \in Y_2 \\[2mm] \langle \quad \max_{y'_b = x''_c \in Y_1 \cap X_2} \min(\mu_R(x'_a, y'_b), \mu_S(x''_c, y''_d)), \\[2mm] \quad \min_{y'_b = x''_c \in Y_1 \cap X_2} \max(\nu_R(x'_a, y'_b), \nu_S(x''_c, y''_d)) \rangle, \\[2mm] \quad \text{if } x_i = x'_a \in X_1 \text{ and } y_j = y''_d \in Y_2 \\[2mm] \langle 0, 1 \rangle, \text{ otherwise} \end{cases}$$

Therefore,

$$X_1 \cup (X_2 - Y_1) = \{x_1, x_2, \ldots, x_M\}$$

and

$$Y_2 \cup (Y_1 - X_2) = \{y_1, y_2, \ldots, y_N\}.$$

8.4 Intuitionistic Fuzzy Graphs (IFGs)

Now, we consider the applications of IFSs, IFRs and IMs to graph theory. Following [31, 33, 45, 424, 425] the concept of an *Intuitionistic Fuzzy Graph (IFG)* are introduced.

Let E_1 and E_2 be two sets. In this Section we assume that $x \in E_1$ and $y \in E_2$ and operation $\times$ denotes the standard Cartesian product operation. Therefore $\langle x, y \rangle \in E_1 \times E_2$. Let the operation $o \in \{\times_1, \times_2, \ldots, \times_6\}$.

The set

$$G^* = \{\langle \langle x, y \rangle, \mu_G(x, y), \nu_G(x, y) \rangle \, \langle x, y \rangle \in E_1 \times E_2\}$$

is called an *o*-IFG (or briefly, an IFG) if the functions $\mu_G : E_1 \times E_2 \to [0, 1]$ and $\nu_G : E_1 \times E_2 \to [0, 1]$ define the degree of membership and the degree of non-membership, respectively, of the element $\langle x, y \rangle \in E_1 \times E_2$ to the set

$G \subseteq E_1 \times E_2$; these functions have the forms of the corresponding components of the o-Cartesian product over IFSs; and for all $\langle x, y \rangle \in E_1 \times E_2$,

$$0 \leq \mu_G(x, y) + \nu_G(x, y) \leq 1.$$

For simplicity, we write G instead of G^*.

As in [301], we illustrate the above definition by an example of a Berge's graph (see Fig. 8.1; the labels of the arcs show the corresponding degrees). Let the following two tables giving μ- and ν-values be defined for it (for example, the data can be obtained as a result of some observations).

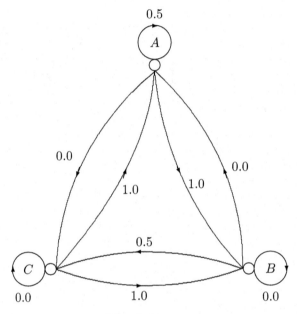

Fig. 8.1

μ_G	A	B	C
A	0.5	1	0
B	0	0	0.5
C	1	1	0

ν_G	A	B	C
A	0.3	0	1
B	1	0.4	0.2
C	0	0	0.7

The data for $\mu_G(x, y)$ are taken from [301]. On the other hand, the IFG G has the form shown in Fig. 8.2.

Let the oriented graph $G = (V, A)$ be given, where V is a set of vertices and A is a set of arcs. Every graph arc connects two graph vertices. Therefore, $A \subseteq V \times V$ and hence A can be described as a $(1, 0)$-IM. If the graph is fuzzy, the IM has elements from the set $[0, 1]$; if the graph is an IFG, the IM has elements from the set $[0, 1] \times [0, 1]$.

The IM of the graph G is given by

$$A = \begin{array}{c|cccc} & v_1 & v_2 & \ldots & v_n \\ \hline v_1 & a_{1,1} & a_{1,2} & \ldots & a_{1,n} \\ v_2 & a_{2,1} & a_{2,2} & \ldots & a_{2,n} \\ \vdots & \ldots & \ldots & \ldots & \ldots \\ v_n & a_{n,1} & a_{n,2} & \ldots & a_{n,n} \end{array}$$

where

$$a_{i,j} = \langle \mu_{i,j}, \nu_{i,j} \rangle \in [0,1] \times [0,1] (1 \le i, j \le n),$$

$$0 \le \mu_G(x,y) + \nu_G(x,y) \le 1,$$

$$V = \{v_1, v_2, \ldots, v_n\}.$$

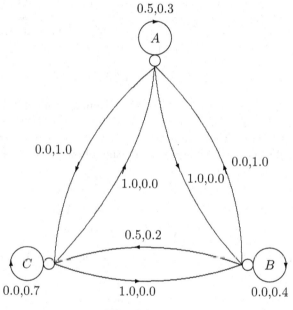

Fig. 8.2

We write briefly,

$$G = [V, V, \{a_{i,j}\}].$$

It can be easily seen that the above IM can be modified to the following form:

$$G = [V_I \cup \overline{V}, \overline{V} \cup V_O, \{a_{i,j}\}],$$

where V_I, V_O and $\overline{V}$ are respectively the sets of the input, output and internal vertices of the graph. At least one arc leaves every vertex of the first type, but none enters; at least one arc enters each vertex of the second type but none leaves it; every vertex of the third type has at least one arc ending in it and at least one arc starting from it.

Obviously, the graph matrix (in the sense of IM) now will be of a smaller dimension than the ordinary graph matrix. Moreover, it can be nonsquare, unlike the ordinary graph, matrices.

As in the ordinary case, the vertex $v_p \in \overline{V}$ has a loop iff $a_{p,p} = \langle \mu_{p,p}, \nu_{p,p} \rangle$ for the vertex v_p and $\mu_{p,p} > 0$ and $\nu_{p,p} < 1$.

Let the graphs G_1 and G_2 be given and let $G_s = [V'_s, V''_s, \{a^s_{i,j}\}]$, where $s = 1, 2$ and V'_s and V''_s are the sets of the graph vertices (input and internal, and output and internal, respectively).

Then, using the apparatus of the IMs, we construct the graph which is a union of the graphs G_1 and G_2. The new graph has the description

$$G = G_1 \cup G_2 = [V'_1 \cup V'_2, V''_1 \cup V''_2, \{\overline{a}_{i,j}\}],$$

where $\overline{a}_{i,j}$ is determined by the above IM-formulae, using min-max operations between its elements, for the case of operation "+" between IMs.

Analogously, we can construct a graph which is the intersection of the two given graphs G_1 and G_2. It would have the form

$$G = G_1 \cap G_2 = [V'_1 \cap V'_2, V''_1 \cap V''_2, \{\overline{\overline{a}}_{i,j}\}],$$

where $\overline{\overline{a}}_{i,j}$ is determined by the above IM-formulae, using min-max operations between its elements, for the case of operation "." between IMs.

Following the definitions from Section **6.1**, for some given $\alpha, \beta \in [0, 1]$ and for a given IFG $G = [V, V, A]$, we define the following three IFGs:

$$G_1 = N_\alpha(G) = [V', V'', A_1]$$

$$G_2 = N^\beta(G) = [V', V'', A_2]$$

$$G_3 = N_{\alpha,\beta}(G) = [V', V'', A_3]$$

For the first graph the arc between the vertices $v_i \in V'$ and $v_j \in V''$ is indexed by $\langle a_{i,j}, b_{i,j} \rangle$, where,

$$a_{i,j} = \begin{cases} \mu(v_i, v_j), & \text{if } \mu(v_i, v_j) \geq \alpha \\ 0, & \text{otherwise} \end{cases}$$

$$b_{i,j} = \begin{cases} \nu(v_i, v_j), & \text{if } \mu(v_i, v_j) \geq \alpha \\ 1, & \text{otherwise} \end{cases}$$

for the second graph - the same pair of numbers, but now having the values:

$$a_{i,j} = \begin{cases} \mu(v_i, v_j), & \text{if } \nu(v_i, v_j) \leq \beta \\ 0, & \text{otherwise} \end{cases}$$

$$b_{i,j} = \begin{cases} \nu(v_i, v_j), & \text{if } \nu(v_i, v_j) \leq \beta \\ 1, & \text{otherwise} \end{cases}$$

for the third graph - the same pair of numbers, but having the values:

$$a_{i,j} = \begin{cases} \mu(v_i, v_j), & \text{if } \mu(v_i, v_j) \geq \alpha \text{ and } \nu(v_i, v_j) \leq \beta \\ 0, & \text{otherwise} \end{cases}$$

$$b_{i,j} = \begin{cases} \nu(v_i, v_j), & \text{if } \mu(v_i, v_j) \geq \alpha \text{ and } \nu(v_i, v_j) \leq \beta \\ 1, & \text{otherwise} \end{cases}.$$

We must note that $v_i \in V'$ and $v_j \in V''$, iff $v_i, v_j \in V$ and in the first and in the third cases $a_{i,j} \geq \alpha$; in the second and in the third cases $b_{i,j} \leq \beta$.

Therefore, in this way we transform a given IFG to a new one whose arcs have high enough degrees of truth and low enough degrees of falsity.

For example, if we apply the operator $N_{\alpha,\beta}$ for $\alpha = 0.5, \beta = 0.25$ to the IFG in Fig. 8.1, we obtain the IFG as in Fig. 8.3.

The following statements hold for every two IFGs A and B and every two numbers $\alpha, \beta \in [0,1]$:

(a) $N_{\alpha,\beta}(A) = N_\alpha(A) \cap N^\beta(A)$,
(b) $N_{\alpha,\beta}(A \cap B) = N_{\alpha,\beta}(A) \cap N_{\alpha,\beta}(B)$,
(c) $N_{\alpha,\beta}(A \cup B) = N_{\alpha,\beta}(A) \cup N_{\alpha,\beta}(B)$.

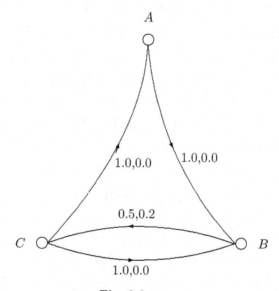

Fig. 8.3

Let us have a (fixed) set of vertices $\mathcal{V}$. An IFTree T (over $\mathcal{V}$) will be the ordered pair $T = (V^*, A^*)$ (see [169, 171, 432]), where

$$V \subset \mathcal{V},$$

$$V^* = \{\langle v, \mu_V(v), \nu_V(v) \rangle | v \in V\},$$

$$A \subset V \times V,$$

$$A^* = \{\langle g, \mu_A(g), \nu_A(g) \rangle | (\exists v, w \in V)(g = \langle v, w \rangle \in A)\},$$

where $\mu_V(v)$ and $\nu_V(v)$ are degrees of membership and non-membership of the element $v \in \mathcal{V}$ to V and

$$0 \le \mu_V(v) + \nu_V(v) \le 1.$$

The IFTree $T = (V^*, A^*)$ is:

a) *weak well constructed (WWC-IFTree)* if

$$(\forall v, w \in V)((\exists g \in A)(g = \langle v, w \rangle) \to (\mu_V(v) \ge \mu_V(w) \ \& \ \nu_V(v) \le \nu_V(w));$$

b) *strong well constructed (SWC-IFTree)* if

$$(\forall v, w \in V)((\exists g \in A)(g = \langle v, w \rangle)$$

$$\to (\mu_V(v) \ge \max(\mu_V(w), \mu_A(g)) \ \& \ \nu_V(v) \le \min(\nu_V(w), \nu_A(g)));$$

c) *average well constructed (AWC-IFTree)* if

$$(\forall v, w \in V)((\exists g \in A)$$

$$(g = \langle v, w \rangle) \to (\mu_V(v) \ge \frac{\mu_V(w) + \mu_A(g)}{2} \ \& \ \nu_V(v) \le \frac{\nu_V(w) + \nu_A(g)}{2}).$$

Let two IFTrees $T_1 = (V_1^*, G_1^*)$ and $T_2 = (V_2^*, G_2^*)$ be given. We define:

$$T_1 \cup T_2 = (V_1^*, A_1^*) \cup (V_2^*, A_2^*) = (V_1^* \cup V_2^*, A_1^* \cup A_2^*),$$

$$T_1 \cap T_2 = (V_1^*, A_1^*) \cap (V_2^*, A_2^*) = (V_1^* \cap V_2^*, A_1^* \cap A_2^*).$$

Let

$$\mathcal{P}(X) = \{Y | Y \subset X\},$$

and let for $T = (V^*, A^*)$

$$T_{full} = (E(V), E(A)),$$

$$T_{empty} = (O(V), O(A)),$$

where

$$E(V) = \{\langle v, 1, 0 \rangle | v \in \mathcal{V}\},$$

$$O(V) = \{\langle v, 0, 1 \rangle | v \in \mathcal{V}\},$$

$$E(A) = \{\langle g, 1, 0 \rangle | (\exists v, w \in V)(g = \langle v, w \rangle \in \mathcal{V} \times \mathcal{V})\},$$

$$O(A) = \{\langle g, 0, 1 \rangle | (\exists v, w \in V)(g = \langle v, w \rangle \in \mathcal{V} \times \mathcal{V})\}.$$

Theorem 8.5: $(\mathcal{P}(\mathcal{V}), \cup, T_{empty})$ and $(\mathcal{P}(\mathcal{V}), \cap, T_{full})$ are commutative monoids.

Let $G = (V, A)$ be a given IFTree. We construct its standard incidence matrix. After this, we change the elements of the matrix with their degrees of membership and non-membership. Finally, numbering the rows and columns of the matrix with the identifiers of the IFTree vertices, will result an IM.

For example, if we have the IFTree as in Fig. 8.4, we can construct the IM that corresponds to its incidence matrix:

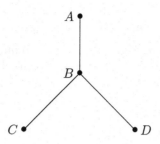

Fig. 8.4

$$[\{A, B, C, D\}, \{A, B, C, D\},$$

	A	B
A	$\langle\mu(A, A), \nu(A, A)\rangle$	$\langle\mu(A, B), \nu(A, B)\rangle$
B	$\langle\mu(B, A), \nu(B, A)\rangle$	$\langle\mu(B, B), \nu(B, B)\rangle$
C	$\langle\mu(C, A), \nu(C, A)\rangle$	$\langle\mu(C, B), \nu(C, B)\rangle$
D	$\langle\mu(D, A), \nu(D, A)\rangle$	$\langle\mu(D, B), \nu(D, B)\rangle$

...

	C	D
A	$\langle\mu(A, C), \nu(A, C)\rangle$	$\langle\mu(A, D), \nu(A, D)\rangle$
B	$\langle\mu(B, C), \nu(B, C)\rangle$	$\langle\mu(B, D), \nu(B, D)\rangle$
C	$\langle\mu(C, C), \nu(C, C)\rangle$	$\langle\mu(C, D), \nu(C, D)\rangle$
D	$\langle\mu(D, C), \nu(D, C)\rangle$	$\langle\mu(D, D), \nu(D, D)\rangle$

...   ],

where here and below by "..." we note the fact that the IM from the first row continues on the second row.

Having in mind that arcs AA, AC, AD, BB, CC, CD and DD do not exist, we can modify the above IM to the form:

$$[\{A, B, C, D\}, \{A, B, C, D\},$$

	A	B	C	D
A	$\langle 0, 1\rangle$	$\langle\mu(A, B), \nu(A, B)\rangle$	$\langle 0, 1\rangle$	$\langle 0, 1\rangle$
B	$\langle 0, 1\rangle$	$\langle 0, 1\rangle$	$\langle\mu(B, C), \nu(B, C)\rangle$	$\langle\mu(B, D), \nu(B, D)\rangle$
C	$\langle 0, 1\rangle$	$\langle 0, 1\rangle$	$\langle 0, 1\rangle$	$\langle 0, 1\rangle$
D	$\langle 0, 1\rangle$	$\langle 0, 1\rangle$	$\langle 0, 1\rangle$	$\langle 0, 1\rangle$

].

Now, it is seen, that all elements of the column indexed with A and all elements of the rows indexed with C and D are $\langle 0, 1\rangle$. Therefore, we can omit these two rows and the column and we obtain the simpler IM as

$$[\{A, B, C, D\}, \{A, B, C, D\},$$

	B	C	D
A	$\langle\mu(A,B),\nu(A,B)\rangle$	$\langle 0,1\rangle$	$\langle 0,1\rangle$
B	$\langle 0,1\rangle$	$\langle\mu(B,C),\nu(B,C)\rangle$	$\langle\mu(B,D),\nu(B,D)\rangle$

].

Finally, having in mind that there is no more a column indexed with A and rows indexed with C and D, we obtain a final form of the IM as

$$[\{A,B\},\{B,C,D\},$$

	B	C	D
A	$\langle\mu(A,B),\nu(A,B)\rangle$	$\langle 0,1\rangle$	$\langle 0,1\rangle$
B	$\langle 0,1\rangle$	$\langle\mu(B,C),\nu(B,C)\rangle$	$\langle\mu(B,D),\nu(B,D)\rangle$

].

Let us have an IFTree $G = (V,A)$ and let L be one of its leaves. Let $F = (W,B)$ be another IFTree so that

$$V \cap W = \{L\},$$

$$A \cup B = \emptyset.$$

Now, we describe the result of operation "substitution of an IFTree's leaf L with the IFTree F. The result will have the form of the IFTree $(V \cup W, A \cup B)$.

For example, if G is the IFTree as in Fig. 8.4 and if we substitute its leaf D with the IFTree F as in Fig. 8.5 that has the shorter IM-representation as

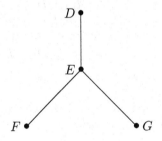

Fig. 8.5

$$[\{D,E\},\{E,F,G\},$$

	E	F	G
D	$\langle\mu(D,E),\nu(D,E)\rangle$	$\langle 0,1\rangle$	$\langle 0,1\rangle$
E	$\langle 0,1\rangle$	$\langle\mu(E,F),\nu(E,F)\rangle$	$\langle\mu(E,G),\nu(E,G)\rangle$

then, the result will be the IFTree as in Fig. 8.6 and it has the IM-representation as

$$[\{A,B,D,E\},\{B,C,D,E,F,G\},$$

	B	C	D
A	$\langle\mu(A,B),\nu(A,B)\rangle$	$\langle 0,1\rangle$	$\langle 0,1\rangle$
B	$\langle 0,1\rangle$	$\langle\mu(B,C),\nu(B,C)\rangle$	$\langle\mu(B,D),\nu(B,D)\rangle$
D	$\langle 0,1\rangle$	$\langle 0,1\rangle$	$\langle 0,1\rangle$
E	$\langle 0,1\rangle$	$\langle 0,1\rangle$	$\langle 0,1\rangle$

$\ldots$

	E	F	G
A	$\langle 0,1\rangle$	$\langle 0,1\rangle$	$\langle 0,1\rangle$
B	$\langle 0,1\rangle$	$\langle 0,1\rangle$	$\langle 0,1\rangle$
D	$\langle\mu(D,E),\nu(D,E)\rangle$	$\langle 0,1\rangle$	$\langle 0,1\rangle$
E	$\langle 0,1\rangle$	$\langle\mu(E,F),\nu(E,F)\rangle$	$\langle\mu(E,G),\nu(E,G)\rangle$

$\ldots$

$$= [\{A,B,D,E\},\{B,C,D,E,F,G\},$$

	B	C	D
A	$\langle\mu(A,B),\nu(A,B)\rangle$	$\langle 0,1\rangle$	$\langle 0,1\rangle$
B	$\langle 0,1\rangle$	$\langle\mu(B,C),\nu(B,C)\rangle$	$\langle\mu(B,D),\nu(B,D)\rangle$

$\oplus$

	E	F	G
D	$\langle\mu(D,E),\nu(D,E)\rangle$	$\langle 0,1\rangle$	$\langle 0,1\rangle$
E	$\langle 0,1\rangle$	$\langle\mu(E,F),\nu(E,F)\rangle$	$\langle\mu(E,G),\nu(E,G)\rangle$

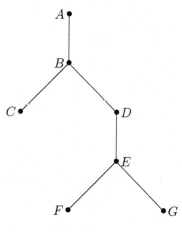

Fig. 8.6

Let the IFTree $T = [V,A]$ be given, where V is the set of its vertices and A is the set of its arcs, and let it has the following IM-form

$$T = [V,V,\{a_{k_i,l_j}\}].$$

Let its vertex w be fixed and let the subtree with source vertex w be

$$U = [W, W, \{b_{k_i, l_j}\}],$$

where
$$W = \{w, w_1, w_2, ..., w_s\} \subseteq V.$$

Let P be the new IFTree to be inserted at the vertex w of the IFTree T and has the IM-form
$$P = [Q, Q, \{c_{k_i, l_j}\}],$$

for which $w \in Q$ and $\{q_1, q_2, ..., q_r\} \subset Q$ are destination vertices.

Then the IM-form of the new IFTree T^* is

$$T^* = ([V, V, \{a_{k_i, l_j}\}] \ominus [W, W, \{\overline{a}_{k_i, l_j}\}]) \oplus [Q, Q, \{c_{k_i, l_j}\}]$$

$$\oplus \sum_{i=1}^{r} \begin{bmatrix} q_i \\ w \end{bmatrix} \begin{bmatrix} w_{i,1} \\ w_1 \end{bmatrix} ... \begin{bmatrix} w_{i,s} \\ w_s \end{bmatrix} [W, W, \{b_{k_i, l_j}\}].$$

We illustrate these definitions by two examples.

Let the ordered IFTree T_1, in Fig. 8.7, be given, and let P (see Fig. 8.8) be the new ordered IFTree to be inserted at vertex w of T_1. The resultant IFTree T_1^* is given in Fig. 8.9.

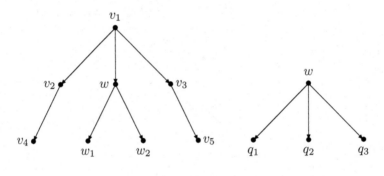

Fig. 8.7   Fig. 8.8

Let the two IFTrees have representations, respectively

$$T_1 = \begin{array}{c|cccccccc}
 & v_1 & v_2 & w & v_3 & v_4 & w_1 & w_2 & v_5 \\
\hline
v_1 & \alpha_{v_1,v_1} & \alpha_{v_1,v_2} & \alpha_{v_1,w} & \alpha_{v_1,v_3} & \alpha_{v_1,v_4} & \alpha_{v_1,w_1} & \alpha_{v_1,w_2} & \alpha_{v_1,v_5} \\
v_2 & \alpha_{v_2,v_1} & \alpha_{v_2,v_2} & \alpha_{v_2,w} & \alpha_{v_2,v_3} & \alpha_{v_2,v_4} & \alpha_{v_2,w_1} & \alpha_{v_2,w_2} & \alpha_{v_2,v_5} \\
w & \alpha_{w,v_1} & \alpha_{w,v_2} & \alpha_{w,w} & \alpha_{w,v_3} & \alpha_{w,v_4} & \alpha_{w,w_1} & \alpha_{w,w_2} & \alpha_{w,v_5} \\
v_3 & \alpha_{v_3,v_1} & \alpha_{v_3,v_2} & \alpha_{v_3,w} & \alpha_{v_3,v_3} & \alpha_{v_3,v_4} & \alpha_{v_3,w_1} & \alpha_{v_3,w_2} & \alpha_{v_3,v_5} \\
v_4 & \alpha_{v_4,v_1} & \alpha_{v_4,v_2} & \alpha_{v_4,w} & \alpha_{v_4,v_3} & \alpha_{v_4,v_4} & \alpha_{v_4,w_1} & \alpha_{v_4,w_2} & \alpha_{v_4,v_5} \\
w_1 & \alpha_{w_1,v_1} & \alpha_{w_1,v_2} & \alpha_{w_1,w} & \alpha_{w_1,v_3} & \alpha_{w_1,v_4} & \alpha_{w_1,w_1} & \alpha_{w_1,w_2} & \alpha_{w_1,v_5} \\
w_2 & \alpha_{w_2,v_1} & \alpha_{w_2,v_2} & \alpha_{w_2,w} & \alpha_{w_2,v_3} & \alpha_{w_2,v_4} & \alpha_{w_2,w_1} & \alpha_{w_2,w_2} & \alpha_{w_2,v_5} \\
v_5 & \alpha_{v_5,v_1} & \alpha_{v_5,v_2} & \alpha_{v_5,w} & \alpha_{v_5,v_3} & \alpha_{v_5,v_4} & \alpha_{v_5,w_1} & \alpha_{v_5,w_2} & \alpha_{v_5,v_5}
\end{array},$$

where $\alpha_{a,b} = \langle \mu_{a,b}, \nu_{a,b} \rangle$ for $a, b \in \{v_1, v_2, w, v_3, v_4, w_1, w_2, v_5\}$ and

$$
P = \begin{array}{c|cccc}
 & w & q_1 & q_2 & q_3 \\
\hline
w & \beta_{w,w} & \beta_{w,q_1} & \beta_{w,q_2} & \beta_{w,q_3} \\
q_1 & \beta_{q_1,w} & \beta_{q_1,q_1} & \beta_{q_1,q_2} & \beta_{q_1,q_3} \\
q_2 & \beta_{q_2,w} & \beta_{q_2,q_1} & \beta_{q_2,q_2} & \beta_{q_2,q_3} \\
q_3 & \beta_{q_3,w} & \beta_{q_3,q_1} & \beta_{q_3,q_2} & \beta_{q_3,q_3}
\end{array},
$$

where $\beta_{a,b} = \langle \mu_{a,b}, \nu_{a,b} \rangle$ for $a, b \in \{w, q_1, q_2, q_3\}$.

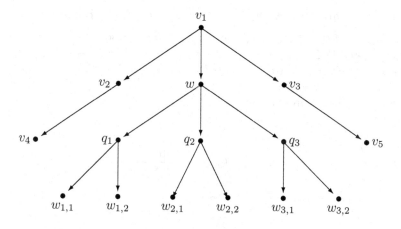

Fig. 8.9

Hawing in mind the above remark for reduction of the IM-representation of a graph and the fact that the IFTree is ordered, we can rewrite the IMs T_1 and P to the (equivalent) forms

$$
T_1 = \begin{array}{c|ccccccc}
 & v_2 & w & v_3 & v_4 & w_1 & w_2 & v_5 \\
\hline
v_1 & \alpha_{v_1,v_2} & \alpha_{v_1,w} & \alpha_{v_1,v_3} & \overline{O} & \overline{O} & \overline{O} & \overline{O} \\
v_2 & \overline{O} & \overline{O} & \overline{O} & \alpha_{v_2,v_4} & \overline{O} & \overline{O} & \overline{O} \\
w & \overline{O} & \overline{O} & \overline{O} & \overline{O} & \alpha_{w,w_1} & \alpha_{w,w_2} & \overline{O} \\
v_3 & \overline{O} & \overline{O} & \overline{O} & \overline{O} & \overline{O} & \overline{O} & \alpha_{v_3,v_5}
\end{array}
$$

and

$$
P = \begin{array}{c|ccc}
 & q_1 & q_2 & q_3 \\
\hline
w & \beta_{w,q_1} & \beta_{w,q_2} & \beta_{w,q_3}
\end{array},
$$

where $\overline{O} = \langle 0, 1 \rangle$.

The IM-form of the IFTree T_1^* is

$$
T_1^* = ([\{v_1, v_2, w, v_3, v_4, w_1, w_2, v_5\}, \{v_1, v_2, w, v_3, v_4, w_1, w_2, v_5\}, \{a_{k_i, l_j}\}]
$$

$$\ominus [\{w, w_1, w_2\}, \{w, w_1, w_2\}, \{\overline{a}_{k_i, l_j}\}])$$

$$\oplus [\{w, q_1, q_2, q_3\}, \{w, q_1, q_2, q_3\}, \{c_{k_i, l_j}\}]$$

$$\oplus \sum_{i=1}^{3} \left[\frac{q_i}{w}\right] \left[\frac{w_{i,1}}{w_1}\right] \left[\frac{w_{i,2}}{w_2}\right] [\{w, w_1, w_2\}, \{w, w_1, w_2\}, \{b_{k_i, l_j}\}]$$

$$= [\{v_1, v_2, w, v_3, v_4, q_1, q_2, q_3, v_5, w_{1,1}, w_{1,2}, w_{2,1}, w_{2,2}, w_{3,1}, w_{3,2}\},$$

$$\{v_1, v_2, w, v_3, v_4, q_1, q_2, q_3, v_5, w_{1,1}, w_{1,2}, w_{2,1}, w_{2,2}, w_{3,1}, w_{3,2}\}, \{d_{a,b}\}]$$

$$= [\{v_1, v_2, w, v_3, q_1, q_2, q_3\},$$

$$\{v_2, w, v_3, v_4, q_1, q_2, q_3, v_5, w_{1,1}, w_{1,2}, w_{2,1}, w_{2,2}, w_{3,1}, w_{3,2}, \}, \{d_{a,b}\}],$$

where the values of the elements $d_{a,b}$ are determined as above.

Let the ordered IFTree T_2, in Fig. 8.10, be given, and let P (see Fig. 8.8) be the new ordered IFTree to be inserted at vertex w of T_2. The resultant IFTree T_2^* is given in Fig. 8.11.

The IFTree T_2 has representation

$$T_2 = \begin{array}{c|cccc} & w & w_1 & w_2 & w_3 \\ \hline w & \beta_{w,w} & \beta_{w,w_1} & \beta_{w,w_2} & \beta_{w,w_3} \\ w_1 & \beta_{w_1,w} & \beta_{w_1,w_1} & \beta_{w_1,w_2} & \beta_{w_1,w_3} \\ w_2 & \beta_{w_2,w} & \beta_{w_2,w_1} & \beta_{w_2,w_2} & \beta_{w_2,w_3} \\ w_3 & \beta_{w_3,w} & \beta_{w_3,w_1} & \beta_{w_3,w_2} & \beta_{w_3,w_3} \end{array} = \begin{array}{c|ccc} & w_1 & w_2 & w_3 \\ \hline w & \beta_{w,w_1} & \beta_{w,w_2} & \beta_{w,w_3} \end{array}.$$

Then

$$T_2^* = ([\{w, w_1, w_2, w_3\}, \{w, w_1, w_2, w_3\}, \{a_{k_i, l_j}\}]$$

$$\ominus [\{w, w_1, w_2, w_3\}, \{w, w_1, w_2, w_3\}, \{a_{k_i, l_j}\}])$$

$$\oplus [\{w, q_1, q_2, q_3\}, \{w, q_1, q_2, q_3\}, \{c_{k_i, l_j}\}]$$

$$\oplus \sum_{i=1}^{3} \left[\frac{q_i}{w}\right] \left[\frac{w_{i,1}}{w_1}\right] \left[\frac{w_{i,2}}{w_2}\right] \left[\frac{w_{i,3}}{w_3}\right] [\{w, w_1, w_2, w_3\}, \{w, w_1, w_2, w_3\}, \{b_{k_i, l_j}\}]$$

$$= [\{w, q_1, q_2, q_3\}, \{w, q_1, q_2, q_3\}, \{c_{k_i, l_j}\}]$$

$$\oplus \sum_{i=1}^{3} \left[\frac{q_i}{w}\right] \left[\frac{w_{i,1}}{w_1}\right] \left[\frac{w_{i,2}}{w_2}\right] \left[\frac{w_{i,3}}{w_3}\right] [\{w, w_1, w_2, w_3\}, \{w, w_1, w_2, w_3\}, \{b_{k_i, l_j}\}]$$

$$= [\{w, q_1, q_2, q_3, w_{1,1}, w_{1,2}, w_{1,3}, w_{2,1}, w_{2,2}, w_{2,3}, w_{3,1}, w_{3,2}, w_{3,3}\},$$

$$\{w, q_1, q_2, q_3, w_{1,1}, w_{1,2}, w_{1,3}, w_{2,1}, w_{2,2}, w_{2,3}, w_{3,1}, w_{3,2}, w_{3,3}\}, \{c_{k_i, l_j}\}].$$

$$= [\{w, q_1, q_2, q_3\}, \{q_1, q_2, q_3, w_{1,1}, w_{1,2}, w_{1,3}, w_{2,1}, w_{2,2}, w_{2,3}, w_{3,1}, w_{3,2}, w_{3,3}\},$$

$$\{c_{k_i, l_j}\}].$$

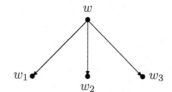

Fig. 8.10

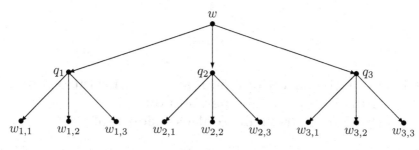

Fig. 8.11

Finally, let the ordered IFTree T_3, in Fig. 8.12, be given, and let P (see Fig. 8.8) be the new ordered IFTree to be inserted at vertex w of T_3. The resultant IFTree T_2^* is given in Fig. 8.13.

The IFTree T_3 has representation

$$T_3 = \begin{array}{c|cccc} & v_1 & v_2 & w & v_3 \\ \hline v_1 & \beta_{v_1,v_1} & \beta_{v_1,v_2} & \beta_{v_1,w} & \beta_{v_1,v_3} \\ v_2 & \beta_{v_2,v_1} & \beta_{v_2,v_2} & \beta_{v_2,w} & \beta_{v_2,v_3} \\ w & \beta_{w,v_1} & \beta_{w,v_2} & \beta_{w,w} & \beta_{w,v_3} \\ v_3 & \beta_{v_3,v_1} & \beta_{v_3,v_2} & \beta_{v_3,w} & \beta_{v_3,v_3} \end{array} = \begin{array}{c|ccc} & v_2 & w & v_3 \\ \hline v_1 & \beta_{v,v_2} & \beta_{v_1,w} & \beta_{v_1,v_3} \end{array}.$$

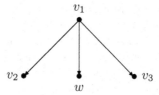

Fig. 8.12

The IM-form of the IFTree T_3^* is

$$T_3^* = (\lfloor \{v_1, v_2, w, v_3\}, \{v_1, v_2, w, v_3\}, \{a_{k_i,l_j}\}] \ominus [\{w\}, \{w\}, \{0\}])$$

$$\oplus [\{w, q_1, q_2, q_3\}, \{w, q_1, q_2, q_3\}, \{c_{k_i,l_j}\}]$$

$$= [\{v_1, v_2, w, v_3, q_1, q_2, q_3\}, \{v_1, v_2, w, v_3, q_1, q_2, q_3\}, \{d_{a,b}\}]$$

$$= [\{v_1, w,\}, \{v_2, w, v_3, q_1, q_2, q_3\}, \{d_{a,b}\}].$$

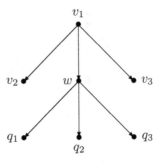

Fig. 8.13

8.5   Example: Intuitionistic Fuzzy Interpretations of Multi-criteria Multi-person and Multi-measurement Tool Decision Making

In Group Decision Making (GDM) a set of experts in a given field is involved in a decision process concerning the selection of the best alternative(s) among a set of predefined ones. An evaluation of the alternatives is performed independently by each decision maker: the experts express their evaluations on the basis of some decision scheme, which can be either implicitly assumed or explicitly specified in the form of a set of predefined criteria [228, 284]. In both cases, the aim is to obtain an evaluation (performance judgment or rating) of the alternatives by each expert. In the case in which a set of predefined criteria is specified, a performance judgment is expressed by each expert for each criterion; this kind of decision problem is called multi-person multi-criteria decision making [284]. Its aim is to compute a consensual judgment and a consensus degree for a majority of the experts on each alternative. As the main actors in a multi-person multi-criteria decision making activity are individuals with their inherent subjectivity, it often happens that performance judgments cannot be precisely assessed; the imprecision may arise from the nature of the characteristics of the alternatives, which can be either unquantifiable or unavailable. It may also be derived from the inability of the experts to formulate a precise evaluation [229, 230, 582]. Several works in the literature have approached the problem of simplifying the experts' formulation of evaluations. To this aim some fuzzy models of GDM have been proposed which relieve experts from quantifying qualitative concepts [140, 165, 230, 258, 294]. This objective has been pursued by dealing directly with performance or preference judgments expressed linguistically.

The second phase of a group decision process is the definition of a collective evaluation for each alternative: once the alternatives have been evaluated, the main problem is to aggregate the experts' performance judgments to obtain an overall rating for each alternative. A consequent problem is to compare the experts' judgments to verify the consensus among them. In the case of unanimous consensus, the evaluation process ends with the selection of the best alternative(s). As in real situation humans rarely come to an unanimous agreement, in the literature some fuzzy approaches to evaluate a "soft" degree of consensus have been proposed. It is important to notice that full consensus (degree $= 1$) is not necessarily the result of an unanimous agreement, but it can be obtained [140, 295, 294]. Each expert is asked to evaluate at least a part of the alternatives in terms of their performance with respect to each predefined criterion: the experts evaluations are expressed as a pair of numeric values, interpreted in the intuitionistic fuzzy framework: these numbers express a "positive" and a "negative" evaluations, respectively. With each expert a pair of values is associated, which express the expert's reliability (confidence in her/his evaluation with respect to each criterion). Distinct reliability values are associated with distinct criteria. The *proposed formulation is based on the assumption of alternatives' independence.*

The contents of this Section is based on paper [98] written by Gabriella Pasi, Ronald Yager and the author.

Here, the described procedure gives the possibility to use partial orders, i.e., orders represented by oriented graphs.

The following basic notation is adopted below:

$E = \{E_1, E_2, ..., E_e\}$ is the set of experts involved in the decision process;

$M = \{M_1, M_2, ..., M_m\}$ is the set of measurement tools employed in the decision process;

$A = \{A_1, A_2, ..., A_p\}$ is the set of considered alternatives;

$C = \{C_1, C_2, ..., C_q\}$ is the set of criteria used for evaluating the alternatives.

Using the apparatus of the IFSs, we discuss the possibility of constructing an overall performance judgment, related to the following distinct, although similar, problems.

Problem 1. *Let alternatives $A_1, ...A_p$ be given and let experts $E_1, ..., E_e$ have to order the alternatives with respect to criteria $C_1, C_2, ..., C_q$. Produce an aggregated order of the objects based on experts' opinions.*

Problem 2. *Let alternatives $A_1, ...A_p$ be given and let us have the measurement tools $M_1, ..., M_m$, which estimate the alternatives with respect to the criteria $C_1, C_2, ..., C_q$. The problem consists in producing an aggregated estimation of the objects on the basis of the measurement tool estimations.*

Each measurement tool can work using (at a given time) exactly one criterion, but in distinct times it can be tuned to use different criteria. The quality of the estimation of each measurement tool with respect to the other criteria is subjective. The following basic assumptions are considered:

- at each moment the tools use only one criterion;
- we determine the order of criteria;
- we determine for each moment which tool and which criteria will be used.

First, the proposed method of multi-person multi-criterion decision making will be described and then, the proposed method of multi-measurement tools multi-criteria decision making is given. Finally, some examples of the proposed method in the context of public relation and mass communication are discussed.

8.5.1 Experts Who Order Alternatives

Let there be m experts, $E_1, E_2, \ldots, E_m$, p alternatives which have to be evaluated by the experts $A_1, A_2, \ldots, A_p$ and q evaluation criteria $C_1, C_2, \ldots, C_q$. Let i-th expert have his/her own (current) reliability score $\langle \delta_i, \varepsilon_i \rangle \in [0,1]^2$ and his/her own (current) number of participations in expert investigations γ_i (these two values correspond to her/his last evaluation). Expert's reliability scores can be interpreted, as

$$
\begin{cases}
\delta_i = \dfrac{\sum\limits_{j=1}^{q} \delta_{i,j}}{q} \\[2em]
\varepsilon_i = \dfrac{\sum\limits_{j=1}^{q} \varepsilon_{i,j}}{q}
\end{cases},
$$

where $\langle \delta_{i,j}, \varepsilon_{i,j} \rangle$ are elements of the IM

$$
T = \begin{array}{c|cccc}
 & C_1 & C_2 & \ldots & C_q \\
\hline
E_1 & & & & \\
 & & \langle \delta_{i,j}, \varepsilon_{i,j} \rangle & & \\
E_2 & & & & \\
 & & (1 \le i \le m, & & \\
\vdots & & & & \\
 & & 1 \le j \le q) & & \\
E_m & & & &
\end{array}
$$

and $\langle \delta_{i,j}, \varepsilon_{i,j} \rangle$ is the rating of the i-th expert with respect to the j-th criterion (assume that the i-th expert's knowledge reliability may differ over different criteria).

To illustrate the expert's reliability score, we give the following example: a sport journalist gave 10 prognoses for the results of 10 football matches. In 5 of the cases he/she guessed the winner, in 3 of the cases he failed and in the rest two cases he did not engage with a final opinion about the result. That is why, we determine his reliability score as $\langle 0.5, 0.3 \rangle$.

Let each of the experts show which criteria they shall use for a concrete evaluation. We use the set of all criteria provided by the experts. For example, each expert will obtain cards with the different criteria written on them. Each expert ranks these criteria (or a part of them, if he/she deems some of them unnecessary), on the vertices of a graph. The highest vertices of this graph corresponds to the most relevant criteria according to the respective expert. The second top-down vertices interpret the criteria that are "an idea" weaker than the first ones. There are no arcs between vertices which are incomparable due to some criterion. Therefore, each of the experts not only ranks the criteria that he/she uses (it is possible, omitting some of them), but his/her order is not linear one. As a result, we obtain m different graphs. Now, transform these graphs to IFGs, labelling each arc of the i-th expert's graph with a pair of values, corresponding to his/her expert's reliability score.

Using operation "$\cup$" over the IFGs, we obtain a new IFG, say G. It represents all expert opinions about the criteria ordering. Now, its arcs have intuitionistic fuzzy weights being the disjunctions of the weights, of the same arcs in the separate IFGs. Of course, the new graph may not be well ordered, while the expert graphs are well orderd. Now, we reconfigure IFG G as follows. If there is a loop between two vertices V_1 and V_2, i.e., there are vertices $U_1, U_2, ..., U_u$ and vertices $W_1, W_2, ..., W_w$, such that $V_1, U_1, U_2, ..., U_u, V_2$ and $V_2, W_1, W_2, ..., W_w, V_1$ are simple paths in the graph, then we calculate the weights of both paths as conjunctions of the weights of the arcs which take part in the respective paths. The path that has smaller weight must be cut into two, removing its arc with smallest weight. If both arcs have equal weights, these arcs will be removed. Therefore, the new graph is already loop-free. Now, determine the priorities of the vertices of the IFG, i.e., the priorities of the criteria. Let them be $\varphi_1, \varphi_2, ..., \varphi_q$. For example, they have values $\frac{s-1}{t}$ for the vertices from the s-th level bottom-top of the IFG with $t+1$ levels. We use these values below.

This procedure will be used in a next authors' research, but with another form of the algorithm for decision making, using the so constructed IFGs more actively. Here, we use the above construction only to propose the experts' possibility to work with non-linearly ordered criteria and to obtain priorities of these criteria.

Having in mind that the i-th expert can use only a part of the criteria and can estimate only a part of the alternatives, we can construct the IM of his/her estimations in the form

$$S_i = \begin{array}{c|cccc} & A_{l_1} & A_{l_2} & \cdots & A_{l_{p_i}} \\ \hline C_{i_1} & & & & \\ & & \langle \alpha^i_{i_j,l_k} \beta^i_{i_j,l_k} \rangle & & \\ C_{i_2} & & & & \\ & & (1 \le i_j \le q_i \le q, & & \\ \vdots & & & & \\ & & 1 \le l_k \le p_i \le p) & & \\ C_{i_{q_i}} & & & & \end{array}$$

where $\alpha^i_{i_j,k}, \beta^i_{i_j,k} \in [0,1]$, $\alpha^i_{i_j,k} + \beta^i_{i_j,k} \le 1$ and $\langle \alpha^i_{j,k}, \beta^i_{j,k} \rangle$ is the i-th expert estimation for the k-th alternative about the j-th criterion; $C_{i_1}, ..., C_{i_{q_i}}$ and $A_{l_1}, ..., A_{l_{p_i}}$ are only those of the criteria and alternatives which the i-th expert prefers. Let us assume that in cases when pair $\langle \alpha^i_{j,k}, \beta^i_{j,k} \rangle$ does not exist, we work with pair $\langle 0, 1 \rangle$.

Now, construct an IM containing the aggregated estimations of the form

$$
S = \begin{array}{c|cccc}
 & A_1 & A_2 & \ldots & A_p \\
\hline
C_1 & & & & \\
 & & \langle \alpha_{j,k} \beta_{j,k} \rangle & & \\
C_2 & & & & \\
 & & (1 \leq j \leq q, & & \\
\vdots & & & & \\
 & & 1 \leq k \leq p) & & \\
C_q & & & &
\end{array}
$$

where $\alpha_{j,k}$ and $\beta_{j,k}$ can be calculated by different formulae, with respect to some specific aims. For example, the formulae are

$$
\left\{
\begin{array}{l}
\alpha_{j,k} = \dfrac{\sum\limits_{i=1}^{m} \delta_i . \alpha^i_{j,k}}{m} \\[3ex]
\beta_{j,k} = \dfrac{\sum\limits_{i=1}^{m} \varepsilon_i . \beta^i_{j,k}}{m}
\end{array}
\right.
$$

(here, only the average degrees of experts' reliability participate),

$$
\left\{
\begin{array}{l}
\alpha_{j,k} = \dfrac{\sum\limits_{i=1}^{m} \delta_{i,j} . \alpha^i_{j,k}}{m} \\[3ex]
\beta_{j,k} = \dfrac{\sum\limits_{i=1}^{m} \varepsilon_{i,j} . \beta^i_{j,k}}{m}
\end{array}
\right.
$$

(here estimated by the corresponding criteria, only the experts' degrees of reliability participate),

$$
\left\{
\begin{array}{l}
\alpha_{j,k} = \dfrac{\sum\limits_{i=1}^{m} \overline{\alpha}^i_{j,k}}{m} \\[3ex]
\beta_{j,k} = \dfrac{\sum\limits_{i=1}^{m} \overline{\beta}^i_{j,k}}{m}
\end{array}
\right.
$$

(here only the experts' degrees of reliability estimated by the corresponding criteria participate), where $\overline{\alpha}^i_{j,k}$ and $\overline{\beta}^i_{j,k}$ can be calculated by various formulae, according to the particular goals and the experts' knowledge. For example, the formulae are

$$
\begin{cases}
\overline{\alpha}^i_{j,k} = \gamma_i \cdot \dfrac{\alpha^i_{j,k} \cdot \delta_{i,j} + \beta^i_{j,k} \cdot \varepsilon_{i,j}}{\gamma_i + 1} \\[3mm]
\overline{\beta}^i_{j,k} = \gamma_i \cdot \dfrac{\alpha^i_{j,k} \cdot \varepsilon_{i,j} + \beta^i_{j,k} \cdot \delta_{i,j}}{\gamma_i + 1}
\end{cases}
$$

or

$$
\begin{cases}
\overline{\alpha}^i_{j,k} = \alpha^i_{j,k} \cdot \dfrac{\delta_{i,j} + 1 - \varepsilon_{i,j}}{2} \\[3mm]
\overline{\beta}^i_{j,k} = \beta^i_{j,k} \cdot \dfrac{\varepsilon_{i,j} + 1 - \delta_{i,j}}{2}
\end{cases}.
$$

The first formula takes into account not only the rating of each expert by the different criteria, but also the number of times he has made a prognosis (his first time is neglected, for the lack of previous experience). Obviously, the so constructed elements of the IM satisfy the inequality: $\alpha_{j,k} + \beta_{j,k} \leq 1$. This IM contains the average experts' estimations taking into account the experts' ratings. As we noted above, each of the criteria $C_j (1 \leq j \leq q)$ has itself a priority, denoted by $\varphi_j \in [0,1]$. For every alternative A_k, we determine the global estimation $\langle \alpha_k, \beta_k \rangle$, where

$$
\begin{cases}
\alpha_k = \dfrac{\sum\limits_{j=1}^{q} \varphi_j \cdot \alpha_{j,k}}{q} \\[5mm]
\beta_k = \dfrac{\sum\limits_{j=1}^{q} \varphi_j \cdot \beta_{j,k}}{q}
\end{cases}.
$$

Let alternatives (processes) have the following (objective) values with regard to the different criteria after the end of the expert estimations:

	A_1	A_2	...	A_p
C_1				
C_2		$\langle a_{j,k} b_{j,k} \rangle$		
		$(1 \leq j \leq q,$		
⋮				
		$1 \leq k \leq p)$		
C_q				

where $a_{j,k}, b_{j,k} \in [0,1]$ and $a_{j,k} + b_{j,k} \leq 1$. Then the i-th expert's new rating, $\langle \delta_i, \varepsilon_i \rangle$, and new number of participations in expert investigations, γ'_i will be:

$$
\gamma'_i = \gamma_i + 1,
$$

and

$$
\begin{cases}
\delta_i' = \dfrac{\gamma_i.\delta_i + \frac{c_M - c_i}{2}}{\gamma_i'}, \\[3mm]
\varepsilon_i' = \dfrac{\gamma_i.\varepsilon_i - \frac{c_M - c_i}{2}}{\gamma_i'},
\end{cases}
$$

where

$$
c_i = \frac{\sum\limits_{j=1}^{q} \sum\limits_{k=1}^{p} ((\alpha_{j,k} - a_{j,k})^2 + (\beta_{j,k} - b_{j,k})^2)^{1/2}}{p.q},
$$

and

$$
c_M = \frac{\sum\limits_{i=1}^{n} c_i}{n}.
$$

8.5.2 Measurement Tools That Evaluate Alternatives

Assume that we have m measurement tools $M_1, M_2, \ldots, M_m$, p alternatives $A_1, A_2, \ldots, A_p$ that have to be evaluated by the m measurement tools, and q evaluation criteria $C_1, C_2, \ldots, C_q$. Also, assume that the j-th criterion has a given preliminary score $\varphi_j \in [0, 1]$, which denotes the importance of the criterion in the evaluation strategy. This score can be determined, by some experts.

Let us assume that each measurement tool has its own (current) reliability score $\langle \delta_i, \varepsilon_i \rangle \in [0, 1]^2$, and its own (current) number of use in the measurement investigations γ_i. These two values correspond to the measurement tool's last using. The measurement tool reliability scores can be obtained, e.g., by the elements of the IM

$$
T =
\begin{array}{c|cccc}
 & C_1 & C_2 & \ldots & C_q \\
\hline
M_1 & & & & \\
M_2 & & \langle \delta_{i,j}, \varepsilon_{i,j} \rangle & & \\
\vdots & & (1 \leq i \leq m, & & \\
 & & 1 \leq j \leq q) & & \\
M_m & & & &
\end{array}
$$

where $\langle \delta_{i,j}, \varepsilon_{i,j} \rangle$ is the rating of the i-th measurement tool with respect to the j-th criterion. Here, we assume that each measurement tool can be used to evaluate only one criterion. Hence, it is important that we must find the most suitable measurement tool for each criterion.

The following is the procedure aimed at determining the different couples $\langle \delta_{i,j}, \varepsilon_{i,j} \rangle$ with the highest values with respect to formulae:

$$\langle a, b \rangle \leq \langle c, d \rangle \quad \text{iff} \quad a \leq c \text{ and } b \geq d, \tag{8.7}$$

where $a, b, c, d \in [0, 1]$ and $a + b \leq 1, c + d \leq 1$.

1 Define the empty IM $U = [\emptyset, \emptyset, *]$, where "$*$" denotes the lack of elements in a matrix (i.e., a matrix with a dimension 0×0).

2 Choose $\langle \delta_{i,j}, \varepsilon_{i,j} \rangle$ as the maximal element with respect to the order $\geq$ by using (8.7).

3 We construct the reduced IM $T_{(M_i, C_j)}$ and construct the IM

$$U := U + [\{M_i\}, \{C_j\}, A_{i,j}],$$

where $A_{i,j}$ is an ordinary matrix of dimension 1×1 and with a unique element $\langle \delta_{i,j}, \varepsilon_{i,j} \rangle$.

4 We check whether some of the index sets of IM $T_{(M_i, C_j)}$ are not already empty. If yes - end; if not - go to **1**.

As a result of the above procedure, we obtain a list of measurement tool scores. For brevity, write

$$\langle \delta_{i,j}, \varepsilon_{i,j} \rangle = \langle \delta_i, \varepsilon_i \rangle,$$

because for the concrete sitting the i-th measurement tool uses only these values of its score. The procedure shows that we determine the most suitable measurement tools for the criteria with the highest sense for the estimated alternatives. Perhaps, a part of the criteria or a part of the measurement tools may not be used for the current sitting. Of course, the procedure shows that for the i-th measurement tool (if it is in the list with the most suitable measurement tools), there exists a j-th criterion and therefore, the present value of j is function of i, i.e., $j = j(i)$.

Briefly, let the i-th measurement tool M_i ($1 \leq i \leq m$), that uses $j = j(i)$-th criterion, have given the following estimations, which are described by the IM

$$S_i = \frac{\begin{array}{c|cccc} & A_1 & A_2 & \cdots & A_p \end{array}}{C_j \begin{array}{|cccc} \langle \alpha^i_{j,1} \beta^i_{j,1} \rangle & \langle \alpha^i_{j,2} \beta^i_{j,2} \rangle & \cdots & \langle \alpha^i_{j,k} \beta^i_{j,k} \rangle \end{array}},$$

where: $\alpha^i_{j,k}, \beta^i_{j,k} \in [0, 1]$ and $\alpha^i_{j,k} + \beta^i_{j,k} \leq 1$ for $1 \leq k \leq p$.

Then, we construct the IM

$$S = \begin{array}{c|cccc} & A_1 & A_2 & \cdots & A_p \\ \hline C_1 & & & & \\ & & \langle \alpha_{j,k} \beta_{j,k} \rangle & & \\ C_2 & & & & \\ & & (1 \leq j \leq q, & & \\ \vdots & & & & \\ & & 1 \leq k \leq p) & & \\ C_q & & & & \end{array}$$

where $\alpha_{j,k}$ and $\beta_{j,k}$ can be calculated by using

$$
\begin{cases}
\alpha_{j,k} = \dfrac{\sum\limits_{i=1}^{m} \delta_i \cdot \alpha_{j,k}^i}{m} \\[3ex]
\beta_{j,k} = \dfrac{\sum\limits_{i=1}^{m} \varepsilon_i \cdot \beta_{j,k}^i}{m}
\end{cases}
$$

(here only the average degrees of measurement tool reliability is taken),

$$
\begin{cases}
\alpha_{j,k} = \dfrac{\sum\limits_{i=1}^{m} \delta_{i,j} \cdot \alpha_{j,k}^i}{m} \\[3ex]
\beta_{j,k} = \dfrac{\sum\limits_{i=1}^{m} \varepsilon_{i,j} \cdot \beta_{j,k}^i}{m}
\end{cases}
$$

(what participates here is only the measurement tool degrees of reliability, estimated according to the corresponding criteria),

$$
\begin{cases}
\alpha_{j,k} = \dfrac{\sum\limits_{i=1}^{m} \overline{\alpha}_{j,k}^i}{m} \\[3ex]
\beta_{j,k} = \dfrac{\sum\limits_{i=1}^{m} \overline{\beta}_{j,k}^i}{m}
\end{cases}
,
$$

where $\overline{\alpha}_{j,k}^i$ and $\overline{\beta}_{j,k}^i$ can also be calculated by various formulae, according to particular goals and measurement tool estimations, by using the formulae

$$
\begin{cases}
\overline{\alpha}_{j,k}^i = \gamma_i \cdot \dfrac{\alpha_{j,k}^i \cdot \delta_{i,j} + \beta_{j,k}^i \cdot \varepsilon_{i,j}}{\gamma_i + 1} \\[3ex]
\overline{\beta}_{j,k}^i = \gamma_i \cdot \dfrac{\alpha_{j,k}^i \cdot \varepsilon_{i,j} + \beta_{j,k}^i \cdot \delta_{i,j}}{\gamma_i + 1}
\end{cases}
$$

or

$$
\begin{cases}
\overline{\alpha}_{j,k}^i = \alpha_{j,k}^i \cdot \dfrac{\delta_{i,j} + 1 - \varepsilon_{i,j}}{2} \\[3ex]
\overline{\beta}_{j,k}^i = \beta_{j,k}^i \cdot \dfrac{\varepsilon_{i,j} + 1 - \delta_{i,j}}{2}
\end{cases}
.
$$

The first formula takes into account not only the score of each measurement tool by the different criteria, but also the number of times it has been used so far. Obviously, the so constructed elements of the IM satisfy the inequality $\alpha_{j,k} + \beta_{j,k} \leq 1$.

Now, we discuss another possibility to use measurement tool estimations, accounting our opinion about the separate tools. On the basis of the measurement tool estimations and the measurement tool scores, we deform the measurement tool estimations, as follows:

• optimistic estimation:

$$\langle \alpha^i_{j,1} \beta^i_{j,1} \rangle = J_{\delta_i, \varepsilon_i}(\langle \alpha^i_{j,1} \beta^i_{j,1} \rangle)$$

or

$$\langle \alpha^i_{j,1} \beta^i_{j,1} \rangle = J^*_{\delta_i, \varepsilon_i}(\langle \alpha^i_{j,1} \beta^i_{j,1} \rangle);$$

• optimistic estimation with restrictions:

$$\langle \alpha^i_{j,1} \beta^i_{j,1} \rangle = P_{\alpha, \beta}(\langle \alpha^i_{j,1} \beta^i_{j,1} \rangle),$$

where $\alpha, \beta \in [0,1]$ are fixed levels and $\alpha + \beta \le 1$;

• pessimistic estimation:

$$\langle \alpha^i_{j,1} \beta^i_{j,1} \rangle = H_{\delta_i, \varepsilon_i}(\langle \alpha^i_{j,1} \beta^i_{j,1} \rangle)$$

or

$$\langle \alpha^i_{j,1} \beta^i_{j,1} \rangle = H^*_{\delta_i, \varepsilon_i}(\langle \alpha^i_{j,1} \beta^i_{j,1} \rangle);$$

• pessimistic estimation with restrictions:

$$\langle \alpha^i_{j,1} \beta^i_{j,1} \rangle = Q_{\alpha, \beta}(\langle \alpha^i_{j,1} \beta^i_{j,1} \rangle),$$

where $\alpha, \beta \in [0,1]$ are fixed levels and $\alpha + \beta \le 1$;

• estimation with decreasing uncertainty:

$$\langle \alpha^i_{j,1} \beta^i_{j,1} \rangle = F_{\delta_i, \varepsilon_i}(\langle \alpha^i_{j,1} \beta^i_{j,1} \rangle)$$

(the condition $\delta_i + \varepsilon_i \le 1$ is obviously valid);

• estimation with increasing uncertainty:

$$\langle \alpha^i_{j,1} \beta^i_{j,1} \rangle = G_{\delta_i, \varepsilon_i}(\langle \alpha^i_{j,1} \beta^i_{j,1} \rangle).$$

After calculating new values of $\langle \alpha^i_{j,1} \beta^i_{j,1} \rangle$, they are used in the above formulae.

We determine, for every alternative A_k, the global estimation $\langle \alpha_k, \beta_k \rangle$, where

$$\begin{cases} \alpha_k = \dfrac{\sum\limits_{j=1}^{q} \varphi_j . \alpha_{j,k}}{q} \\[3em] \beta_k = \dfrac{\sum\limits_{j=1}^{q} \varphi_j . \beta_{j,k}}{q} \end{cases}.$$

Let alternatives (processes) have the following (objective) values with regard to the different criteria after the end of the evaluations measurement tools:

$$\begin{array}{c|cccc}
 & A_1 & A_2 & \dots & A_p \\
\hline
C_1 & & & & \\
 & & \langle a_{j,k} b_{j,k} \rangle & & \\
C_2 & & & & \\
 & & (1 \le j \le q, & & \\
\vdots & & & & \\
 & & 1 \le k \le p) & & \\
C_q & & & &
\end{array}$$

where: $a_{j,k}, b_{j,k} \in [0,1]$ and $a_{j,k} + b_{j,k} \le 1$. Then, the measurement tool's new score, $\langle \delta_i, \varepsilon_i \rangle$, and the new number of measurement tools usage, γ_i', is calculated similar to Section **8.5.1**.

The present algorithm should be quite useful when searching for an objective answer on the basis of subjective initial data. Having in mind the experts' reliability scores with respect to their successful prognoses hitherto (objective data), and their present evaluations (subjective data), we try to derive an objective estimation about the current event, so that it would cover the estimations of the widest possible circle of people involved. This formulation of the problem implies that the areas, which will find the proposed algorithm a suitable tool for analysis and representation of the data, are the areas involving evaluation of the public opinion about currently flowing processes and tendencies in the society, or evaluating the ratings of the politicians, the media and other similar phenomena. If there exists some causal relation between two of the chosen parameters in our evaluation, it seems natural to grade them from top to bottom in the graph representation of the problem. If both factors are indepedent but equal in weight, their place is next to one another on the same hierarchy level in the graph. The experts themselves must definitely be specialists in the area they are giving estimations about.

8.6   Some Ways of Determining Membership and Non-membership Functions

In the theory of fuzzy sets, various methods are discussed for the generation of values of the membership function (for instance, see [144, 145, 146, 301, 147, 593, 592]).

Here, we discuss a way of generation of the two degrees – of membership and of non-membership that exist in the intuitionistic fuzzy sets (IFSs). For other approaches of assigning membership and non-membership functions of IFSs see [456].

Let us have k different generators $G_1, G_2, ..., G_k$ of fuzzy estimations for n different objects $O_1, O_2, ..., O_n$. In [144, 145, 146] these generators are called "estimators".

Let the estimations be collected in the IM

$$\begin{array}{c|cccccc}
 & O_1 & O_2 & \dots & O_j & \dots & O_n \\
\hline
G_1 & \alpha_{1,1} & \alpha_{1,2} & \dots & \alpha_{1,j} & \dots & \alpha_{1,n} \\
G_2 & \alpha_{2,1} & \alpha_{2,2} & \dots & \alpha_{2,j} & \dots & \alpha_{2,n} \\
\vdots & \vdots & \vdots & & \vdots & & \vdots \\
G_i & \alpha_{i,1} & \alpha_{i,2} & \dots & \alpha_{i,j} & \dots & \alpha_{i,n} \\
\vdots & \vdots & \vdots & & \vdots & & \vdots \\
G_k & \alpha_{k,1} & \alpha_{k,2} & \dots & \alpha_{k,j} & \dots & \alpha_{k,n}
\end{array}$$

On the basis of the values of the IM, we can construct the following two types of fuzzy sets:

$$O_1^* = \{\langle G_i, \alpha_{i,1}\rangle | 1 \leq i \leq k\},$$
$$O_2^* = \{\langle G_i, \alpha_{i,2}\rangle | 1 \leq i \leq k\},$$

$$\dots$$

$$O_n^* = \{\langle G_i, \alpha_{i,n}\rangle | 1 \leq i \leq k\},$$

and

$$G_1^* = \{\langle O_j \alpha_{1,j}\rangle | 1 \leq j \leq n\},$$
$$G_2^* = \{\langle O_j \alpha_{2,j}\rangle | 1 \leq j \leq n\},$$

$$\dots$$

$$G_k^* = \{\langle O_j \alpha_{k,j}\rangle | 1 \leq j \leq n\}.$$

Now, using these sets we construct new different sets, already IFSs. First, we construct the IFSs:

$$O_1^I = \{\langle G_i, \alpha_{i,1}, \sum_{2 \leq s \leq n} \alpha_{i,s}\rangle | 1 \leq i \leq k\},$$

$$O_2^I = \{\langle G_i, \alpha_{i,2}, \sum_{1 \leq s \leq n;\ s \neq 2} \alpha_{i,s}\rangle | 1 \leq i \leq k\},$$

$$\dots$$

$$O_n^I = \{\langle G_i, \alpha_{i,n}, \sum_{1 \leq s \leq n-1} \alpha_{i,s}\rangle | 1 \leq i \leq k\},$$

or

$$O_j^I = \{\langle G_i, \alpha_{i,j}, \sum_{1 \leq s \leq n;\ s \neq j} \alpha_{i,s}\rangle | 1 \leq i \leq k\}, \text{ for } j = 1, 2, ..., n;$$

and

$$G_1^I = \{\langle O_j, \alpha_{1,j}, \sum_{2 \leq s \leq n} \alpha_{s,j}\rangle | 1 \leq j \leq n\},$$

$$G_2^I = \{\langle O_j, \alpha_{2,j}, \sum_{1 \leq s \leq n;\ s \neq 2} \alpha_{s,j}\rangle | 1 \leq j \leq n\},$$

$$\dots$$

$$G_k^I = \{\langle O_j, \alpha_{k,j}, \sum_{1 \le s \le n-1} \alpha_{j,s}\rangle | 1 \le j \le n\},$$

or

$$G_i^I = \{\langle O_j, \alpha_{i,j}, \sum_{1 \le s \le n;\ s \ne i} \alpha_{j,s}\rangle | 1 \le j \le n\}, \text{ for } j = 1, 2, ..., k;$$

Second, we construct the IFSs:

$$G_{\max,\min}^I = \{\langle O_j, \max_{1 \le i \le n} \alpha_{i,j}, \min_{1 \le i \le n} \alpha_{i,j}\rangle | 1 \le j \le n\},$$

$$G_{av}^I = \{\langle O_j, \frac{1}{k}\sum_{i=1}^{k}\alpha_{i,j}, \frac{1}{k}\sum_{1 \le s \le n; s \ne j}\sum_{i=1}^{k}\alpha_{i,s}\rangle | 1 \le j \le n\},$$

$$G_{\min,\max}^I = \{\langle O_j, \min_{1 \le i \le n} \alpha_{i,j}, \max_{1 \le i \le n} \alpha_{i,j}\rangle | 1 \le j \le n\}.$$

Now, we illustrate the above constructions.

Let five experts E_1, E_2, E_3, E_4 and E_5 offer their evaluations of the percentage of votes, obtained by the political parties P_1, P_2 and P_3:

	P_1	P_2	P_3
E_1	32%	9%	37%
E_2	27%	7%	39%
E_3	26%	11%	35%
E_4	31%	8%	39%
E_5	29%	9%	41%

Now, we are able to generate the fuzzy sets

$$P_1^* = \{\langle E_1, 0.32\rangle, \langle E_2, 0.27\rangle, \langle E_3, 0.26\rangle, \langle E_4, 0.31\rangle, \langle E_5, 0.29\rangle\},$$

$$P_2^* = \{\langle E_1, 0.09\rangle, \langle E_2, 0.07\rangle, \langle E_3, 0.11\rangle, \langle E_4, 0.08\rangle, \langle E_5, 0.09\rangle\},$$

$$P_3^* = \{\langle E_1, 0.37\rangle, \langle E_2, 0.39\rangle, \langle E_3, 0.35\rangle, \langle E_4, 0.39\rangle, \langle E_5, 0.41\rangle\},$$

$$E_1^* = \{\langle P_1, 0.32\rangle, \langle P_2, 0.09\rangle, \langle P_3, 0.37\rangle\},$$

$$E_2^* = \{\langle P_1, 0.27\rangle, \langle P_2, 0.07\rangle, \langle P_3, 0.39\rangle\},$$

$$E_3^* = \{\langle P_1, 0.26\rangle, \langle P_2, 0.11\rangle, \langle P_3, 0.35\rangle\},$$

$$E_4^* = \{\langle P_1, 0.31\rangle, \langle P_2, 0.08\rangle, \langle P_3, 0.39\rangle\},$$

$$E_5^* = \{\langle P_1, 0.29\rangle, \langle P_2, 0.09\rangle, \langle P_3, 0.41\rangle\}.$$

We can aggregate the last five sets, e.g., by operation @ and obtain the fuzzy set

$$E_{FS} = \{\langle P_1, 0.29\rangle, \langle P_2, 0.088\rangle, \langle P_3, 0.382\rangle\}.$$

Below, we show why we use the above information for constructing IFSs.

It is easy to figure out that if expert E_1 believes that party P_1 would obtain 32% of the election votes, then he deems that 68% of the voters are against this party. If we take for granted that all the five experts are equally competent, i.e. their opinions are of equal worth, then we may conclude that party P_1 will receive between 26% and 32% of the votes, therefore, the opposers of this party will count between 68% and 74% of the voters. Now, an IFS can be constructed for the universe $\{P_1, P_2, P_3\}$ that would have the form:

$$E_{IFS,1} = \{\langle P_1, 0.26, 0.68\rangle, \langle P_2, 0.07, 0.89\rangle, \langle P_3, 0.35, 0.59\rangle\}.$$

This shows that at least 26% of the voters would support party P_1 and at least 68% would oppose it.

Another possible IFS that we can construct on the basis of the above data, is

$$E_{IFS,2} = \{\langle P_1, 0.29, 0.47\rangle, \langle P_2, 0.088, 0.672\rangle, \langle P_3, 0.382, 0.378\rangle\}.$$

The μ-components of this IFS are obtained directly from E_{FS}, while the ν-components are sums of the μ-components of the other two parties.

Following the above formulae, we can construct the next IFSs:

$$P_1^* = \{\langle E_1, 0.32, 0.46\rangle, \langle E_2, 0.27, 0.46\rangle, \langle E_3, 0.26, 0.46\rangle, \langle E_4, 0.31, 0.47\rangle,$$
$$\langle E_5, 0.29, 0.50\rangle\},$$

$$P_2^* = \{\langle E_1, 0.09, 0.59\rangle, \langle E_2, 0.07, 0.66\rangle, \langle E_3, 0.11, 0.61\rangle, \langle E_4, 0.08, 0.70\rangle,$$
$$\langle E_5, 0.09, 0.70\rangle\},$$

$$P_3^* = \{\langle E_1, 0.37, 0.41\rangle, \langle E_2, 0.39, 0.34\rangle, \langle E_3, 0.35, 0.37\rangle, \langle E_4, 0.39, 0.39\rangle,$$
$$\langle E_5, 0.41, 0.38\rangle\},$$

$$E_1^* = \{\langle P_1, 0.32, 0.46\rangle, \langle P_2, 0.09, 0.59\rangle, \langle P_3, 0.37, 0.41\rangle\},$$
$$E_2^* = \{\langle P_1, 0.27, 0.46\rangle, \langle P_2, 0.07, 0.66\rangle, \langle P_3, 0.39, 0.34\rangle\},$$
$$E_3^* = \{\langle P_1, 0.26, 0.46\rangle, \langle P_2, 0.11, 0.61\rangle, \langle P_3, 0.35, 0.37\rangle\},$$
$$E_4^* = \{\langle P_1, 0.31, 0.47\rangle, \langle P_2, 0.08, 0.70\rangle, \langle P_3, 0.39, 0.39\rangle\},$$
$$E_5^* = \{\langle P_1, 0.29, 0.50\rangle, \langle P_2, 0.09, 0.70\rangle, \langle P_3, 0.41, 0.38\rangle\}.$$

When some of the estimators are incorrect, we can use the algorithms from Section 1.7 for correction of their estimations.

New Intuitionistic Fuzzy Operations

When discussions for the name "IFS" started and I understood that the general critique is that over the IFSs we can define operation "negation" that is not classical, I started searching the implications and negations that have non-classical nature. My first step (naturally, not in the most suitable direction) was to introduce the operators, defined in Section **5.8**. As it is mentioned there, one of them satisfies axiom $A \to \neg\neg A$, but does not satisfy axiom $\neg\neg A \to A$, and another satisfies axiom $\neg\neg A \to A$, but does not satisfy axiom $A \to \neg\neg A$, where operation $\neg$ is changed with one of these operators. On one hand, these operators are difficult to use, and on another – they are not close to the idea for the operation "negation". So, I started searching for another form of implication and of negation-type of operators, and in a series of papers [51, 52, 54, 55, 56, 57, 58, 59, 60, 62, 67, 81, 89, 92, 97, 109, 110, 543] more than 140 different operations for implication and more than 35 different operations for negation are defined. The research is in two aspects: logical and set-theoretical. Firstly, we give short remarks on the logical aspect. I hope that in future these results will be a basis for an independent book. After this, the basic results related to implication and negation operations over IFSs will be discussed and their basic properties will be described.

In this chapter we stop using the overline symbol for operation "negation" and instead we use symbol $\neg$. Since,a lot of different negations are introduce here, they are indexed $\neg_1$, $\neg_2$,... . Practically, the overline symbol coincides with the classical negation $\neg_1$.

9.1 Remarks on IFL. Part 4

During the last years a discussion started, relating to the name "intuitionistic" of the extension of the fuzzy set introduced by me in 1983 in [11]. I hope, after [108], that the concept of "intuitionistic fuzzy set", introduced by Takeuti and Titani in [510] one and a half years later to mine, is a particular case of the ordinary fuzzy sets, while my concept of an IFS essentially

K.T. Atanassov: On Intuitionistic Fuzzy Sets Theory, STUDFUZZ 283, pp. 195–257.
springerlink.com   © Springer-Verlag Berlin Heidelberg 2012

extends fuzzy sets. It is also clear that the construction of the IFSs contains Brauwer's idea for intuitionism [141]. After all, the discussion is very useful for IFS theory.

Firstly, as it is mentioned in Chapter 1, one of the most important author's mistakes is that more than 20 years in IFS theory only classical negation (denoted by $\neg_1$), which is given by the formula

$$\neg x = x \to 0,$$

or in the intuitionistic fuzzy case

$$\neg x = x \to \langle 0, 1 \rangle,$$

and a classical implication (denoted by $\to_1$), which satisfies the formula

$$x \to y = \neg x \vee y$$

have been used. Really, another implication (denoted by $\to_{11}$) and an intuitionistic fuzzy negation (denoted by $\neg_2$), were introduced in IFS theory [17], but for a long period of time, I had not estimated their merits as an intuitionistic implication and negation and I had not used them as they are more complex than the standard ones. Now, I realize this mistake (better late than never!) and the properties of different versions of both operations are discussed below.

Secondly, searching for an argument for defending the name of the IFS, I and some colleagues of me, found a lot of implications and negations. A part of them have classical, other - intuitionistic and third - even more nonstandard properties. These new operations are the object of discussion for the present Section.

9.1.1 Definitions and Properties of Some Intuitionistic Fuzzy Implications and Negations

The first ten variants of intuitionistic fuzzy implications are discussed in [51, 55], using the book [311] by George Klir and Bo Yuan as a basis, where the conventional fuzzy implications are given. Other five implications, defined by the author and his colleagues Boyan Kolev and Trifon Trifonov, are introduced in [52, 56, 58, 97, 109, 110].

Let us denote each one of these implications by $\mathcal{I}(x, y)$.

In IFL, if x is a variable then its truth-value is represented by the ordered couple

$$V(x) = \langle a, b \rangle,$$

so that $a, b, a + b \in [0, 1]$, where a and b are degrees of validity and of non-validity of x. Any other formula is estimated by analogy.

Throughout this Section, let us assume that for the three variables x, y and z there hold the equalities: $V(x) = \langle a, b \rangle, V(y) = \langle c, d \rangle, V(z) = \langle e, f \rangle$ where $a, b, c, d, e, f, a + b, c + d, e + f \in [0, 1]$.

For discussion, we use the notions of Intuitionistic Fuzzy Tautology (IFT)

$$\langle a, b \rangle \text{ iff } a \geq b$$

and tautology

$$\langle a, b \rangle \text{ iff } a = 1 \text{ and } b = 0.$$

Also, let

$$V(x) \leq V(y) \text{ iff } a \leq c \text{ and } b \geq d.$$

In some definitions, we use functions sg and $\overline{sg}$ defined by,

$$\text{sg}(x) = \begin{cases} 1 & \text{if } x > 0 \\ 0 & \text{if } x \leq 0 \end{cases},$$

$$\overline{sg}(x) = \begin{cases} 0 & \text{if } x > 0 \\ 1 & \text{if } x \leq 0 \end{cases}$$

In Table 9.1, we include the first 15 implications. There are a lot of other implications, defined by Dimiter Dimitrov [209], Lilija Atanassova [111, 113] and the author, that will be discussed in the next Section. New implications are introduced by Piotr Dworniczak in [222, 223, 224, 225].

Table 9.1 List of intuitionistic fuzzy implications

Notation	Name	Form of implication
$\rightarrow_1$	Zadeh	$\langle \max(b, \min(a, c)), \min(a, d) \rangle$
$\rightarrow_2$	Gaines-Rescher	$\langle 1 - \text{sg}(a - c), d.\text{sg}(a - c) \rangle$
$\rightarrow_3$	Gödel	$\langle 1 - (1 - c).\text{sg}(a - c), d.\text{sg}(a - c) \rangle$
$\rightarrow_4$	Kleene-Dienes	$\langle \max(b, c), \min(a, d) \rangle$
$\rightarrow_5$	Lukasiewicz	$\langle \min(1, b + c), \max(0, a + d - 1) \rangle$
$\rightarrow_6$	Reichenbach	$\langle b + ac, ad \rangle$
$\rightarrow_7$	Willmott	$\langle \min(\max(b, c), \max(a, b), \max(c, d)), \max(\min(a, d), \min(a, b), \min(c, d)) \rangle$
$\rightarrow_8$	Wu	$\langle 1 - (1 - \min(b, c)).\text{sg}(a - c), \max(a, d).\text{sg}(a - c).\text{sg}(d - b) \rangle$
$\rightarrow_9$	Klir and Yuan 1	$\langle b + a^2 c, ab + a^2 d \rangle$
$\rightarrow_{10}$	Klir and Yuan 2	$\langle c.\overline{sg}(1 - a) + \text{sg}(1 - a).(\overline{sg}(1 - c) + b.\text{sg}(1 - c)), d.\overline{sg}(1 - a) + a.\text{sg}(1 - a).\text{sg}(1 - c) \rangle$
$\rightarrow_{11}$	Atanassov 1	$\langle 1 - (1 - c).\text{sg}(a - c), d.\text{sg}(a - c).\text{sg}(d - b) \rangle$

Table 9.1 (*continued*)

$\rightarrow_{12}$	Atanassov 2	$\langle \max(b,c), 1 - \max(b,c) \rangle$
$\rightarrow_{13}$	Atanassov and Kolev	$\langle b + c - b.c, a.d \rangle$
$\rightarrow_{14}$	Atanassov and Trifonov	$\langle 1 - (1-c).\mathrm{sg}(a-c) - d.\overline{sg}(a-c)$ $.\mathrm{sg}(d-b), \; d.\mathrm{sg}(d-b) \rangle$
$\rightarrow_{15}$	Atanassov 3	$\langle 1 - (1 - \min(b,c)).\mathrm{sg}(\mathrm{sg}(a-c)$ $+\mathrm{sg}(d-b)) - \min(b,c).\mathrm{sg}(a-c).\mathrm{sg}(d-b),$ $1 - (1 - \max(a,d)).\mathrm{sg}(\overline{sg}(a-c) + \overline{sg}(d-b))$ $- \max(a,d).\overline{sg}(a-c).\overline{sg}(d-b) \rangle$

The correctness of the definitions of the above implications is directly checked. For example, to check the validity of the definition of $\rightarrow_{15}$, we check that for every $a, b, c, d \in [0,1]$ such that $a+b \leq 1$ and $c+d \leq 1$ for expression,

$$X \equiv 1 - (1 - \min(b,c)).\mathrm{sg}(\mathrm{sg}(a-c) + \mathrm{sg}(d-b)) - \min(b,c).\mathrm{sg}(a-c).\mathrm{sg}(d-b)$$

$$+1 - (1 - \max(a,d)).\mathrm{sg}(\overline{sg}(a-c) + \overline{sg}(d-b)) - \max(a,d).\overline{sg}(a-c).\overline{sg}(d-b)$$

we obtain the following.
If $a \leq c$ and $b \geq d$, then,

$$X = 1 - (1 - \min(b,c)).\mathrm{sg}(0+0) - \min(b,c).0 + 1$$

$$-(\max(a,d)).\mathrm{sg}(1+1) - \max(a,d).1$$

$$= 1 + 1 - (1 - \max(a,d)) - \max(a,d) = 1.$$

If $a \leq c$ and $b < d$, then,

$$X = 1 - (1 - \min(b,c)).\mathrm{sg}(0+1) - \min(b,c).0.1$$

$$+1 - (1 - \max(a,d)).\mathrm{sg}(1+0) - \max(a,d).1.0$$

$$= 1 - 1 + \min(b,c) + 1 - 1 + \max(a,d)$$

$$= \min(b,c) + \max(a,d) \leq 1.$$

If $a > c$ and $b \geq d$, then,

$$X = 1 - (1 - \min(b,c)).\mathrm{sg}(1+0) - \min(b,c).1.0$$

$$+1 - (1 - \max(a,d)).\mathrm{sg}(0+1) - \max(a,d).0.1$$

$$= 1 - 1 + \min(b,c) + 1 - 1 + \max(a,d)$$

$$= \min(b,c) + \max(a,d) \leq 1.$$

If $a > c$ and $b < d$ then,

$$X = 1 - (1 - \min(b,c)).\mathrm{sg}(1+1) - \min(b,c).1.1$$

$$+1 - (1 - \max(a, d)).\text{sg}(0 + 0) - \max(a, d).0.0$$

$$= 1 - (1 - \min(b, c)).1 - \min(b, c) + 1 - (1 - \max(a, d)).0$$

$$= 1 - 1 + \min(b, c)) - \min(b, c) + 1 = 1.$$

Therefore, the definition of the implication $\to_{15}$ is valid.

Now, we introduce an implication, inspired by G. Takeuti and S. Titani's paper [510] and its form given by T. Trifonov and the author in [108].

In [510] G. Takeuti and S.Titani introduced the following implication. For $p, q \in [0, 1]$

$$p \to q = \bigvee \{ r \in [0, 1] \mid p \wedge r \leq q \} = \begin{cases} 1, & \text{if } p \leq q \\ q, & \text{if } p > q \end{cases}$$

In [108], its intuitionistic fuzzy extension is given in the form

$$\langle a, b \rangle \to \langle c, d \rangle = \langle \max(c, \overline{sg}(a - c)), \min(d, \text{sg}(a - c)) \rangle$$

$$= \begin{cases} \langle 1, 0 \rangle, & \text{if } a \leq c \text{ and } b \geq d \\ \langle 1, 0 \rangle, & \text{if } a \leq c \text{ and } b < d \\ \langle c, d \rangle, & \text{if } a > c \text{ and } b \geq d \\ \langle c, d \rangle, & \text{if } a > c \text{ and } b < d \end{cases}$$

and it has been proved that the latter implication coincides with Gödel's implication $(\to_3)$, because if

$$X \equiv 1 - (1 - c).\text{sg}(a - c) - \max(c, \overline{sg}(a - c)),$$

$$Y \equiv d.\text{sg}(a - c) - \min(d, \text{sg}(a - c)),$$

then we obtain the following. If $a > c$,

$$X - 1 - (1 - c) - \max(c, 0) = c - c = 0,$$

$$Y = d.1 - \min(d, 1) = d - d = 0.$$

If $a \leq c$,

$$X = 1 - (1 - c).0 - \max(c, 1) = 1 - 1 = 0,$$

$$Y = d.0 - \min(d, 0) = 0 - 0 = 0,$$

i.e.

$$\langle 1 - (1 - c).\text{sg}(a - c), d.\text{sg}(a - c) \rangle = \langle \max(c, \overline{sg}(a - c)), \min(d, \text{sg}(a - c)) \rangle.$$

All constructions in [510] are re-written in [108] for the intuitionistic fuzzy case with the aim to show that Takeuti and Titani's sets are particular cases of intuitionistic fuzzy sets in the present sense. Practically, the set constructed by them is an ordinary fuzzy set with elements, satisfying intuitionistic logic axioms. In [108], it is shown that there is an *intuitionistic fuzzy set* with elements, satisfying intuitionistic logic axioms.

Let I be an arbitrary one of the above implications and has the form

$$I(x, y) = x \to y,$$

so that

$$I(\langle 0, 1 \rangle, \langle 0, 1 \rangle) = \langle 1, 0 \rangle,$$

$$I(\langle 0, 1 \rangle, \langle 1, 0 \rangle) = \langle 1, 0 \rangle,$$

$$I(\langle 1, 0 \rangle, \langle 1, 0 \rangle) = \langle 1, 0 \rangle,$$

$$I(\langle 1, 0 \rangle, \langle 0, 1 \rangle) = \langle 0, 1 \rangle.$$

Therefore, the restriction of each of these implications coincides over constants *false* and *true* with the implication from the ordinary propositional calculus.

We introduce the expression

$$Z_i = x \to_i y,$$

where $1 \le i \le 15$. We say that Z_i is more powerful than Z_j for $1 \le i, j \le 15$, if

$$V(Z_i) \ge V(Z_j).$$

We construct Table 9.2 in which the lack of relation between two implications is denoted by " $*$ ".

Table 9.2 Relation between elements of set $\{Z_i | 1 \le i \le 15\}$

	Z_1	Z_2	Z_3	Z_4	Z_5	Z_6	Z_7	Z_8	Z_9	Z_{10}	Z_{11}	Z_{12}	Z_{13}	Z_{14}	Z_{15}
Z_1	$=$	$*$	$*$	$\le$	$\le$	$*$	$\ge$	$*$	$*$	$*$	$*$	$*$	$\le$	$*$	$*$
Z_2	$*$	$=$	$\le$	$*$	$*$	$*$	$*$	$*$	$*$	$*$	$\le$	$*$	$*$	$*$	$*$
Z_3	$*$	$\ge$	$=$	$*$	$*$	$*$	$*$	$*$	$*$	$*$	$\le$	$*$	$*$	$*$	$\ge$
Z_4	$\ge$	$*$	$*$	$=$	$\le$	$*$	$\ge$	$*$	$*$	$\ge$	$*$	$\ge$	$\le$	$*$	$*$
Z_5	$\ge$	$*$	$*$	$\ge$	$=$	$\ge$	$\ge$	$*$	$*$	$\ge$	$*$	$\ge$	$\ge$	$*$	$*$
Z_6	$*$	$*$	$*$	$*$	$\le$	$=$	$*$	$*$	$*$	$*$	$*$	$*$	$\le$	$*$	$*$
Z_7	$\le$	$*$	$*$	$\le$	$\le$	$*$	$=$	$*$	$*$	$*$	$*$	$*$	$\le$	$*$	$*$
Z_8	$*$	$*$	$*$	$*$	$*$	$*$	$*$	$=$	$*$	$*$	$\le$	$*$	$*$	$*$	$\ge$
Z_9	$*$	$*$	$*$	$*$	$*$	$*$	$*$	$*$	$=$	$*$	$*$	$*$	$*$	$*$	$*$
Z_{10}	$*$	$*$	$*$	$\le$	$\le$	$*$	$*$	$*$	$*$	$=$	$*$	$*$	$\le$	$*$	$*$
Z_{11}	$*$	$\ge$	$\ge$	$*$	$*$	$*$	$*$	$\ge$	$*$	$*$	$=$	$*$	$*$	$\ge$	$\ge$
Z_{12}	$*$	$*$	$*$	$\le$	$\le$	$*$	$*$	$*$	$*$	$*$	$*$	$=$	$\le$	$*$	$*$
Z_{13}	$\ge$	$*$	$*$	$\ge$	$\le$	$\ge$	$\ge$	$*$	$*$	$\ge$	$*$	$\ge$	$=$	$*$	$*$
Z_{14}	$*$	$*$	$*$	$*$	$*$	$*$	$*$	$*$	$*$	$*$	$\le$	$*$	$*$	$=$	$\ge$
Z_{15}	$*$	$*$	$\le$	$*$	$*$	$*$	$*$	$\le$	$*$	$*$	$\le$	$*$	$*$	$\le$	$=$

We construct the oriented graph in Fig. 9.1, where the arcs are directed top-down.

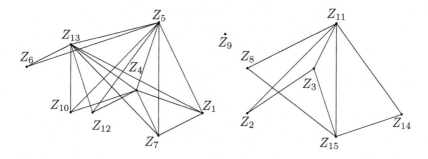

Fig. 9.1

Now, corresponding negations to each of the first 15 of the above implications are constructed. These negations are given in Table 9.3.

We see that the negations are from the five different types and they are given in Table 9.4.

Table 9.3 List of intuitionistic fuzzy negations

Name	Form of negation
Zadeh	$\langle b, a \rangle$
Gaines-Rescher	$\langle 1 - \mathrm{sg}(a), \mathrm{sg}(a) \rangle$
Gödel	$\langle 1 - \mathrm{sg}(a), \mathrm{sg}(a) \rangle$
Kleene-Dienes	$\langle b, a \rangle$
Lukasiewicz	$\langle b, a \rangle$
Reichenbach	$\langle b, a \rangle$
Willmott	$\langle b, a \rangle$
Wu	$\langle 1 - \mathrm{sg}(a), \mathrm{sg}(a) \rangle$
Klir and Yuan 1	$\langle b, a.b + a^2 \rangle$
Klir and Yuan 2	$\langle b, a \rangle$
Atanassov 1	$\langle 1 - \mathrm{sg}(a), \mathrm{sg}(a) \rangle$
Atanassov 2	$\langle b, 1 - b \rangle$
Atanassov and Kolev	$\langle b, a \rangle$
Atanassov and Trifonov	$\langle \overline{sg}(1 - b), \mathrm{sg}(1 - b) \rangle$
Atanassov 3	$\langle \overline{sg}(1 - b), \mathrm{sg}(1 - b) \rangle$

The following three properties are checked In [55] for the separate negations:
Property P1: $A \rightarrow \neg\neg A$,
Property P2: $\neg\neg A \rightarrow A$,
Property P3: $\neg\neg\neg A = \neg A$.

Table 9.4 List of the distinct intuitionistic fuzzy negations

Notation	Form of negation
$\neg_1$	$\langle b, a \rangle$
$\neg_2$	$\langle 1 - \mathrm{sg}(a), \mathrm{sg}(a) \rangle$
$\neg_3$	$\langle b, a.b + a^2 \rangle$
$\neg_4$	$\langle b, 1 - b \rangle$
$\neg_5$	$\langle \overline{sg}(1 - b), \mathrm{sg}(1 - b) \rangle$

It is proved that each of the negations in Table 9.4 satisfies Property 1 and Property 3; only negation $\neg_1$ satisfies Property 2, while the other negations do not satisfy it. For example, the check of the latest assertion for the case of negation $\neg_5$ is:

$$\neg_5 \neg_5 \neg_5 \langle a, b \rangle = \neg_5 \neg_5 \langle \overline{sg}(1 - b), \mathrm{sg}(1 - b) \rangle$$

$$= \neg_5 \langle \overline{sg}(1 - \mathrm{sg}(1 - b)), \mathrm{sg}(1 - \mathrm{sg}(1 - b)) \rangle$$

$$= \langle \overline{sg}(1 - \mathrm{sg}(1 - \mathrm{sg}(1 - b))), \mathrm{sg}(1 - \mathrm{sg}(1 - \mathrm{sg}(1 - b))) \rangle.$$

Let

$$X \equiv \overline{sg}(1 - \mathrm{sg}(1 - \mathrm{sg}(1 - b))) - \overline{sg}(1 - b).$$

If $b = 1$, then

$$X = \overline{sg}(1 - \mathrm{sg}(1)) - \overline{sg}(0) = \overline{sg}(0) - 1 = 0.$$

If $b < 1$, then

$$X = \overline{sg}(1 - \mathrm{sg}(1 - 1)) - 0 = \overline{sg}(1) = 0.$$

Following [55, 81], we study the relations between the distinct negations, by checking the validity of Table 9.5.

The lack of relation between two negations is noted in Table 9.5 by symbol " $*$ ".

Table 9.5 List of the relations between the distinct intuitionistic fuzzy negations

	$\neg_1$	$\neg_2$	$\neg_3$	$\neg_4$	$\neg_5$
$\neg_1$	$=$	$*$	$\leq$	$\geq$	$\geq$
$\neg_2$	$*$	$=$	$*$	$*$	$\geq$
$\neg_3$	$\geq$	$*$	$=$	$\geq$	$\geq$
$\neg_4$	$\leq$	$*$	$\leq$	$=$	$\geq$
$\neg_5$	$\leq$	$\leq$	$*$	$\leq$	$=$

Therefore, the ordered graph (see Fig. 9.2) with vertices $\neg_1, ..., \neg_5$, can be constructed.

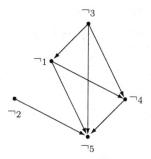

Fig. 9.2

In [311], George Klir and Bo Yuan discussed the following 9 axioms related to fuzzy implications.

Axiom 1 $(\forall x, y)(x \leq y \to (\forall z)(I(x, z) \geq I(y, z)))$.
Axiom 2 $(\forall x, y)(x \leq y \to (\forall z)(I(z, x) \leq I(z, y)))$.
Axiom 3 $(\forall y)(I(0, y) = 1)$.
Axiom 4 $(\forall y)(I(1, y) = y)$.
Axiom 5 $(\forall x)(I(x, x) = 1)$.
Axiom 6 $(\forall x, y, z)(I(x, I(y, z)) = I(y, I(x, z)))$.
Axiom 7 $(\forall x, y)(I(x, y) = 1$ iff $x \leq y)$.
Axiom 8 $(\forall x, y)(I(x, y) = I(N(y), N(x)))$, where N is an operation for negation.
Axiom 9 I is a continuous function.

We must note that Table 9.6 summarizes the axioms of Klir and Yuan which are satisfied by the 15 implications.

If the axiom is valid as an IFT, the number of the axiom is marked with an asterisk (*). These axioms are

Axiom 3* $(\forall y)(I(0, y)$ is an IFT).
Axiom 5* $(\forall x)(I(x, x)$ is an IFT).
Axiom 7* $(\forall x, y)(I(x, y)$ is an IFT iff $x \leq y)$.

We should note that Axiom 8 is checked using the classical intuitionistic fuzzy negation ($\neg_1$); if it is valid using the respective generated negation as $N(x)$, then the axiom is marked as 8^N. We should also note that the validity of Axiom 7 does not imply the validity of Axiom 7*.

The validity of each of these assertions can be directly checked. In some cases, this is a long and tedious procedure, which can be automatically performed by a proof-checking program suggested in [210, 542] by Dimiter Dimitrov and Trifon Trifonov.

Table 9.6 List of axioms of Klir and Yuan that are satisfied by intuitionistic fuzzy implications

Notation	Axioms
$\to_1$	2,3,4,5*,9
$\to_2$	1,2,3,5

Table 9.6 (*continued*)

$\rightarrow_3$	1,2,3,4,5,6
$\rightarrow_4$	1,2,3,4,5*,6,8,9
$\rightarrow_5$	1,2,3,4,5*,6,8,9
$\rightarrow_6$	2,3,4,5*,9
$\rightarrow_7$	3*,4,5*,8,9
$\rightarrow_8$	1,2,3,5
$\rightarrow_9$	2,3,4,5*
$\rightarrow_{10}$	2,3,4
$\rightarrow_{11}$	1,2,3,4,5,6
$\rightarrow_{12}$	1,2,3,6,8,9
$\rightarrow_{13}$	1,2,3,4,5*,6,8,9
$\rightarrow_{14}$	1,2,3,4,5,6,7
$\rightarrow_{15}$	1,2,3,5,7,7*,8

9.1.2   On Law of Excluded Middle and Its Modifications

Following [52, 81], we study the validity of the Law of Excluded Middle (LEM) in the forms

$$\langle a, b\rangle \vee \neg\langle a, b\rangle = \langle 1, 0\rangle$$

(tautology-form) and

$$\langle a, b\rangle \vee \neg\langle a, b\rangle = \langle p, q\rangle,$$

(IFT-form), and a Modified LEM in the forms

$$\neg\neg\langle a, b\rangle \vee \neg\langle a, b\rangle = \langle 1, 0\rangle$$

(tautology-form) and

$$\neg\neg\langle a, b\rangle \vee \neg\langle a, b\rangle = \langle p, q\rangle,$$

(IFT-form), where $1 \geq p \geq q \geq 0$ and $i = 1, 2, ..., 5$.

In [52], the following assertions are proved:

- no negation satisfies the LEM in the tautological form;
- negations $\neg_1, \neg_3$ and $\neg_4$ satisfy the LEM in the IFT-form;
- only $\neg_2$ and $\neg_5$ satisfy the Modified LEM in the tautological form;
- all negations satisfy the Modified LEM in the IFT-form.

As illustration, the third and fourth assertions for the case of implication $\neg_5$ are proved here.

$$\neg_5\neg_5\langle a, b\rangle \vee \neg_5\langle a, b\rangle$$

$$= \langle \overline{sg}(1 - \text{sg}(1 - b)), \text{sg}(1 - \text{sg}(1 - b))\rangle \vee \langle \overline{sg}(1 - b), \text{sg}(1 - b)\rangle$$

$$= \langle \max(\overline{sg}(1 - \text{sg}(1 - b)), \overline{sg}(1 - b)), \min(\text{sg}(1 - \text{sg}(1 - b)), \text{sg}(1 - b))\rangle.$$

If $b = 1$, then

$$\max(\overline{sg}(1 - sg(1 - b)), \overline{sg}(1 - b)) = \max(\overline{sg}(1), 1) = \max(0, 1) = 1.$$

and

$$\min(sg(1 - sg(1 - b)), sg(1 - b)) = \min(sg(1), 0) = \min(1, 0) = 0.$$

If $b < 1$, then

$$\max(\overline{sg}(1 - sg(1 - b)), \overline{sg}(1 - b)) = \max(\overline{sg}(0), 0) = \max(1, 0) = 1.$$

and

$$\min(sg(1 - sg(1 - b)), sg(1 - b)) = \min(sg(0), 1) = \min(0, 1) = 0.$$

Therefore, negation $\neg_5$ satisfies the Modified LEM in the IFT-form. On the other hand, in all cases the evaluation of the expression is equal to $\langle 1, 0 \rangle$, i.e., this negation satisfies the Modified LEM in the tautological form.

9.1.3   On De Morgan's Laws and Their Modifications

Following [57], some forms of De Morgan's Laws are discussed in this Section.
Usually, De Morgan's Laws have the forms

$$\neg x \wedge \neg y = \neg(x \vee y),$$

$$\neg x \vee \neg y = \neg(x \wedge y).$$

For every two propositional forms x and y,

$$\neg_i x \wedge \neg_i y = \neg_i(x \vee y),$$

$$\neg_i x \vee \neg_i y = \neg_i(x \wedge y)$$

for $i = 1, 2, 4, 5$, while negation $\neg_3$ does not satisfy these equalities.
We illustrate only the fact that the De Morgan's Laws are not valid for $i = 3$. For example, if $a = b = 0.5, c = 0.1, d = 0$, then

$$V(\neg_3 x \wedge \neg_3 y) = 0.5$$

$$V(\neg_3(x \vee y)) = 0.25.$$

The above mentioned change of the LEM inspired the idea to study the validity of De Morgan's Laws that the classical negation $\neg$ (here it is negation $\neg_1$) satisfies. It can easily be proved that the expressions

$$\neg_1(\neg_1 x \vee \neg_1 y) = x \wedge y$$

and

$$\neg_1(\neg_1 x \wedge \neg_1 y) = x \vee y$$

are IFTs, but the other negations do not satisfy these equalities. On the other hand, for every two propositional forms x and y,

$$\neg_I(\neg_i x \vee \neg_i y) = \neg_i \neg_i x \wedge \neg_i \neg_i y$$

and

$$\neg_I(\neg_i x \wedge \neg_i y) = \neg_i \neg_i x \vee \neg_i \neg_i y$$

for $i = 1, 2, 4, 5$, while negation $\neg_3$ does not satisfy these equalities.

Now, we discuss another form of the LEM that in propositional calculus is equivalent to the standard LEM, if De Morgan's Laws are valid. It is the following:

$$\neg(x \wedge \neg x).$$

Negations $\neg_2$ and $\neg_5$ satisfy it as tautologies, negations $\neg_1$, $\neg_3$ and $\neg_4$ satisfy it as IFTs.

9.1.4　On the Axioms of Propositional Intuitionistic Logic

The next and more important question is, which of the introduced implications and negations satisfy all the axioms of Propositional Intuitionistic Logic (IL, see, e.g., [401])? We use the following list of axioms for propositional IL:

(a) $A \to A$,

(b) $A \to (B \to A)$,

(c) $A \to (B \to (A\&B))$,

(d) $(A \to (B \to C)) \to (B \to (A \to C))$,

(e) $(A \to (B \to C)) \to ((A \to B) \to (A \to C))$,

(f) $A \to \neg\neg A$,

(g) $\neg(A\&\neg A)$,

(h) $(\neg A \vee B) \to (A \to B)$,

(i) $\neg(A \vee B) \to (\neg A\&\neg B)$,

(j) $(\neg A\&\neg B) \to \neg(A \vee B)$,

(k) $(\neg A \vee \neg B) \to \neg(A\&B)$,

(l) $(A \to B) \to (\neg B \to \neg A)$,

(m) $(A \to \neg B) \to (B \to \neg A)$,

(n) $\neg\neg\neg A \to \neg A$,

(o) $\neg A \to \neg\neg\neg A$,

(p) $\neg\neg(A \to B) \to (A \to \neg\neg B)$,

(q) $(C \to A) \to ((C \to (A \to B)) \to (C \to B))$.

The validity of the IL axioms was already checked for some implications in [543]. Here, following [81, 211], we give a full list of valid axioms for each of the 15 implications. We again verify the validity axioms in two variants - tautological validity (Table 9.7) and IFT validity (Table 9.8).

Table 9.7 List of axioms of the intuitionistic logic that are satisfied by intuitionistic fuzzy implications as tautologies

	a	b	c	d	e	f	g	h	i	j	k	l	m	n	o	p	q
$\rightarrow_1$	-	-	-	-	-	-	-	-	-	-	-	-	-	-	-	-	-
$\rightarrow_2$	+	-	-	-	+	+	+	-	+	+	+	+	+	+	+	+	+
$\rightarrow_3$	+	+	+	+	+	+	+	+	+	+	+	+	+	+	+	+	+
$\rightarrow_4$	-	-	-	-	-	-	-	-	-	-	-	-	-	-	-	-	-
$\rightarrow_5$	-	-	-	-	-	-	-	-	-	-	-	-	-	-	-	-	-
$\rightarrow_6$	-	-	-	-	-	-	-	-	-	-	-	-	-	-	-	-	-
$\rightarrow_7$	-	-	-	-	-	-	-	-	-	-	-	-	-	-	-	-	-
$\rightarrow_8$	+	-	-	-	-	+	+	-	+	+	+	+	+	+	+	+	-
$\rightarrow_9$	-	-	-	-	-	-	-	-	-	-	-	-	-	-	-	-	-
$\rightarrow_{10}$	-	-	-	-	-	-	-	-	-	-	-	-	-	-	-	-	-
$\rightarrow_{11}$	+	+	+	+	+	+	+	+	+	+	+	+	+	+	+	+	+
$\rightarrow_{12}$	-	-	-	-	-	-	-	-	-	-	-	-	-	-	-	-	-
$\rightarrow_{13}$	-	-	-	-	-	-	-	-	-	-	-	-	-	-	-	-	-
$\rightarrow_{14}$	+	+	+	+	+	+	+	+	+	+	+	+	+	+	+	+	+
$\rightarrow_{15}$	+	-	-	-	-	+	+	-	+	+	+	+	+	+	+	+	-

Table 9.7 shows that only implications $\rightarrow_3$, $\rightarrow_{11}$, $\rightarrow_{14}$, satisfy all intuitionistic logic axioms as tautologies.

Table 9.8 List of axioms of the intuitionistic logic that are satisfied by intuitionistic fuzzy implications as IFTs

	a	b	c	d	e	f	g	h	i	j	k	l	m	n	o	p	q
$\rightarrow_1$	+	+	+	+	+	+	+	+	+	+	+	+	+	+	+	+	+
$\rightarrow_2$	+	-	-	-	+	+	+	-	+	+	+	+	+	+	+	+	+
$\rightarrow_3$	+	+	+	+	+	+	+	+	+	+	+	+	+	+	+	+	+
$\rightarrow_4$	+	+	+	+	+	+	+	+	+	+	+	+	+	+	+	+	+
$\rightarrow_5$	+	+	+	+	+	+	+	+	+	+	+	+	+	+	+	+	+
$\rightarrow_6$	+	+	+	+	-	+	+	+	+	+	+	+	+	+	+	+	+
$\rightarrow_7$	+	-	-	-	-	+	+	-	+	+	+	+	+	+	+	+	-
$\rightarrow_8$	+	-	-	+	+	+	+	-	+	+	+	+	+	+	+	+	+
$\rightarrow_9$	+	+	+	+	+	+	+	+	+	+	+	+	+	+	+	+	+
$\rightarrow_{10}$	-	-	-	-	-	-	+	-	-	-	-	-	-	-	-	-	-
$\rightarrow_{11}$	+	+	+	+	+	+	+	+	+	+	+	+	+	+	+	+	+
$\rightarrow_{12}$	-	-	-	+	+	+	+	-	+	+	+	+	+	+	+	+	+
$\rightarrow_{13}$	+	+	+	+	+	+	+	+	+	+	+	+	+	+	+	+	+
$\rightarrow_{14}$	+	+	+	+	+	+	+	+	+	+	+	+	+	+	+	+	+
$\rightarrow_{15}$	+	-	-	-	-	+	+	-	+	+	+	+	+	+	+	+	-

Table 9.8 shows that implications $\rightarrow_1$, $\rightarrow_3$, $\rightarrow_4$, $\rightarrow_5$, $\rightarrow_9$, $\rightarrow_{11}$, $\rightarrow_{13}$, $\rightarrow_{14}$ satisfy all intuitionistic logic as IFTs.

9.1.5   A New Argument That the Intuitionistic Fuzzy Sets Have Intuitionistic Nature

The above assertions show that, all negations except the first one satisfy the conditions of the IL, but not of a classical logic. A part of these negations were generated by implications, that were generated by fuzzy implications. Now, let us return from the intuitionistic fuzzy negations to ordinary fuzzy negations. The result is shown in Table 9.9, where $b = 1 - a$.

Table 9.9 List of the fuzzy negations, generated by intuitionistic fuzzy negations

Notation	Form of the intuitionistic fuzzy negation	Form of the fuzzy negation
$\neg_1$	$\langle b, a \rangle$	$1 - a$
$\neg_2$	$\langle 1 - \mathrm{sg}(a), \mathrm{sg}(a) \rangle$	$1 - \mathrm{sg}(a)$
$\neg_3$	$\langle b, a.b + a^2 \rangle$	$1 - a$
$\neg_4$	$\langle b, 1 - b \rangle$	$1 - a$
$\neg_5$	$\langle \overline{sg}(1 - b), \mathrm{sg}(1 - b) \rangle$	$1 - \mathrm{sg}(a)$

Therefore, from the intuitionistic fuzzy negations we can generate fuzzy negations, so that two of them ($\neg_3$ and $\neg_4$) coincide with the standard fuzzy negation ($\neg_1$). Therefore, there are intuitionistic fuzzy negations that lose their properties when they are restricted to ordinary fuzzy case. In other words, the construction of the intuitionistic fuzzy estimation

$$\langle \text{degree of membership/validity}, \text{degree of non-membership/non-validity} \rangle$$

that is specific for the intuitionistic fuzzy sets, is the reason for the intuitionistic behaviour of these sets. Intuitionistic as well as classical negations can be defined over these sets.

In fuzzy case, negations $\neg_2$ and $\neg_5$ coincide, generating a fuzzy negation that satisfies Properties 1 and 3 and does not satisfy Property 2, i.e., it is intuitionistic in nature.

9.2   IF Implications and Negations over IFSs

9.2.1   Definitions and Properties of IF Implications and Negations

After the above remarks, where the first steps of introducing intuitionistic fuzzy implications and negation in the case of IFL are described, we start directly with the same definitions, but for the case of IFSs.

The forms of the 138 implications over IFSs are given in Table 9.10, while the 34 negations are described on Table 9.11.

In Table 9.12 the relations between implications and negations are shown.

Table 9.10 List of the intuitionistic fuzzy implications

$\rightarrow_1$	$\{\langle x, \max(\nu_A(x), \min(\mu_A(x), \mu_B(x))), \min(\mu_A(x), \nu_B(x))\rangle \mid x \in E\}$
$\rightarrow_2$	$\{\langle x, \overline{sg}(\mu_A(x) - \mu_B(x)), \nu_B(x).sg(\mu_A(x) - \mu_B(x))\rangle \mid x \in E\}$
$\rightarrow_3$	$\{\langle x, 1 - (1 - \mu_B(x)).sg(\mu_A(x) - \mu_B(x)))),$ $\nu_B(x).sg(\mu_A(x) - \mu_B(x))\rangle \mid x \in E\}$
$\rightarrow_4$	$\{\langle x, \max(\nu_A(x), \mu_B(x)), \min(\mu_A(x), \nu_B(x))\rangle \mid x \in E\}$
$\rightarrow_5$	$\{\langle x, \min(1, \nu_A(x) + \mu_B(x)), \max(0, \mu_A(x) + \nu_B(x) - 1)\rangle \mid x \in E\}$
$\rightarrow_6$	$\{\langle x, \nu_A(x) + \mu_A(x)\mu_B(x), \mu_A(x)\nu_B(x)\rangle \mid x \in E\}$
$\rightarrow_7$	$\{\langle x, \min(\max(\nu_A(x), \mu_B(x)), \max(\mu_A(x), \nu_A(x)),$ $\max(\mu_B(x), \nu_B(x))), \ \max(\min(\mu_A(x), \nu_B(x)),$ $\min(\mu_A(x), \nu_A(x)), \min(\mu_B(x), \nu_B(x)))\rangle \mid x \in E\}$
$\rightarrow_8$	$\{\langle x, 1 - (1 - \min(\nu_A(x), \mu_B(x))).sg(\mu_A(x) - \mu_B(x)),$ $\max(\mu_A(x), \nu_B(x)).sg(\mu_A(x) - \mu_B(x)),$ $sg(\nu_B(x) - \nu_A(x))\rangle \mid x \in E\}$
$\rightarrow_9$	$\{\langle x, \nu_A(x) + \mu_A(x)^2\mu_B(x), \mu_A(x)\nu_A(x) + \mu_A(x)^2\nu_B(x)\rangle \mid x \in E\}$
$\rightarrow_{10}$	$\{\langle x, \mu_B(x).\overline{sg}(1 - \mu_A(x))$ $+ sg(1 - \mu_A(x)).(\overline{sg}(1 - \mu_B(x)) + \nu_A(x).sg(1 - \mu_B(x))),$ $\nu_B(x).\overline{sg}(1 - \mu_A(x)) + \mu_A(x).sg(1 - \mu_A(x)).sg(1 - \mu_B(x))\rangle \mid x \in E\}$
$\rightarrow_{11}$	$\{\langle x, 1 - (1 - \mu_B(x)).sg(\mu_A(x) - \mu_B(x)),$ $\nu_B(x).sg(\mu_A(x) - \mu_B(x)).sg(\nu_B(x) - \nu_A(x))\rangle \mid x \in E\}$
$\rightarrow_{12}$	$\{\langle x, \max(\nu_A(x), \mu_B(x)), 1 - \max(\nu_A(x), \mu_B(x))\rangle \mid x \in E\}$
$\rightarrow_{13}$	$\{\langle x, \nu_A(x) + \mu_B(x) - \nu_A(x).\mu_B(x), \mu_A(x).\nu_B(x)\rangle \mid x \in E\}$
$\rightarrow_{14}$	$\{\langle x, 1 - (1 - \mu_B(x)).sg(\mu_A(x) - \mu_B(x))$ $- \nu_B(x).\overline{sg}(\mu_A(x) - \mu_B(x)).sg(\nu_B(x) - \nu_A(x)),$ $\nu_B(x).sg(\nu_B(x) - \nu_A(x))\rangle \mid x \in E\}$
$\rightarrow_{15}$	$\{\langle x, 1 - (1 - \min(\nu_A(x), \mu_B(x)))$ $.sg(sg(\mu_A(x)) - \mu_B(x)) + sg(\nu_B(x)) - \nu_A(x)))$ $\min(\nu_A(x)), \mu_B(x)) \ sg(\mu_A(x) - \mu_B(x)).sg(\nu_B(x) - \nu_A(x)),$ $1 - (1 - \max(\mu_A(x), \nu_B(x))).sg(\overline{sg}(\mu_A(x) - \mu_B(x))$ $+ \overline{sg}(\nu_B(x) - \nu_A(x))) - \max(\mu_A(x), \nu_B(x)).\overline{sg}(\mu_A(x) - \mu_B(x))$ $.\overline{sg}(\nu_B(x) - \nu_A(x))\rangle \mid x \in E\}$
$\rightarrow_{16}$	$\{\langle x, \max(\overline{sg}(\mu_A(x)), \mu_B(x)), \min(sg(\mu_A(x)), \nu_B(x))\rangle \mid x \in E\}$
$\rightarrow_{17}$	$\{\langle x, \max(\nu_A(x), \mu_B(x)), \min(\mu_A(x).\nu_A(x) + \mu_A(x)^2, \nu_B(x))\rangle \mid x \in E\}$
$\rightarrow_{18}$	$\{\langle x, \max(\nu_A(x), \mu_B(x)), \min(1 - \nu_A(x), \nu_B(x))\rangle \mid x \in E\}$
$\rightarrow_{19}$	$\{\langle x, \max(1 - sg(sg(\mu_A(x)) + sg(1 - \nu_A(x))), \mu_B(x)),$ $\min(sg(1 - \nu_A(x)), \nu_B(x))\rangle \mid x \in E\}$
$\rightarrow_{20}$	$\{\langle x, \max(\overline{sg}(\mu_A(x)), sg(\mu_B(x))), \min(sg(\mu_A(x)), \overline{sg}(\mu_B(x)))\rangle$ $\mid x \in E\}$
$\rightarrow_{21}$	$\{\langle x, \max(\nu_A(x), \mu_B(x).(\mu_B(x) + \nu_B(x))),$ $\min(\mu_A(x).(\mu_A(x) + \nu_A(x)), \nu_B(x).$ $(\mu_B(x)^2 + \nu_B(x) + \mu_B(x).\nu_B(x)))\rangle \mid x \in E\}$
$\rightarrow_{22}$	$\{\langle x, \max(\nu_A(x), 1 - \nu_B(x)), \min(1 - \nu_A(x), \nu_B(x))\rangle \mid x \in E\}$

Table 9.10 (*continued*)

$\rightarrow_{23}$	$\{\langle x, 1 - \min(\mathrm{sg}(1 - \nu_A(x)), \overline{sg}(1 - \nu_B(x))),$ $\min(\mathrm{sg}(1 - \nu_A(x)), \overline{sg}(1 - \nu_B(x)))\rangle \mid x \in E\}$
$\rightarrow_{24}$	$\{\langle x, \overline{sg}(\mu_A(x) - \mu_B(x)).\overline{sg}(\nu_B(x) - \nu_A(x)),$ $\mathrm{sg}(\mu_A(x) - \mu_B(x)).\mathrm{sg}(\nu_B(x) - \nu_A(x))\rangle \mid x \in E\}$
$\rightarrow_{25}$	$\{\langle x, \max(\nu_A(x), \overline{sg}(\mu_A(x)).\overline{sg}(1 - \nu_A(x)),$ $\mu_B(x).\overline{sg}(\nu_B(x)).\overline{sg}(1 - \mu_B(x))),$ $\min(\mu_A(x), \nu_B(x))\rangle \mid x \in E\}$
$\rightarrow_{26}$	$\{\langle x, \max(\overline{sg}(1 - \nu_A(x)), \mu_B(x)), \min(\mathrm{sg}(\mu_A(x)), \nu_B(x))\rangle \mid x \in E\}$
$\rightarrow_{27}$	$\{\langle x, \max(\overline{sg}(1 - \nu_A(x)), \mathrm{sg}(\mu_B(x))),$ $\min(\mathrm{sg}(\mu_A(x)), \overline{sg}(1 - \nu_B(x)))\rangle \mid x \in E\}$
$\rightarrow_{28}$	$\{\langle x, \max(\overline{sg}(1 - \nu_A(x)), \mu_B(x)), \min(\mu_A(x), \nu_B(x))\rangle \mid x \in E\}$
$\rightarrow_{29}$	$\{\langle x, \max(\overline{sg}(1 - \nu_A(x)), \overline{sg}(1 - \mu_B(x))),$ $\min(\mu_A(x), \overline{sg}(1 - \nu_B(x)))\rangle \mid x \in E\}$
$\rightarrow_{30}$	$\{\langle x, \max(1 - \mu_A(x), \min(\mu_A(x), 1 - \nu_B(x))),$ $\min(\mu_A(x), \nu_B(x))\rangle \mid x \in E\}$
$\rightarrow_{31}$	$\{\langle x, \overline{sg}(\mu_A(x) + \nu_B(x) - 1), \nu_B(x).\mathrm{sg}(\mu_A(x) + \nu_B(x) - 1)\rangle \mid x \in E\}$
$\rightarrow_{32}$	$\{\langle x, 1 - \nu_B(x).\mathrm{sg}(\mu_A(x) + \nu_B(x) - 1),$ $\nu_B(x).\mathrm{sg}(\mu_A(x) + \nu_B(x) - 1)\rangle \mid x \in E\}$
$\rightarrow_{33}$	$\{\langle x, 1 - \min(\mu_A(x), \nu_B(x)), \min(\mu_A(x), \nu_B(x))\rangle \mid x \in E\}$
$\rightarrow_{34}$	$\{\langle x, \min(1, 2 - \mu_A(x) - \nu_B(x)), \max(0, \mu_A(x) + \nu_B(x) - 1)\rangle \mid x \in E\}$
$\rightarrow_{35}$	$\{\langle x, 1 - \mu_A(x).\nu_B(x), \mu_A(x).\nu_B(x)\rangle \mid x \in E\}$
$\rightarrow_{36}$	$\{\langle x, \min(1 - \min(\mu_A(x), \nu_B(x)),$ $\max(\mu_A(x), 1 - \mu_A(x)), \max(1 - \nu_B(x), \nu_B(x))),$ $\max(\min(\mu_A(x), \nu_B(x)), \min(\mu_A(x), 1 - \mu_A(x)),$ $\min(1 - \nu_B(x), \nu_B(x)))\rangle \mid x \in E\}$
$\rightarrow_{37}$	$\{\langle x, 1 - \max(\mu_A(x), \nu_B(x)).\mathrm{sg}(\mu_A(x) + \nu_B(x) - 1),$ $\max(\mu_A(x), \nu_B(x)).\mathrm{sg}(\mu_A(x) + \nu_B(x) - 1)\rangle \mid x \in E\}$
$\rightarrow_{38}$	$\{\langle x, 1 - \mu_A(x) + (\mu_A(x)^2.(1 - \nu_B(x))),$ $\mu_A(x).(1 - \mu_A(x)) + \mu_A(x)^2.\nu_B(x)\rangle \mid x \in E\}$
$\rightarrow_{39}$	$\{\langle x, (1 - \nu_B(x)).\overline{sg}(1 - \mu_A(x))$ $+ \mathrm{sg}(1 - \mu_A(x)).(\overline{sg}(\nu_B(x)) + (1 - \mu_A(x)).\mathrm{sg}(\nu_B(x))),$ $\nu_B(x).\overline{sg}(1 - \mu_A(x)) + \mu_A(x).\mathrm{sg}(1 - \mu_A(x))$ $.\mathrm{sg}(\nu_B(x))\rangle \mid x \in E\}$
$\rightarrow_{40}$	$\{\langle x, 1 - \mathrm{sg}(\mu_A(x) + \nu_B(x) - 1), 1 - \overline{sg}(\mu_A(x) + \nu_B(x) - 1)\rangle \mid x \in E\}$
$\rightarrow_{41}$	$\{\langle x, \max(\overline{sg}(\mu_A(x)), 1 - \nu_B(x)), \min(\mathrm{sg}(\mu_A(x)), \nu_B(x))\rangle \mid x \in E\}$
$\rightarrow_{42}$	$\{\langle x, \max(\overline{sg}(\mu_A(x)), \mathrm{sg}(1 - \nu_B(x))),$ $\min(\mathrm{sg}(\mu_A(x)), \overline{sg}(1 - \nu_B(x)))\rangle \mid x \in E\}$
$\rightarrow_{43}$	$\{\langle x, \max(\overline{sg}(\mu_A(x)), 1 - \nu_B(x)), \min(\mathrm{sg}(\mu_A(x)), \nu_B(x))\rangle \mid x \in E\}$
$\rightarrow_{44}$	$\{\langle x, \max(\overline{sg}(\mu_A(x)), 1 - \nu_B(x)), \min(\mu_A(x), \nu_B(x))\rangle \mid x \in E\}$
$\rightarrow_{45}$	$\{\langle x, \max(\overline{sg}(\mu_A(x)), \overline{sg}(\nu_B(x))),$ $\min(\mu_A(x), \overline{sg}(1 - \nu_B(x)))\rangle \mid x \in E\}$
$\rightarrow_{46}$	$\{\langle x, \max(\nu_A(x), \min(1 - \nu_A(x), \mu_B(x))),$ $1 - \max(\nu_A(x), \mu_B(x))\rangle \mid x \in E\}$

Table 9.10 (*continued*)

$\rightarrow_{47}$	$\{\langle x, \overline{sg}(1 - \nu_A(x) - \mu_B(x)),$ $(1 - \mu_B(x)).\mathrm{sg}(1 - \nu_A(x) - \mu_B(x))\rangle \| x \in E\}$
$\rightarrow_{48}$	$\{\langle x, 1 - (1 - \mu_B(x)).\mathrm{sg}(1 - \nu_A(x) - \mu_B(x)),$ $(1 - \mu_B(x)).\mathrm{sg}(1 - \nu_A(x) - \mu_B(x))\rangle \| x \in E\}$
$\rightarrow_{49}$	$\{\langle x, \min(1, \nu_A(x) + \mu_B(x)), \max(0, 1 - \nu_A(x) - \mu_B(x))\rangle \| x \in E\}$
$\rightarrow_{50}$	$\{\langle x, \nu_A(x) + \mu_B(x) - \nu_A(x).\mu_B(x),$ $1 - \nu_A(x) - \mu_B(x) + \nu_A(x).\mu_B(x)\rangle \| x \in E\}$
$\rightarrow_{51}$	$\{\langle x, \min(\max(\nu_A(x), \mu_B(x)),$ $\max(1 - \nu_A(x), \nu_A(x)), \max(\mu_B(x), 1 - \mu_B(x))),$ $\max(1 - \max(\nu_A(x), \mu_B(x)), \min(1 - \nu_A(x), \nu_A(x)),$ $\min(\mu_B(x), 1 - \mu_B(x)))\rangle \| x \in E\}$
$\rightarrow_{52}$	$\{\langle x, 1 - (1 - \min(\nu_A(x), \mu_B(x))).\mathrm{sg}(1 - \nu_A(x) - \mu_B(x)),$ $1 - \min(\nu_A(x), \mu_B(x)).\mathrm{sg}(1 - \nu_A(x) - \mu_B(x))\rangle \| x \in E\}$
$\rightarrow_{53}$	$\{\langle x, \nu_A(x) + (1 - \nu_A(x))^2.\mu_B(x),$ $(1 - \nu_A(x)).\nu_A(x) + (1 - \nu_A(x))^2.(1 - \mu_B(x))\rangle \| x \in E\}$
$\rightarrow_{54}$	$\{\langle x, \mu_B(x).\overline{sg}(\nu_A(x))$ $+ \mathrm{sg}(\nu_A(x)).(\overline{sg}(1 - \mu_B(x)) + \nu_A(x).\mathrm{sg}(1 - \mu_B(x))),$ $(1 - \mu_B(x)).\overline{sg}(\nu_A(x)) + (1 - \nu_A(x)).\mathrm{sg}(\nu_A(x)).\mathrm{sg}(1 - \mu_B(x))\rangle \| x \in E\}$
$\rightarrow_{55}$	$\{\langle x, 1 - \mathrm{sg}(1 - \nu_A(x) - \mu_B(x)), 1 - \overline{sg}(1 - \nu_A(x) - \mu_B(x))\rangle \| x \in E\}$
$\rightarrow_{56}$	$\{\langle x, \max(\overline{sg}(1 - \nu_A(x)), \mu_B(x)),$ $\min(\mathrm{sg}(1 - \nu_A(x)), 1 - \mu_B(x))\rangle \| x \in E\}$
$\rightarrow_{57}$	$\{\langle x, \max(\overline{sg}(1 - \nu_A(x)), \mathrm{sg}(\mu_B(x))),$ $\min(\mathrm{sg}(1 - \nu_A(x)), \overline{sg}(\mu_B(x)))\rangle \| x \in E\}$
$\rightarrow_{58}$	$\{\langle x, \max(\overline{sg}(1 - \nu_A(x)), \overline{sg}(1 - \mu_B(x))),$ $1 - \max(\nu_A(x), \mu_B(x))\rangle \| x \in E\}$
$\rightarrow_{59}$	$\{\langle x, \max(\overline{sg}(1 - \nu_A(x)), \mu_B(x)),$ $(1 - \max(\nu_A(x), \mu_B(x)))\rangle \| x \in E\}$
$\rightarrow_{60}$	$\{\langle x, \max(\overline{sg}(1 - \nu_A(x)), \overline{sg}(1 - \mu_B(x))),$ $\min(1 - \nu_A(x), \overline{sg}(\mu_B(x)))\rangle \| x \in E\}$
$\rightarrow_{61}$	$\{\langle x, \max(\mu_B(x), \min(\nu_B(x), \nu_A(x))), \min(\nu_B(x), \mu_A(x))\rangle \| x \in E\}$
$\rightarrow_{62}$	$\{\langle x, \overline{sg}(\nu_B(x) - \nu_A(x)), \mu_A(x).\mathrm{sg}(\nu_B(x) - \nu_A(x))\rangle \| x \in E\}$
$\rightarrow_{63}$	$\{\langle x, 1 - (1 - \nu_A(x)).\mathrm{sg}(\nu_B(x) - \nu_A(x)),$ $\mu_A(x).\mathrm{sg}(\nu_B(x) - \nu_A(x))\rangle \| x \in E\}$
$\rightarrow_{64}$	$\{\langle x, \mu_B(x) + \nu_B(x).\nu_A(x), \nu_B(x).\mu_A(x)\rangle \| x \in E\}$
$\rightarrow_{65}$	$\{\langle x, 1 - (1 - \min(\mu_B(x), \nu_A(x))).\mathrm{sg}(\nu_B(x) - \nu_A(x)),$ $\max(\nu_B(x), \mu_A(x)).\mathrm{sg}(\nu_B(x) - \nu_A(x)).\mathrm{sg}(\mu_A(x) - \mu_B(x))\rangle \| x \in E\}$
$\rightarrow_{66}$	$\{\langle x, \mu_B(x) + \nu_B(x)^2.\nu_A(x),$ $\nu_B(x).\mu_B(x) + \nu_B(x)^2.\mu_A(x)\rangle \| x \in E\}$
$\rightarrow_{67}$	$\{\langle x, \nu_A(x).\overline{sg}(1 - \nu_B(x)) + \mathrm{sg}(1 - \nu_B(x)).(\overline{sg}(1 - \nu_A(x))$ $+ \mu_B(x).\mathrm{sg}(1 - \nu_A(x))),$ $\mu_A(x).\overline{sg}(1 - \nu_B(x)) + \nu_B(x).\mathrm{sg}(1 - \nu_B(x)).\mathrm{sg}(1 - \nu_A(x))\rangle \| x \in E\}$
$\rightarrow_{68}$	$\{\langle x, 1 - (1 - \nu_A(x)).\mathrm{sg}(\nu_B(x) - \nu_A(x)),$ $\mu_A(x).\mathrm{sg}(\nu_B(x) - \nu_A(x)).\mathrm{sg}(\mu_A(x) - \mu_B(x))\rangle \| x \in E\}$

Table 9.10 (*continued*)

$\to_{69}$	$\{\langle x, 1 - (1 - \nu_A(x)).\mathrm{sg}(\nu_B(x) - \nu_A(x))$ $-\mu_A(x).\overline{sg}(\nu_B(x) - \nu_A(x)).\mathrm{sg}(\mu_A(x) - \mu_B(x)),$ $\mu_A(x).\mathrm{sg}(\mu_A(x) - \mu_B(x))\rangle | x \in E\}$
$\to_{70}$	$\{\langle x, \max(\overline{sg}(\nu_B(x)), \nu_A(x)), \min(\mathrm{sg}(\nu_B(x)), \mu_A(x))\rangle | x \in E\}$
$\to_{71}$	$\{\langle x, \max(\mu_B(x), \nu_A(x)), \min(\nu_B(x).\mu_B(x) + \nu_B(x)^2, \mu_A(x))\rangle | x \in E\}$
$\to_{72}$	$\{\langle x, \max(\mu_B(x), \nu_A(x)), \min(1 - \mu_B(x), \mu_A(x))\rangle | x \in E\}$
$\to_{73}$	$\{\langle x, \max(1 - \max(\mathrm{sg}(\nu_B(x)), \mathrm{sg}(1 - \mu_B(x))), \nu_A(x)),$ $\min(\mathrm{sg}(1 - \mu_B(x)), \mu_A(x))\rangle | x \in E\}$
$\to_{74}$	$\{\langle x, \max(\overline{sg}(\nu_B(x)), \mathrm{sg}(\nu_A(x))), \min(\mathrm{sg}(\nu_B(x)), \overline{sg}(\nu_A(x)))\rangle | x \in E\}$
$\to_{75}$	$\{\langle x, \max(\mu_B(x), \nu_A(x).(\nu_A(x) + \mu_A(x))),$ $\min(\nu_B(x).(\nu_B(x) + \mu_B(x)), \mu_A(x).(\nu_A(x)^2 + \mu_A(x))$ $+\nu_A(x).\mu_A(x))\rangle | x \in E\}$
$\to_{76}$	$\{\langle x, \max(\mu_B(x), 1 - \mu_A(x)), \min(1 - \mu_B(x), \mu_A(x))\rangle | x \in E\}$
$\to_{77}$	$\{\langle x, 1 - \min(\mathrm{sg}(1 - \mu_B(x)), \overline{sg}(1 - \mu_A(x))),$ $\min(\mathrm{sg}(1 - \mu_B(x)), \overline{sg}(1 - \mu_A(x)))\rangle | x \in E\}$
$\to_{78}$	$\{\langle x, \max(\overline{sg}(1 - \mu_B(x)), \nu_A(x)), \min(\mathrm{sg}(\nu_B(x)), \mu_A(x))\rangle | x \in E\}$
$\to_{79}$	$\{\langle x, \max(\overline{sg}(1 - \mu_B(x)), \mathrm{sg}(\nu_A(x))),$ $\min(\mathrm{sg}(\nu_B(x)), \overline{sg}(1 - \mu_A(x)))\rangle | x \in E\}$
$\to_{80}$	$\{\langle x, \max(\overline{sg}(1 - \mu_B(x)), \nu_A(x)), \min(\nu_B(x), \mu_A(x))\rangle | x \in E\}$
$\to_{81}$	$\{\langle x, \max(\overline{sg}(1 - \mu_B(x)), \overline{sg}(1 - \nu_A(x))),$ $\min(\nu_B(x), \overline{sg}(1 - \mu_A(x)))\rangle | x \in E\}$
$\to_{82}$	$\{\langle x, \max(1 - \nu_B(x), \min(\nu_B(x), 1 - \mu_A(x))),$ $\min(\nu_B(x), \mu_A(x))\rangle | x \in E\}$
$\to_{83}$	$\{\langle x, \overline{sg}(\nu_B(x) + \mu_A(x) - 1), \mu_A(x).\mathrm{sg}(\nu_B(x) + \mu_A(x) - 1)\rangle | x \in E\}$
$\to_{84}$	$\{\langle x, 1 - \mu_A(x).\mathrm{sg}(\nu_B(x) + \mu_A(x) + 1),$ $\mu_A(x).\mathrm{sg}(\nu_B(x) + \mu_A(x) + 1)\rangle | x \in E\}$
$\to_{85}$	$\{\langle x, 1 - \nu_B(x) + \nu_B(x)^2.(1 - \mu_A(x)),$ $\nu_B(x).(1 - \nu_B(x)) + \nu_B(x)^2\rangle | x \in E\}$
$\to_{86}$	$\{\langle x, (1 - \mu_A(x)).\overline{sg}(1 - \nu_B(x))$ $+\mathrm{sg}(1 - \nu_B(x)).\overline{sg}(\mu_A(x) + \min(1 - \nu_B(x), \mathrm{sg}(\mu_A(x)))),$ $\mu_A(x).\overline{sg}(1 - \nu_B(x)) + \nu_B(x).\mathrm{sg}(1 - \nu_B(x)).\mathrm{sg}(\mu_A(x))\rangle | x \in E\}$
$\to_{87}$	$\{\langle x, \max(\overline{sg}(\nu_B(x)), 1 - \mu_A(x)), \min(\mathrm{sg}(\nu_B(x)), \mu_A(x))\rangle | x \in E\}$
$\to_{88}$	$\{\langle x, \max(\overline{sg}(\nu_B(x)), \mathrm{sg}(1 - \mu_A(x))),$ $\min(\mathrm{sg}(\nu_B(x)), \overline{sg}(1 - \mu_A(x)))\rangle | x \in E\}$
$\to_{89}$	$\{\langle x, \max(\overline{sg}(\nu_B(x)), 1 - \mu_A(x)), \min(\nu_B(x), \mu_A(x))\rangle | x \in E\}$
$\to_{90}$	$\{\langle x, \max(\overline{sg}(\nu_B(x)), \overline{sg}(\mu_A(x))),$ $\min(\nu_B(x), \overline{sg}(1 - \mu_A(x)))\rangle | x \in E\}$
$\to_{91}$	$\{\langle x, \max(\mu_B(x), \min(1 - \mu_B(x), \nu_A(x))),$ $1 - \max(\mu_B(x), \nu_A(x))\rangle | x \in E\}$
$\to_{92}$	$\{\langle x, \overline{sg}(1 - \mu_B(x) - \nu_A(x)),$ $\min(1 - \nu_A(x), \mathrm{sg}(1 - \mu_B(x) - \nu_A(x)))\rangle | x \in E\}$
$\to_{93}$	$\{\langle x, 1 - \min(1 - \nu_A(x), \mathrm{sg}(1 - \mu_B(x) - \nu_A(x))),$ $\min(1 - \nu_A(x), \mathrm{sg}(1 - \mu_B(x) - \nu_A(x)))\rangle | x \in E\}$

Table 9.10 (*continued*)

$\rightarrow_{94}$	$\{\langle x, \mu_B(x) + (1 - \mu_B(x))^2.\nu_A(x),$ $(1 - \mu_B(x)).\mu_B(x) + (1 - \mu_B(x))^2.(1 - \nu_A(x))\rangle | x \in E\}$
$\rightarrow_{95}$	$\{\langle x, \min(\nu_A(x), \overline{sg}(\mu_B(x))) + sg(\mu_B(x)).(\overline{sg}(1 - \nu_A(x))$ $+ \min(\mu_B(x), sg(1 - \nu_A(x)))),$ $\min(1 - \nu_A(x), \overline{sg}(\mu_B(x))) + \min(\min(1 - \mu_B(x), sg(\mu_B(x))),$ $sg(1 - \nu_A(x)))\rangle | x \in E\}$
$\rightarrow_{96}$	$\{\langle x, \max(\overline{sg}(1 - \mu_B(x)), \nu_A(x)),$ $\min(sg(1 - \mu_B(x)), 1 - \nu_A(x))\rangle | x \in E\}$
$\rightarrow_{97}$	$\{\langle x, \max(\overline{sg}(1 - \mu_B(x)), sg(\nu_A(x))),$ $\min(sg(1 - \mu_B(x)), \overline{sg}(\nu_A(x)))\rangle | x \in E\}$
$\rightarrow_{98}$	$\{\langle x, \max(\overline{sg}(1 - \mu_B(x)), \nu_A(x)),$ $1 - \max(\mu_B(x), \nu_A(x))\rangle | x \in E\}$
$\rightarrow_{99}$	$\{\langle x, \max(\overline{sg}(1 - \mu_B(x)), \overline{sg}(1 - \nu_A(x))),$ $\min(1 - \mu_B(x), \overline{sg}(\nu_A(x)))\rangle | x \in E\}$
$\rightarrow_{100}$	$\{\langle x, \max(\min(\nu_A(x), sg(\mu_A(x))), \mu_B(x)),$ $\min(\min(\mu_A(x), sg(\nu_A(x))), \nu_B(x))\rangle | x \in E\}$
$\rightarrow_{101}$	$\{\langle x, \max(\min(\nu_A(x), sg(\mu_A(x))), \min(\mu_B(x), sg(\nu_B(x)))),$ $\min(\min(\mu_A(x), sg(\nu_A(x))), \min(\nu_B(x), sg(\mu_B(x))))\rangle | x \in E\}$
$\rightarrow_{102}$	$\{\langle x, \max(\nu_A(x), \min(\mu_B(x), sg(\nu_B(x)))),$ $\min(\mu_A(x), \min(\nu_B(x), sg(\mu_B(x))))\rangle | x \in E\}$
$\rightarrow_{103}$	$\{\langle x, \max(\min(1 - \mu_A(x), sg(\mu_A(x))), 1 - \nu_B(x)),$ $\min(\mu_A(x), sg(1 - \mu_A(x)), \nu_B(x))\rangle | x \in E\}$
$\rightarrow_{104}$	$\{\langle x, \max(\min(1 - \mu_A(x), sg(\mu_A(x))),$ $\min(1 - \nu_B(x), sg(\nu_B(x)))),$ $\min(\min(\mu_A(x), sg(1 - \mu_A(x))),$ $\min(\nu_B(x), sg(1 - \nu_B(x))))\rangle | x \in E\}$
$\rightarrow_{105}$	$\{\langle x, \max(1 - \mu_A(x), \min(1 - \nu_B(x), sg(\nu_B(x)))),$ $\min(\mu_A(x), \min(\nu_B(x), sg(1 - \nu_B(x))))\rangle | x \in E\}$
$\rightarrow_{106}$	$\{\langle x, \max(\min(\nu_A(x), sg(1 - \nu_A(x))), \mu_B(x)),$ $\min(\min(1 - \nu_A(x), sg(\nu_A(x))), 1 - \mu_B(x))\rangle | x \in E\}$
$\rightarrow_{107}$	$\{\langle x, \max(\min(\nu_A(x), sg(1 - \nu_A(x))),$ $\min(\mu_B(x), sg(1 - \mu_B(x)))),$ $\min(\min(1 - \nu_A(x), sg(\nu_A(x))),$ $\min(1 - \mu_B(x), sg(\mu_B(x))))\rangle | x \in E\}$
$\rightarrow_{108}$	$\{\langle x, \max(\nu_A(x), \min(\mu_B(x), sg(1 - \mu_B(x)))),$ $\min(1 - \nu_A(x), \min(1 - \mu_B(x), sg(\mu_B(x))))\rangle | x \in E\}$
$\rightarrow_{109}$	$\{\langle x, \nu_A(x) + \min(\overline{sg}(1 - \mu_A(x)), \mu_B(x)),$ $\mu_A(x).\nu_A(x) + \min(\overline{sg}(1 - \mu_A(x)), \nu_B(x))\rangle | x \in E\}$
$\rightarrow_{110}$	$\{\langle x, \max(\nu_A(x), \mu_B(x)),$ $\min(\mu_A(x).\nu_A(x) + \overline{sg}(1 - \mu_A(x)), \nu_B(x))\rangle | x \in E\}$
$\rightarrow_{111}$	$\{\langle x, \max(\nu_A(x), \mu_B(x).\nu_B(x) + \overline{sg}(1 - \mu_B(x))),$ $\min(\mu_A(x).\nu_A(x) + \overline{sg}(1 - \mu_A(x)), \nu_B(x).(\mu_B(x).\nu_B(x)$ $+\overline{sg}(1 - \mu_B(x))) + \overline{sg}(1 - \nu_B(x)))\rangle | x \in E\}$

Table 9.10 (*continued*)

$\rightarrow_{112}$	$\{\langle x, \nu_A(x) + \mu_B(x) - \nu_A(x).\mu_B(x),$ $\mu_A(x).\nu_A(x) + \overline{sg}(1 - \mu_A(x)).\nu_B(x)\rangle | x \in E\}$
$\rightarrow_{113}$	$\{\langle x, \nu_A(x) + (\mu_B(x).\nu_B(x)) - \nu_A(x).$ $(\mu_B(x).\nu_B(x) + \overline{sg}(1 - \mu_B(x))),$ $(\mu_A(x).\nu_A(x) + \overline{sg}(1 - \mu_A(x))).(\nu_B(x).(\mu_B(x).\nu_B(x)$ $+\overline{sg}(1 - \mu_B(x))) + \overline{sg}(1 - \nu_B(x)))\rangle | x \in E\}$
$\rightarrow_{114}$	$\{\langle x, 1 - \mu_A(x) + \min(\overline{sg}(1 - \mu_A(x)), 1 - \nu_B(x)),$ $\mu_A(x).(1 - \mu_A(x)) + \min(\overline{sg}(1 - \mu_A(x)), \nu_B(x))\rangle | x \in E\}$
$\rightarrow_{115}$	$\{\langle x, 1 - \min(\mu_A(x), \nu_B(x)),$ $\min(\mu_A(x).(1 - \mu_A(x)) + \overline{sg}(1 - \mu_A(x)), \nu_B(x))\rangle | x \in E\}$
$\rightarrow_{116}$	$\{\langle x, \max(1 - \mu_A(x), (1 - \nu_B(x)).\nu_B(x) + \overline{sg}(\nu_B(x))),$ $\min(\mu_A(x).(1 - \mu_A(x)) + \overline{sg}(1 - \mu_A(x)),$ $\nu_B(x).((1 - \nu_B(x)).\nu_B(x)$ $+\overline{sg}(\nu_B(x))) + \overline{sg}(1 - \nu_B(x)))\rangle | x \in E\}$
$\rightarrow_{117}$	$\{\langle x, 1 - \mu_A(x) - \nu_B(x) + \mu_A(x).\nu_B(x)$ $(\mu_A(x).(1 - \mu_A(x)) + \overline{sg}(1 - \mu_A(x))).\nu_B(x)\rangle | x \in E\}$
$\rightarrow_{118}$	$\{\langle x, (1 - \mu_A(x)).\mathrm{sg}(\nu_B(x)) + \mu_A(x).\nu_B(x).(1 - \nu_B(x)),$ $(\mu_A(x) - \mu_A(x)^2 + \overline{sg}(1 - \mu_A(x))).((1 - \nu_B(x)).\nu_B(x)^2$ $+\overline{sg}(1 - \nu_B(x))) + \overline{sg}(1 - \nu_B(x))\rangle | x \in E\}$
$\rightarrow_{119}$	$\{\langle x, \nu_A(x) + \min(\overline{sg}(\nu_A(x)), \mu_B(x)),$ $(1 - \nu_A(x)).\nu_A(x) + \min(\overline{sg}(\nu_A(x)), 1 - \mu_B(x))\rangle | x \in E\}$
$\rightarrow_{120}$	$\{\langle x, \max(\nu_A(x), \mu_B(x)),$ $\min((1 - \nu_A(x)).\nu_A(x) + \overline{sg}(\nu_A(x)), 1 - \mu_B(x))\rangle | x \in E\}$
$\rightarrow_{121}$	$\{\langle x, \max(\nu_A(x), \mu_B(x).(1 - \mu_B(x)) + \overline{sg}(1 - \mu_B(x))),$ $\min((1 - \nu_A(x)).\nu_A(x) + \overline{sg}(\nu_A(x)), (1 - \mu_B(x)).(\mu_B(x)$ $.(1 - \mu_B(x)) + \overline{sg}(1 - \mu_B(x))) + \overline{sg}(\mu_B(x)))\rangle | x \in E\}$
$\rightarrow_{122}$	$\{\langle x, \nu_A(x) + \mu_B(x) - \nu_A(x).\mu_B(x),$ $((1 - \nu_A(x)).\nu_A(x) + \overline{sg}(\nu_A(x))).(1 - \mu_B(x))\rangle | x \in E\}$
$\rightarrow_{123}$	$\{\langle x, \nu_A(x) + \mu_B(x).(1 - \mu_B(x)) - \nu_A(x)$ $.(\mu_B(x).(1 - \mu_B(x)) + \overline{sg}(1 - \mu_B(x))),$ $((1 - \nu_A(x)).\nu_A(x) + \overline{sg}(\nu_A(x))).(((1 - \mu_B(x)).(\mu_B(x).(1 - \mu_B(x))$ $+\overline{sg}(1 - \mu_B(x)))) + \overline{sg}(\mu_B(x)))\rangle | x \in E\}$
$\rightarrow_{124}$	$\{\langle x, \mu_B(x) + \min(\overline{sg}(1 - \nu_B(x)), \nu_A(x)),$ $\nu_B(x).\mu_B(x) + \min(\overline{sg}(1 - \nu_B(x)), \mu_A(x))\rangle | x \in E\}$
$\rightarrow_{125}$	$\{\langle x, \max(\mu_B(x), \nu_A(x)),$ $\min(\nu_B(x).\mu_B(x) + \overline{sg}(1 - \nu_B(x)), \mu_A(x))\rangle | x \in E\}$
$\rightarrow_{126}$	$\{\langle x, \max(\mu_B(x), \nu_A(x).\mu_A(x) + \overline{sg}(1 - \nu_A(x))),$ $\min(\nu_B(x).\mu_B(x) + \overline{sg}(1 - \nu_B(x)), \mu_A(x).$ $(\nu_A(x).\mu_A(x) + \overline{sg}(1 - \nu_A(x))) + \overline{sg}(1 - \mu_A(x)))\rangle | x \in E\}$
$\rightarrow_{127}$	$\{\langle x, \mu_B(x) + \nu_A(x) - \mu_B(x).\nu_A(x),$ $(\nu_B(x).\mu_B(x) + \overline{sg}(1 - \nu_B(x))).\mu_A(x)\rangle | x \in E\}$

Table 9.10 (*continued*)

$\rightarrow_{128}$	$\{\langle x, \mu_B(x) + \nu_A(x).\mu_A(x) - \mu_B(x).$ $(\nu_A(x).\mu_A(x) + \overline{sg}(1 - \nu_A(x))),$ $(\nu_B(x).\mu_B(x) + \overline{sg}(1 - \nu_B(x))).(\mu_A(x).(\nu_A(x).\mu_A(x)$ $+\overline{sg}(1 - \nu_A(x))) + \overline{sg}(1 - \mu_A(x)))\rangle | x \in E\}$
$\rightarrow_{129}$	$\{\langle x, 1 - \nu_B(x) + \min(\overline{sg}(1 - \nu_B(x)), 1 - \mu_A(x)),$ $\nu_B(x).(1 - \nu_B(x)) + \min(\overline{sg}(1 - \nu_B(x)), \mu_A(x))\rangle | x \in E\}$
$\rightarrow_{130}$	$\{\langle x, 1 - \min(\nu_B(x), \mu_A(x)),$ $\min(\nu_B(x).(1 - \nu_B(x)) + \overline{sg}(1 - \nu_B(x)), \mu_A(x))\rangle | x \in E\}$
$\rightarrow_{131}$	$\{\langle x, \max(1 - \nu_B(x), (1 - \mu_A(x)).\mu_A(x) + \overline{sg}(\mu_A(x))),$ $\min(\nu_B(x).(1 - \nu_B(x)) + \overline{sg}(1 - \nu_B(x)), \mu_A(x).((1 - \mu_A(x))$ $.\mu_A(x) + \overline{sg}(\mu_A(x))) + \overline{sg}(1 - \mu_A(x))))\rangle | x \in E\}$
$\rightarrow_{132}$	$\{\langle x, 1 - \mu_A(x).\nu_B(x),$ $(\nu_B(x).(1 - \nu_B(x)) + \overline{sg}(1 - \nu_B(x))).\mu_A(x)\rangle | x \in E\}$
$\rightarrow_{133}$	$\{\langle x, 1 - \nu_B(x) + (1 - \mu_A(x)).\mu_A(x)$ $-(1 - \nu_B(x)).((1 - \mu_A(x)).\mu_A(x) + \overline{sg}(\mu_A(x))),$ $(\nu_B(x).(1 - \nu_B(x)) + \overline{sg}(1 - \nu_B(x))).(\mu_A(x).((1 - \mu_A(x)).\mu_A(x)$ $+\overline{sg}(\mu_A(x))) + \overline{sg}(1 - \mu_A(x)))\rangle | x \in E\}$
$\rightarrow_{134}$	$\{\langle x, \mu_B(x) + \min(\overline{sg}(\mu_B(x)), \nu_A(x)),$ $(1 - \mu_B(x)).\mu_B(x) + \min(\overline{sg}(\mu_B(x)), 1 - \nu_A(x))\rangle | x \in E\}$
$\rightarrow_{135}$	$\{\langle x, \max(\mu_B(x), \nu_A(x)),$ $\min((1 - \mu_B(x)).\mu_B(x) + \overline{sg}(\mu_B(x)), 1 - \nu_A(x))\rangle | x \in E\}$
$\rightarrow_{136}$	$\{\langle x, \max(\mu_B(x), \nu_A(x).(1 - \nu_A(x)) + \overline{sg}(1 - \nu_A(x))),$ $\min((1 - \mu_B(x)).\mu_B(x) + \overline{sg}(\mu_B(x)), (1 - \nu_A(x))$ $.(\nu_A(x).(1 - \nu_A(x)) + \overline{sg}(1 - \nu_A(x))) + \overline{sg}(\nu_A(x)))\rangle | x \in E\}$
$\rightarrow_{137}$	$\{\langle x, \mu_B(x) + \nu_A(x) - \mu_B(x).\nu_A(x),$ $((1 - \mu_B(x)).\mu_B(x) + \overline{sg}(\mu_B(x))).(1 - \nu_A(x))\rangle | x \in E\}$
$\rightarrow_{138}$	$\{\langle x, \mu_B(x) + \nu_A(x).(1 - \nu_A(x))$ $-\mu_B(x).(\nu_A(x).(1 - \nu_A(x)) + \overline{sg}(1 - \nu_A(x))),$ $((1 - \mu_B(x)).\mu_B(x) + \overline{sg}(\mu_B(x))).(1 - \nu_A(x)).(\nu_A(x).(1 - \nu_A(x))$ $+\overline{sg}(1 - \nu_A(x)) + \overline{sg}(\nu_A(x)))\rangle | x \in E\}$

Table 9.11 List of intuitionistic fuzzy negations

$\neg_1$	$\{\langle x, \nu_A(x), \mu_A(x)\rangle | x \in E\}$
$\neg_2$	$\{\langle x, \overline{sg}(\mu_A(x)), sg(\mu_A(x))\rangle | x \in E\}$
$\neg_3$	$\{\langle x, \nu_A(x), \mu_A(x).\nu_A(x) + \mu_A(x)^2\rangle | x \in E\}$
$\neg_4$	$\{\langle x, \nu_A(x), 1 - \nu_A(x)\rangle | x \in E\}$
$\neg_5$	$\{\langle x, \overline{sg}(1 - \nu_A(x)), sg(1 - \nu_A(x))\rangle | x \subset E\}$
$\neg_6$	$\{\langle x, \overline{sg}(1 - \nu_A(x)), sg(\mu_A(x))\rangle | x \in E\}$
$\neg_7$	$\{\langle x, \overline{sg}(1 - \nu_A(x)), \mu_A(x)\rangle | x \in E\}$
$\neg_8$	$\{\langle x, 1 - \mu_A(x), \mu_A(x)\rangle | x \in E\}$
$\neg_9$	$\{\langle x, \overline{sg}(\mu_A(x)), \mu_A(x)\rangle | x \in E\}$
$\neg_{10}$	$\{\langle x, \overline{sg}(1 - \nu_A(x)), 1 - \nu_A(x)\rangle | x \in E\}$
$\neg_{11}$	$\{\langle x, sg(\nu_A(x)), \overline{sg}(\nu_A(x))\rangle | x \in E\}$

Table 9.11 (*continued*)

$\neg_{12}$	$\{\langle x, \nu_A(x).(\nu_A(x)+\mu_A(x)),$ $\mu_A(x).(\nu_A(x)^2+\mu_A(x)+\nu_A(x).\mu_A(x))\rangle | x \in E\}$
$\neg_{13}$	$\{\langle x, \mathrm{sg}(1 - \mu_A(x)), \overline{sg}(1 - \mu_A(x))\rangle | x \in E\}$
$\neg_{14}$	$\{\langle x, \mathrm{sg}(\nu_A(x)), \overline{sg}(1 - \mu_A(x))\rangle | x \in E\}$
$\neg_{15}$	$\{\langle x, \overline{sg}(1 - \nu_A(x)), \overline{sg}(1 - \mu_A(x))\rangle | x \in E\}$
$\neg_{16}$	$\{\langle x, \overline{sg}(\mu_A(x)), \overline{sg}(1 - \mu_A(x))\rangle | x \in E\}$
$\neg_{17}$	$\{\langle x, \overline{sg}(1 - \nu_A(x)), \overline{sg}(\nu_A(x))\rangle | x \in E\}$
$\neg_{18}$	$\{\langle x, \min(\nu_A(x), \mathrm{sg}(\mu_A(x))), \min(\mu_A(x), \mathrm{sg}(\nu_A(x)))\rangle | x \in E\}$
$\neg_{19}$	$\{\langle x, \min(\nu_A(x), \mathrm{sg}(\mu_A(x))), 0\rangle | x \in E\}$
$\neg_{20}$	$\{\langle x, \nu_A(x), 0\rangle | x \in E\}$
$\neg_{21}$	$\{\langle x, \min(1 - \mu_A(x), \mathrm{sg}(\mu_A(x))), \min(\mu_A(x), \mathrm{sg}(1 - \mu_A(x)))\rangle | x \in E\}$
$\neg_{22}$	$\{\langle x, \min(1 - \mu_A(x), \mathrm{sg}(\mu_A(x))), 0\rangle | x \in E\}$
$\neg_{23}$	$\{\langle x, 1 - \mu_A(x), 0\rangle | x \in E\}$
$\neg_{24}$	$\{\langle x, \min(\nu_A(x), \mathrm{sg}(1 - \nu_A(x))), \min(1 - \nu_A(x), \mathrm{sg}(\nu_A(x)))\rangle | x \in E\}$
$\neg_{25}$	$\{\langle x, \min(\nu_A(x), \mathrm{sg}(1 - \nu_A(x))), 0\rangle | x \in E\}$
$\neg_{26}$	$\{\langle x, \nu_A(x), \mu_A(x).\nu_A(x) + \overline{sg}(1 - \mu_A(x))\rangle | x \in E\}$
$\neg_{27}$	$\{\langle x, 1 - \mu_A(x), \mu_A(x).(1 - \mu_A(x)) + \overline{sg}(1 - \mu_A(x))\rangle | x \in E\}$
$\neg_{28}$	$\{\langle x, \nu_A(x), (1 - \nu_A(x)).\nu_A(x) + \overline{sg}(\nu_A(x))\rangle | x \in E\}$
$\neg_{29}$	$\{\langle x, \nu_A(x).\mu_A(x) + \overline{sg}(1 - \nu_A(x)),$ $\mu_A(x).(\nu_A(x).\mu_A(x) + \overline{sg}(1 - \nu_A(x))) + \overline{sg}(1 - \mu_A(x))\rangle | x \in E\}$
$\neg_{30}$	$\{\langle x, \nu_A(x).\mu_A(x),$ $\mu_A(x).(\nu_A(x).\mu_A(x)+\overline{sg}(1 - \nu_A(x)))+\overline{sg}(1 - \mu_A(x))\rangle | x \in E\}$
$\neg_{31}$	$\{\langle x, (1 - \mu_A(x)).\mu_A(x) + \overline{sg}(\mu_A(x)),$ $\mu_A(x).((1 - \mu_A(x)).\mu_A(x) + \overline{sg}(\mu_A(x))) + \overline{sg}(1 - \mu_A(x))\rangle | x \in E\}$
$\neg_{32}$	$\{\langle x, (1 - \mu_A(x)).\mu_A(x),$ $\mu_A(x).((1 - \mu_A(x)).\mu_A(x) + \overline{sg}(\mu_A(x)))+\overline{sg}(1 - \mu_A(x))\rangle | x \in E\}$
$\neg_{33}$	$\{\langle x, \nu_A(x).(1 - \nu_A(x)) + \overline{sg}(1 - \nu_A(x)),$ $(1 - \nu_A(x)).(\nu_A(x).(1 - \nu_A(x))$ $+\overline{sg}(1 - \nu_A(x))) + \overline{sg}(\nu_A(x))\rangle | x \in E\}$
$\neg_{34}$	$\{\langle x, \nu_A(x).(1 - \nu_A(x)),$ $(1 - \nu_A(x)).(\nu_A(x).(1 - \nu_A(x))+\overline{sg}(1 - \nu_A(x)))+\overline{sg}(\nu_A(x))\rangle | x \in E\}$

Table 9.12 Correspondence between intuitionistic fuzzy negations and implications

$\neg_1$	$\to_1, \to_4, \to_5, \to_6, \to_7, \to_{10}, \to_{13}, \to_{61}, \to_{63}, \to_{64}, \to_{66}, \to_{67},$ $\to_{68}, \to_{69}, \to_{70}, \to_{71}, \to_{72}, \to_{73}, \to_{78}, \to_{80}, \to_{124}, \to_{125}, \to_{127}$
$\neg_2$	$\to_2, \to_3, \to_8, \to_{11}, \to_{16}, \to_{20}, \to_{31}, \to_{32}, \to_{37}, \to_{40}, \to_{41}, \to_{42}$
$\neg_3$	$\to_9, \to_{17}, \to_{21}$
$\neg_4$	$\to_{12}, \to_{18}, \to_{22}, \to_{46}, \to_{49}, \to_{50}, \to_{51}, \to_{53}, \to_{54}, \to_{91}, \to_{93},$ $\to_{94}, \to_{95}, \to_{96}, \to_{98}, \to_{134}, \to_{135}, \to_{137}$
$\neg_5$	$\to_{14}, \to_{15}, \to_{19}, \to_{23}, \to_{47}, \to_{48}, \to_{52}, \to_{55}, \to_{56}, \to_{57}$
$\neg_6$	$\to_{24}, \to_{26}, \to_{27}, \to_{65}$
$\neg_7$	$\to_{25}, \to_{28}, \to_{29}, \to_{62}$

Table 9.12 (*continued*)

$\neg$8	$\to_{30}, \to_{33}, \to_{34}, \to_{35}, \to_{36}, \to_{38}, \to_{39}, \to_{76}, \to_{82}, \to_{84}, \to_{85},$ $\to_{86}, \to_{87}, \to_{89}, \to_{129}, \to_{130}, \to_{132}$
$\neg$9	$\to_{43}, \to_{44}, \to_{45}, \to_{83}$
$\neg$10	$\to_{58}, \to_{59}, \to_{60}, \to_{92}$
$\neg$11	$\to_{74}, \to_{97}$
$\neg$12	$\to_{75}$
$\neg$13	$\to_{77}, \to_{88}$
$\neg$14	$\to_{79}$
$\neg$15	$\to_{81}$
$\neg$16	$\to_{90}$
$\neg$17	$\to_{99}$
$\neg$18	$\to_{100}$
$\neg$19	$\to_{101}$
$\neg$20	$\to_{102}, \to_{108}$
$\neg$21	$\to_{103}$
$\neg$22	$\to_{104}$
$\neg$23	$\to_{105}$
$\neg$24	$\to_{106}$
$\neg$25	$\to_{107}$
$\neg$26	$\to_{109}, \to_{110}, \to_{111}, \to_{112}, \to_{113}$
$\neg$27	$\to_{114}, \to_{115}, \to_{116}, \to_{117}, \to_{118}$
$\neg$28	$\to_{119}, \to_{120}, \to_{121}, \to_{122}, \to_{123}$
$\neg$29	$\to_{126}$
$\neg$30	$\to_{128}$
$\neg$31	$\to_{131}$
$\neg$32	$\to_{133}$
$\neg$33	$\to_{130}$
$\neg$34	$\to_{138}$

Theorem 9.1: For every three IFSs A, B and C the IFSs
(a) $A \to A$,
(b) $A \to (B \to A)$,
(c) $A \to (B \to (A \cap B))$,
(d) $(A \to (B \to C)) \to (B \to (A \to C))$,
(e) $(A \to (B \to C)) \to ((A \to B) \to (A \to C))$,
(f) $A \to \neg\neg A$,
(g) $\neg(A \cap \neg A)$,
(h) $(\neg A \cup B) \to (A \to B)$,
(i) $\neg(A \cup B) \to (\neg A \cap \neg B)$,
(j) $(\neg A \cap \neg B) \to \neg(A \cup B)$,

(k) $(\neg A \cup \neg B) \to \neg (A \cap B)$,

(l) $(A \to B) \to (\neg B \to \neg A)$,

(m) $(A \to \neg B) \to (B \to \neg A)$,

(n) $\neg\neg\neg A \to \neg A$,

(o) $\neg A \to \neg\neg\neg A$,

(p) $\neg\neg(A \to B) \to (A \to \neg\neg B)$,

(q) $(C \to A) \to ((C \to (A \to B)) \to (C \to B))$

are equal to the set E^* (see (2.14)) for the implications $\to_3$, $\to_{11}$, $\to_{14}$, $\to_{20}$, $\to_{23}$, $\to_{74}$, $\to_{77}$.

Theorem 9.2: For every three IFSs A, B and C, the IFSs from (a)-(q) of Theorem 9.1 are IFTSs for the implications $\to_1$, $\to_3$, $\to_4$, $\to_5$, $\to_{11}$, $\to_{13}$, $\to_{14}$, $\to_{17}$, $\to_{18}$, $\to_{20}$, $\to_{22}$, $\to_{23}$, $\to_{27}$, $\to_{28}$, $\to_{29}$, $\to_{61}$, $\to_{66}$, $\to_{71}$, $\to_{74}$, $\to_{76}$, $\to_{77}$, $\to_{79}$, $\to_{81}$, $\to_{100}$, $\to_{101}$, $\to_{102}$, $\to_{109}$, $\to_{110}$, $\to_{111}$, $\to_{112}$, $\to_{113}$, $\to_{118}$.

When an implication satisfies some of the nine axioms from Section **9.1.1**, we write in Table 9.13 " $+$ ", otherwise we write " $-$ ".

Table 9.13

	A1	A2	A3	A3*	A4	A5	A5*	A6	A7	A7*	A8	A8^N	A9
$\to_1$	-	+	+	+	+	-	+	-	-	-	-	-	+
$\to_2$	+	+	+	+	-	+	+	-	-	-	-	-	-
$\to_3$	+	+	+	+	+	+	+	+	-	-	-	-	-
$\to_4$	+	+	+	+	+	-	+	+	-	-	+	+	+
$\to_5$	+	+	+	+	+	-	+	+	-	-	+	+	+
$\to_6$	-	+	+	+	+	-	+	-	-	-	-	-	+
$\to_7$	-	-	-	+	+	-	+	-	-	-	+	+	+
$\to_8$	+	+	+	+	-	+	+	-	-	-	-	-	-
$\to_9$	-	+	+	+	+	-	+	-	-	-	-	-	+
$\to_{10}$	-	+	+	+	+	-	-	-	-	-	-	-	-
$\to_{11}$	+	+	+	+	+	+	+	+	-	-	-	-	-
$\to_{12}$	+	+	+	+	-	-	-	+	-	-	+	-	+
$\to_{13}$	+	+	+	+	+	-	+	+	-	-	+	+	+
$\to_{14}$	+	+	+	+	+	+	+	+	+	-	-	-	-
$\to_{15}$	+	+	+	+	-	+	+	-	+	+	+	-	-
$\to_{16}$	+	+	+	+	+	-	-	+	-	-	-	-	-
$\to_{17}$	-	+	+	+	+	-	+	+	-	-	-	-	+
$\to_{18}$	+	+	+	+	+	-	+	+	-	-	-	-	+
$\to_{19}$	+	+	+	+	+	-	-	+	-	-	-	-	-
$\to_{20}$	+	+	+	+	-	+	+	+	-	-	-	+	-
$\to_{21}$	-	-	+	+	-	-	+	-	-	-	-	-	+
$\to_{22}$	+	+	+	+	-	-	+	+	-	-	-	+	+
$\to_{23}$	+	+	+	+	-	+	+	+	-	-	-	+	-
$\to_{24}$	+	+	+	+	-	+	+	-	+	-	+	-	-

Table 9.13 (*continued*)

	A1	A2	A3	A3*	A4	A5	A5*	A6	A7	A7*	A8	A8^N	A9
$\to_{25}$	+	+	+	+	-	-	-	+	-	-	+	-	-
$\to_{26}$	+	+	+	+	+	-	-	+	-	-	-	-	-
$\to_{27}$	+	+	+	+	-	-	+	+	-	-	-	+	-
$\to_{28}$	+	+	+	+	+	-	+	+	-	-	-	-	-
$\to_{29}$	+	+	+	+	-	-	+	-	-	-	-	-	-
$\to_{30}$	-	+	+	+	-	-	+	-	-	-	-	-	+
$\to_{31}$	+	+	+	+	-	+	+	+	-	-	-	-	-
$\to_{32}$	+	+	+	+	-	+	+	+	-	-	-	-	-
$\to_{33}$	+	+	+	+	-	-	+	+	-	-	+	-	+
$\to_{34}$	+	+	+	+	-	+	+	+	-	-	+	-	+
$\to_{35}$	+	+	+	+	-	-	+	+	-	-	+	-	+
$\to_{36}$	-	-	-	+	-	-	+	+	-	-	+	-	+
$\to_{37}$	+	+	+	+	-	+	+	-	-	-	+	-	-
$\to_{38}$	-	+	+	+	-	-	+	-	-	-	-	-	+
$\to_{39}$	-	+	+	+	-	-	-	-	-	-	-	-	-
$\to_{40}$	+	+	+	+	-	+	+	-	-	-	+	-	-
$\to_{41}$	+	+	+	+	-	-	-	+	-	-	-	-	-
$\to_{42}$	+	+	+	+	-	+	+	+	-	-	-	-	-
$\to_{43}$	+	+	+	+	-	-	-	+	-	-	+	-	-
$\to_{44}$	+	+	+	+	-	-	+	-	-	-	-	-	-
$\to_{45}$	+	+	+	+	-	-	+	-	-	-	-	-	-
$\to_{46}$	-	+	+	+	-	-	-	-	-	-	-	-	+
$\to_{47}$	+	+	+	+	-	-	-	-	-	-	-	-	-
$\to_{48}$	+	+	+	+	-	-	-	+	-	-	-	-	-
$\to_{49}$	+	+	+	+	-	-	-	+	-	-	+	-	+
$\to_{50}$	+	+	+	+	-	-	-	+	-	-	+	-	+
$\to_{51}$	-	-	-	+	-	-	-	+	-	-	+	-	+
$\to_{52}$	+	+	+	+	-	-	-	-	-	-	+	-	-
$\to_{53}$	-	+	+	+	-	-	-	-	-	-	-	-	+
$\to_{54}$	-	+	+	+	-	-	-	-	-	-	-	-	-
$\to_{55}$	+	+	+	+	-	-	-	-	-	-	+	-	-
$\to_{56}$	+	+	+	+	-	-	-	+	-	-	-	-	-
$\to_{57}$	+	+	+	+	-	-	-	+	-	-	-	-	-
$\to_{58}$	+	+	+	+	-	-	-	-	-	-	+	-	-
$\to_{59}$	+	+	+	+	-	-	-	-	-	-	-	-	-
$\to_{00}$	+	+	+	+	-	-	-	-	-	-	-	-	-
$\to_{61}$	+	-	-	+	+	-	+	-	-	-	-	-	+
$\to_{62}$	+	+	+	+	-	+	+	-	-	-	-	-	-
$\to_{63}$	+	+	+	+	-	+	+	-	-	-	-	-	-
$\to_{64}$	+	-	-	+	+	-	+	-	-	-	-	-	+
$\to_{65}$	+	+	+	+	-	+	+	-	-	-	-	-	-
$\to_{66}$	+	-	-	+	-	-	+	-	-	-	-	-	+

Table 9.13 (*continued*)

	A1	A2	A3	A3*	A4	A5	A5*	A6	A7	A7*	A8	A8^N	A9
$\rightarrow_{67}$	+	-	+	+	+	-	-	-	-	-	-	-	-
$\rightarrow_{68}$	+	+	+	+	-	+	+	-	-	-	-	-	-
$\rightarrow_{69}$	+	+	+	+	-	+	+	-	+	-	-	-	-
$\rightarrow_{70}$	+	+	+	+	-	-	-	-	-	-	-	-	-
$\rightarrow_{71}$	+	-	+	+	-	-	+	-	-	-	-	-	+
$\rightarrow_{72}$	+	+	+	+	-	-	+	-	-	-	-	-	+
$\rightarrow_{73}$	+	+	+	+	-	-	-	-	-	-	-	-	-
$\rightarrow_{74}$	+	+	+	+	-	+	+	+	-	-	-	+	-
$\rightarrow_{75}$	-	-	+	+	-	-	+	-	-	-	-	-	+
$\rightarrow_{76}$	+	+	+	+	-	-	+	+	-	-	-	+	+
$\rightarrow_{77}$	+	+	+	+	-	+	+	+	-	-	-	+	-
$\rightarrow_{78}$	+	+	+	+	-	-	-	-	-	-	-	-	-
$\rightarrow_{79}$	+	+	+	+	-	-	+	+	-	-	-	+	-
$\rightarrow_{80}$	+	+	+	+	-	-	+	-	-	-	-	-	-
$\rightarrow_{81}$	+	+	+	+	-	-	+	+	-	-	-	-	-
$\rightarrow_{82}$	+	-	-	+	-	-	+	-	-	-	-	-	+
$\rightarrow_{83}$	+	+	+	+	-	+	+	-	-	-	-	-	-
$\rightarrow_{84}$	+	+	+	+	-	+	+	-	-	-	-	-	-
$\rightarrow_{85}$	+	-	-	+	-	-	+	-	-	-	-	-	+
$\rightarrow_{86}$	+	-	+	+	-	-	-	-	-	-	-	-	-
$\rightarrow_{87}$	+	+	+	+	-	-	-	-	-	-	-	-	-
$\rightarrow_{88}$	+	+	+	+	-	+	+	+	-	-	-	-	-
$\rightarrow_{89}$	+	+	+	+	-	-	+	-	-	-	-	-	-
$\rightarrow_{90}$	+	+	+	+	-	-	+	-	-	-	-	-	-
$\rightarrow_{91}$	+	-	-	+	-	-	-	-	-	-	-	-	+
$\rightarrow_{92}$	+	+	+	+	-	-	-	-	-	-	-	-	-
$\rightarrow_{93}$	+	+	+	+	-	-	-	-	-	-	-	-	-
$\rightarrow_{94}$	+	-	-	+	-	-	-	-	-	-	-	-	+
$\rightarrow_{95}$	+	-	+	+	-	-	-	-	-	-	-	-	-
$\rightarrow_{96}$	+	+	+	+	-	-	-	-	-	-	-	-	-
$\rightarrow_{97}$	+	+	+	+	-	-	-	+	-	-	-	-	-
$\rightarrow_{98}$	+	+	+	+	-	-	-	-	-	-	-	-	-
$\rightarrow_{99}$	+	+	+	+	-	-	-	-	-	-	-	-	-
$\rightarrow_{100}$	-	+	-	+	-	-	+	+	-	-	-	-	-
$\rightarrow_{101}$	-	-	-	+	-	-	+	-	-	-	+	-	-
$\rightarrow_{102}$	+	-	+	+	-	-	+	-	-	-	-	-	-
$\rightarrow_{103}$	-	+	-	+	-	-	+	-	-	-	-	-	-
$\rightarrow_{104}$	-	-	-	+	-	-	+	-	-	-	+	-	-
$\rightarrow_{105}$	+	-	+	+	-	-	+	-	-	-	-	-	-
$\rightarrow_{106}$	-	+	-	+	-	-	-	-	-	-	-	-	-
$\rightarrow_{107}$	-	-	-	+	-	-	-	-	-	-	+	-	-
$\rightarrow_{108}$	+	-	+	+	-	-	-	-	-	-	-	-	-

Table 9.13 (*continued*)

	A1	A2	A3	A3*	A4	A5	A5*	A6	A7	A7*	A8	A8^N	A9
→109	-	+	+	+	+	-	+	-	-	-	-	-	-
→110	-	+	+	+	+	-	+	+	-	-	-	-	-
→111	-	-	+	+	-	-	+	-	-	-	-	-	-
→112	-	+	+	+	+	-	+	+	-	-	-	-	-
→113	-	-	-	+	-	-	+	-	-	-	-	-	-
→114	-	+	+	+	-	-	+	-	-	-	-	-	-
→115	-	+	+	+	-	-	+	-	-	-	-	-	-
→116	-	-	+	+	-	-	+	-	-	-	-	-	-
→117	-	+	+	+	-	-	+	-	-	-	-	-	-
→118	-	-	-	+	-	-	+	-	-	-	-	-	-
→119	-	+	+	+	-	-	-	-	-	-	-	-	-
→120	-	+	+	+	-	-	-	-	-	-	-	-	-
→121	-	-	+	+	-	-	-	-	-	-	-	-	-
→122	-	+	+	+	-	-	-	-	-	-	-	-	-
→123	-	-	-	+	-	-	-	-	-	-	-	-	-
→124	+	-	-	+	-	-	+	-	-	-	-	-	-
→125	+	-	+	+	-	-	+	-	-	-	-	-	-
→126	-	-	+	+	-	-	+	-	-	-	-	-	-
→127	+	-	+	+	-	-	+	-	-	-	-	-	-
→128	-	-	-	+	-	-	+	-	-	-	-	-	-
→129	+	-	-	+	-	-	+	-	-	-	-	-	-
→130	+	-	+	+	-	-	+	-	-	-	-	-	-
→131	-	-	+	+	-	-	+	-	-	-	-	-	-
→132	+	-	+	+	-	-	+	-	-	-	-	-	-
→133	-	-	-	+	-	-	+	-	-	-	-	-	-
→134	+	-	-	+	-	-	-	-	-	-	-	-	-
→135	+	-	+	+	-	-	-	-	-	-	-	-	-
→136	-	-	+	+	-	-	-	-	-	-	-	-	-
→137	+	-	+	+	-	-	-	-	-	-	-	-	-
→138	-	-	-	+	-	-	-	-	-	-	-	-	-

Theorem 9.3 For every three IFSs A, B, C:
(a) equality

$$(A \to B) \cup (B \to C) \cup (C \to A) = (A \to C) \cup (C \to B) \cup (B \to A),$$

is valid for implications $\to_1$, $\to_2$, $\to_3$, $\to_4$, $\to_8$, $\to_{10}$, $\to_{11}$, $\to_{12}$, $\to_{16}$, $\to_{17}$, $\to_{18}$, $\to_{19}$, $\to_{20}$, $\to_{21}$, $\to_{22}$, $\to_{23}$, $\to_{25}$, $\to_{26}$, $\to_{27}$, $\to_{28}$, $\to_{29}$, $\to_{31}$, $\to_{32}$, $\to_{33}$, $\to_{34}$, $\to_{37}$, $\to_{39}$, $\to_{40}$, $\to_{41}$, $\to_{42}$, $\to_{43}$, $\to_{44}$, $\to_{45}$, $\to_{56}$, $\to_{57}$, $\to_{58}$, $\to_{59}$, $\to_{60}$, $\to_{61}$, $\to_{62}$, $\to_{63}$, $\to_{65}$, $\to_{67}$, $\to_{68}$, $\to_{70}$, $\to_{71}$, $\to_{72}$, $\to_{73}$, $\to_{74}$, $\to_{75}$, $\to_{76}$, $\to_{77}$, $\to_{78}$, $\to_{79}$, $\to_{80}$, $\to_{81}$, $\to_{83}$, $\to_{84}$, $\to_{86}$, $\to_{87}$, $\to_{88}$, $\to_{89}$, $\to_{90}$, $\to_{96}$, $\to_{97}$, $\to_{98}$, $\to_{99}$, $\to_{100}$, $\to_{101}$, $\to_{102}$, $\to_{103}$, $\to_{104}$, $\to_{105}$, $\to_{106}$, $\to_{107}$, $\to_{108}$, $\to_{110}$, $\to_{111}$, $\to_{115}$, $\to_{116}$, $\to_{120}$, $\to_{121}$, $\to_{125}$, $\to_{126}$, $\to_{130}$, $\to_{131}$, $\to_{135}$, $\to_{136}$;

(b) equality

$$(A \to B) \cap (B \to C) \cap (C \to A) = (A \to C) \cap (C \to B) \cap (B \to A).$$

is valid for $\to_{14}$, $\to_{19}$, $\to_{20}$, $\to_{23}$, $\to_{47}$, $\to_{48}$, $\to_{52}$, $\to_{55}$, $\to_{56}$, $\to_{57}$, $\to_{73}$, $\to_{74}$, $\to_{77}$, $\to_{92}$, $\to_{93}$, $\to_{96}$, $\to_{97}$.

Note that only 10 implications, namely $\to_{19}$, $\to_{20}$, $\to_{23}$, $\to_{56}$, $\to_{57}$, $\to_{73}$, $\to_{74}$, $\to_{77}$, $\to_{96}$, $\to_{97}$, satisfy both equalities.

Following the definitions of LEM and MLEM in Section **9.1.2**, we can re-formulate them for IFS-case, if for every IFS A over universe E:

$$A \cup \neg A = E^*$$

(tautology-form) and

$$A \cup \neg A = B, \text{ where for each } x \in E : \mu_B(x) \geq \nu_B(x)$$

(an IFTS-form);

$$\neg\neg A \cup \neg A = E^*$$

(tautology-form) and

$$\neg\neg A \cup \neg A = B, \text{ where for each } x \in E : \mu_B(x) \geq \nu_B(x)$$

(an IFTS-form).

Theorem 9.4: Only negation $\neg_{13}$ satisfies the LEM in the tautological form.

Theorem 9.5: Only negations $\neg_2$, $\neg_5$, $\neg_9$, $\neg_{11}$, $\neg_{13}$, $\neg_{16}$ satisfy the MLEM in the tautological form.

Theorem 9.6: Only negations $\neg_2$, $\neg_5$, $\neg_6$, $\neg_{10}$ do not satisfy the LEM in the IFT form.

Theorem 9.7: Only negation $\neg_{10}$, does not satisfy the MLEM in the IFT form.

Now, we study the following properties of an IFS A:

Property $P1_{IFTS}$: $A \to \neg\neg A$ is an IFTS,

Property $P1_{stand}$: $A \to \neg\neg A = E^*$,

Property $P2_{IFTS}$: $\neg\neg A \to A$ is an IFTS,

Property $P2_{stand}$: $\neg\neg A \to A = E^*$,

Property $P3$: $\neg\neg\neg A = \neg A$.

In Table 9.14 we give all couples $(\neg, \to)$ and the list of above properties that they satisfy (marked there by "+").

Table 9.14

Negation	Implication	$P1_{IFTS}$	$P1_{stand}$	$P2_{IFTS}$	$P2_{stand}$	$P3$
$\neg_1$	$\to_1$	+	+			+
$\neg_1$	$\to_4$	+	+			+
$\neg_1$	$\to_5$	+	+			+
$\neg_1$	$\to_6$	+	+			+
$\neg_1$	$\to_7$	+	+			+
$\neg_1$	$\to_{10}$					+

Table 9.14 (*continued*)

Negation	Implication	$P1_{IFTS}$	$P1_{stand}$	$P2_{IFTS}$	$P2_{stand}$	$P3$
$\neg_1$	$\to_{13}$	+	+			+
$\neg_1$	$\to_{61}$	+	+			+
$\neg_1$	$\to_{63}$	+	+	+	+	+
$\neg_1$	$\to_{64}$	+	+			+
$\neg_1$	$\to_{66}$	+	+			+
$\neg_1$	$\to_{67}$					+
$\neg_1$	$\to_{68}$	+	+	+	+	+
$\neg_1$	$\to_{69}$	+	+	+	+	+
$\neg_1$	$\to_{70}$					+
$\neg_1$	$\to_{71}$	+	+			+
$\neg_1$	$\to_{72}$	+	+			+
$\neg_1$	$\to_{73}$					+
$\neg_1$	$\to_{78}$					+
$\neg_1$	$\to_{80}$	+	+			+
$\neg_2$	$\to_{2}$	+		+		+
$\neg_2$	$\to_{3}$	+		+		+
$\neg_2$	$\to_{8}$	+		+		+
$\neg_2$	$\to_{11}$	+		+		+
$\neg_2$	$\to_{16}$	+		+		+
$\neg_2$	$\to_{20}$	+	+	+	+	+
$\neg_2$	$\to_{31}$	+		+		+
$\neg_2$	$\to_{32}$	+		+		+
$\neg_2$	$\to_{37}$	+		+		+
$\neg_2$	$\to_{40}$	+		+		+
$\neg_2$	$\to_{41}$	+		+		+
$\neg_2$	$\to_{42}$	+	+	+	+	+
$\neg_3$	$\to_{9}$	+	+			
$\neg_3$	$\to_{17}$	+	+			
$\neg_3$	$\to_{21}$	+	+			
$\neg_4$	$\to_{12}$	+				+
$\neg_4$	$\to_{18}$	+	+			+
$\neg_4$	$\to_{22}$	+	+			+
$\neg_4$	$\to_{46}$	+				+
$\neg_4$	$\to_{49}$	+		+		+
$\neg_4$	$\to_{50}$	+				+
$\neg_4$	$\to_{51}$	+				+
$\neg_4$	$\to_{53}$	+				+
$\neg_4$	$\to_{54}$					+
$\neg_4$	$\to_{91}$	+				+
$\neg_4$	$\to_{93}$	+		+		+
$\neg_4$	$\to_{94}$	+				+
$\neg_4$	$\to_{95}$					+

Table 9.14 (*continued*)

Negation	Implication	$P1_{IFTS}$	$P1_{stand}$	$P2_{IFTS}$	$P2_{stand}$	$P3$
$\neg_4$	$\rightarrow_{96}$					+
$\neg_4$	$\rightarrow_{98}$	+				+
$\neg_5$	$\rightarrow_{14}$	+		+		+
$\neg_5$	$\rightarrow_{15}$	+		+		+
$\neg_5$	$\rightarrow_{19}$	+		+		+
$\neg_5$	$\rightarrow_{23}$	+	+	+	+	+
$\neg_5$	$\rightarrow_{47}$	+		+		+
$\neg_5$	$\rightarrow_{48}$	+		+		+
$\neg_5$	$\rightarrow_{52}$	+		+		+
$\neg_5$	$\rightarrow_{55}$	+		+		+
$\neg_5$	$\rightarrow_{56}$	+		+		+
$\neg_5$	$\rightarrow_{57}$	+		+		+
$\neg_6$	$\rightarrow_{24}$	+		+		+
$\neg_6$	$\rightarrow_{26}$	+				+
$\neg_6$	$\rightarrow_{27}$	+	+			+
$\neg_6$	$\rightarrow_{65}$	+		+		+
$\neg_7$	$\rightarrow_{25}$	+	+			
$\neg_7$	$\rightarrow_{28}$	+	+			
$\neg_7$	$\rightarrow_{29}$	+	+			
$\neg_7$	$\rightarrow_{62}$	+	+	+		
$\neg_8$	$\rightarrow_{30}$	+	+			+
$\neg_8$	$\rightarrow_{33}$	+	+			+
$\neg_8$	$\rightarrow_{34}$	+	+	+	+	+
$\neg_8$	$\rightarrow_{35}$	+	+			+
$\neg_8$	$\rightarrow_{36}$	+	+			+
$\neg_8$	$\rightarrow_{38}$	+	+			+
$\neg_8$	$\rightarrow_{39}$					+
$\neg_8$	$\rightarrow_{76}$	+	+			+
$\neg_8$	$\rightarrow_{82}$	+	+			+
$\neg_8$	$\rightarrow_{84}$	+	+	+	+	+
$\neg_8$	$\rightarrow_{85}$	+	+			+
$\neg_8$	$\rightarrow_{86}$					+
$\neg_8$	$\rightarrow_{87}$					+
$\neg_8$	$\rightarrow_{89}$	+	+			+
$\neg_9$	$\rightarrow_{43}$	+		+		
$\neg_9$	$\rightarrow_{44}$	+		+		
$\neg_9$	$\rightarrow_{45}$	+	+	+		
$\neg_9$	$\rightarrow_{83}$	+		+		
$\neg_{10}$	$\rightarrow_{58}$					+
$\neg_{10}$	$\rightarrow_{59}$					+
$\neg_{10}$	$\rightarrow_{60}$					+
$\neg_{10}$	$\rightarrow_{92}$					+

Table 9.14 (*continued*)

Negation	Implication	$P1_{IFTS}$	$P1_{stand}$	$P2_{IFTS}$	$P2_{stand}$	$P3$
$\neg11$	$\to_{74}$	+	+	+	+	+
$\neg11$	$\to_{97}$	+		+		+
$\neg12$	$\to_{75}$	+	+			
$\neg13$	$\to_{77}$	+	+	+	+	+
$\neg13$	$\to_{88}$	+	+	+	+	+
$\neg14$	$\to_{79}$	+	+			+
$\neg15$	$\to_{81}$	+	+			+
$\neg16$	$\to_{99}$	+				
$\neg17$	$\to_{90}$	+		+		

9.2.2 On One of Baczynski-Jayaram's Problems

In [124], Michal Baczynski and Balasubramaniam Jayaram formulated some problems related to fuzzy implications I and negations N. Here, we give a solution to one of them:

Problem 1.7.1: *Give examples of fuzzy implications I such that*
(i) I satisfies only property

$$I(x,y) = I(N(y), N(x)) \qquad (CP)$$

(ii) I satisfies only property

$$I(N(x),y) = I(N(y), x) \qquad (L-CP)$$

(iii) I satisfies both (CP) and (L-CP), but not

$$I(x, N(y)) = I(y, N(x)) \qquad (R-CP)$$

with some fuzzy negation N, where $x, y \in [0,1]$.

We must note that in [124] no example is given.

Here, following [90], we give examples of couples of implications and negations that satisfy Problem 1.7.1 (ii) and other problems. Let us denote by the pair (m, n) the expression of m-th implication and n-th negation. First, we formulate the following

Theorem 9.8: The pairs $(4,1)$, $(5,1)$, $(7,1)$, $(12,1)$, $(13,1)$, $(15,1)$, $(24,1)$, $(25,1)$, $(33,1)$, $(34,1)$, $(35,1)$, $(36,1)$, $(37,1)$, $(40,1)$, $(43,1)$, $(49,1)$, $(50,1)$, $(51,1)$, $(52,1)$, $(55,1)$, $(58,1)$, $(101,1)$, $(104,1)$, $(107,1)$, $(20,2)$, $(22,4)$, $(23,5)$, $(27,6)$, $(42,6)$, $(57,6)$, $(76,8)$, $(20,9)$, $(22,10)$, $(74,11)$, $(77,13)$, $(79,14)$, $(88,14)$, $(97,14)$, $(20,16)$, $(74,17)$, $(101,18)$, $(76,23)$, $(76,27)$ satisfy the three axioms (CP), (L-CP) and (R-CP).

Theorem 9.9: The pairs $(52, 7)$, $(55, 7)$, $(52, 15)$, $(55, 15)$, $(88, 19)$, $(33, 20)$, $(34, 20)$, $(35, 20)$, $(37, 20)$, $(40, 20)$, $(43, 20)$, $(88, 22)$, $(88, 25)$ satisfy two axioms and more exactly, they satisfy (L-CP) and (R-CP).

We had not found any couple of implication and negation that are solution of Problem 1.7.1 (iii).

Also, we had not found any couple of implication and negation that satisfy only the first axiom, i.e., we cannot give examples for the case of Problem 1.7.1 (i).

Another result of our search is

Theorem 9.10: The pairs $(2, 2)$, $(3, 2)$, $(8, 2)$, $(11, 2)$, $(16, 2)$, $(31, 2)$, $(32, 2)$, $(37, 2)$, $(40, 2)$, $(41, 2)$, $(42, 2)$, $(12, 3)$, $(17, 3)$, $(49, 3)$, $(50, 3)$, $(51, 3)$, $(52, 3)$, $(55, 3)$, $(58, 3)$, $(107, 3)$, $(12, 4)$, $(18, 4)$, $(49, 4)$, $(50, 4)$, $(51, 4)$, $(52, 4)$, $(55, 4)$, $(58, 4)$, $(107, 4)$, $(14, 5)$, $(15, 5)$, $(19, 5)$, $(47, 5)$, $(48, 5)$, $(52, 5)$, $(55, 5)$, $(56, 5)$, $(57, 5)$, $(24, 6)$, $(26, 6)$, $(31, 6)$, $(32, 6)$, $(37, 6)$, $(40, 6)$, $(41, 6)$, $(47, 6)$, $(48, 6)$, $(52, 6)$, $(55, 6)$, $(56, 6)$, $(25, 7)$, $(28, 7)$, $(33, 7)$, $(34, 7)$, $(35, 7)$, $(36, 7)$, $(37, 7)$, $(40, 7)$, $(43, 7)$, $(47, 7)$, $(48, 7)$, $(56, 7)$, $(57, 7)$, $(104, 7)$, $(33, 8)$, $(34, 8)$, $(35, 8)$, $(36, 8)$, $(37, 8)$, $(40, 8)$, $(43, 8)$, $(104, 8)$, $(33, 9)$, $(34, 9)$, $(35, 9)$, $(36, 9)$, $(37, 9)$, $(40, 9)$, $(43, 9)$, $(104, 9)$, $(47, 10)$, $(48, 10)$, $(52, 10)$, $(55, 10)$, $(56, 10)$, $(57, 10)$, $(97, 11)$, $(88, 13)$, $(47, 15)$, $(48, 15)$, $(56, 15)$, $(57, 15)$, $(81, 15)$, $(88, 15)$, $(88, 16)$, $(47, 17)$, $(48, 17)$, $(52, 17)$, $(55, 17)$, $(56, 17)$, $(57, 17)$, $(23, 18)$, $(42, 18)$, $(100, 18)$, $(22, 19)$, $(23, 19)$, $(31, 19)$, $(32, 19)$, $(33, 19)$, $(34, 19)$, $(35, 19)$, $(37, 19)$, $(39, 19)$, $(40, 19)$, $(41, 19)$, $(42, 19)$, $(43, 19)$, $(44, 19)$, $(45, 19)$, $(62, 19)$, $(63, 19)$, $(65, 19)$, $(68, 19)$, $(70, 19)$, $(74, 19)$, $(82, 19)$, $(83, 19)$, $(84, 19)$, $(85, 19)$, $(86, 19)$, $(87, 19)$, $(89, 19)$, $(90, 19)$, $(100, 19)$, $(103, 19)$, $(115, 19)$, $(116, 19)$, $(117, 19)$, $(118, 19)$, $(129, 19)$, $(130, 19)$, $(131, 19)$, $(132, 19)$, $(4, 20)$, $(5, 20)$, $(12, 20)$, $(13, 20)$, $(17, 20)$, $(18, 20)$, $(22, 20)$, $(23, 20)$, $(25, 20)$, $(29, 20)$, $(31, 20)$, $(32, 20)$, $(39, 20)$, $(41, 20)$, $(42, 20)$, $(44, 20)$, $(45, 20)$, $(49, 20)$, $(50, 20)$, $(51, 20)$, $(52, 20)$, $(55, 20)$, $(58, 20)$, $(62, 20)$, $(63, 20)$, $(65, 20)$, $(68, 20)$, $(70, 20)$, $(71, 20)$, $(74, 20)$, $(81, 20)$, $(82, 20)$, $(83, 20)$, $(84, 20)$, $(85, 20)$, $(86, 20)$, $(87, 20)$, $(88, 20)$, $(89, 20)$, $(90, 20)$, $(103, 20)$, $(107, 20)$, $(110, 20)$, $(112, 20)$, $(115, 20)$, $(116, 20)$, $(117, 20)$, $(118, 20)$, $(125, 20)$, $(127, 20)$, $(129, 20)$, $(130, 20)$, $(131, 20)$, $(132, 20)$, $(23, 21)$, $(42, 21)$, $(104, 21)$, $(22, 22)$, $(23, 22)$, $(31, 22)$, $(32, 22)$, $(33, 22)$, $(34, 22)$, $(35, 22)$, $(37, 22)$, $(39, 22)$, $(40, 22)$, $(41, 22)$, $(42, 22)$, $(43, 22)$, $(44, 22)$, $(45, 22)$, $(62, 22)$, $(63, 22)$, $(65, 22)$, $(68, 22)$, $(70, 22)$, $(74, 22)$, $(82, 22)$, $(83, 22)$, $(84, 22)$, $(85, 22)$, $(86, 22)$, $(87, 22)$, $(89, 22)$, $(90, 22)$, $(103, 22)$, $(115, 22)$, $(116, 22)$, $(117, 22)$, $(118, 22)$, $(129, 22)$, $(130, 22)$, $(131, 22)$, $(132, 22)$, $(2, 23)$, $(22, 23)$, $(23, 23)$, $(24, 23)$, $(31, 23)$, $(32, 23)$, $(33, 23)$, $(34, 23)$, $(35, 23)$, $(37, 23)$, $(39, 23)$, $(40, 23)$, $(41, 23)$, $(42, 23)$, $(43, 23)$, $(44, 23)$, $(45, 23)$, $(62, 23)$, $(63, 23)$, $(65, 23)$, $(68, 23)$, $(70, 23)$, $(74, 23)$, $(82, 23)$, $(83, 23)$, $(84, 23)$, $(85, 23)$, $(86, 23)$, $(87, 23)$, $(88, 23)$, $(89, 23)$, $(90, 23)$, $(103, 23)$, $(115, 23)$, $(116, 23)$, $(117, 23)$, $(118, 23)$, $(129, 23)$, $(130, 23)$, $(131, 23)$, $(132, 23)$, $(23, 24)$, $(42, 24)$, $(107, 24)$, $(22, 25)$, $(23, 25)$, $(31, 25)$, $(32, 25)$, $(33, 25)$, $(34, 25)$, $(35, 25)$, $(37, 25)$, $(39, 25)$, $(40, 25)$, $(41, 25)$, $(42, 25)$, $(43, 25)$, $(44, 25)$, $(45, 25)$, $(62, 25)$, $(63, 25)$, $(65, 25)$, $(68, 25)$, $(70, 25)$, $(74, 25)$, $(82, 25)$, $(83, 25)$, $(84, 25)$, $(85, 25)$, $(86, 25)$, $(87, 25)$, $(89, 25)$, $(90, 25)$, $(103, 25)$,

(107,25), (115,25), (116,25), (117,25), (118,25), (129, 25), (130, 25), (131, 25), (132, 25), (12, 26), (107, 25), (49, 26), (50, 26), (51, 26), (52, 26), (55, 26), (58, 26), (107, 26), (110, 26), (112, 26), (12, 28), (49, 28), (50, 28), (51, 28), (52, 28), (55, 28), (58, 28), (107, 28) satisfy only the Axiom (R-CP).

The most interesting is the following

Theorem 9.11: The pairs (57, 2), (21, 3), (25, 3), (33, 3), (34, 3), (35, 3), (36, 3), (37, 3), (40, 3), (43, 3), (104, 3), (33, 4), (34, 4), (35, 4), (36, 4), (37, 4), (40, 4), (43, 4), (104, 4), (42, 5), (12, 7), (29, 7), (42, 7), (49, 7), (50, 7), (51, 7), (58, 7), (107, 7), (12, 8), (49, 8), (50, 8), (51, 8), (52, 8), (55, 8), (58, 8), (72, 8), (107,8), (12,9), (49,9), (50,9), (51,9), (52, 9), (55, 9), (58, 9), (107, 9), (42, 10), (37, 11), (40, 11), (62, 11), (63, 11), (65, 11), (68, 11), (70, 11), (83, 11), (84, 11), (87, 11), (88, 11), (15, 13), (52, 13), (55, 13), (69, 13), (73, 13), (92, 13), (93, 13), (96, 13), (97, 13), (24, 14), (37, 14), (40, 14), (52, 14), (55, 14), (78, 14), (83, 14), (84, 14), (87, 14), (92, 14), (93, 14), (96, 14), (29, 15), (42, 15), (92, 15), (93, 15), (96, 15), (97, 15), (52, 16), (55, 16), (92, 16), (93, 16), (96, 16), (97, 16), (42, 17), (77, 18), (88, 18), (102, 18), (77, 19), (109, 19), (36, 20), (104, 20), (77, 21), (88, 21), (107, 21), (77, 22), (109, 22), (77, 24), (88, 24), (104, 24), (77, 25), (104,25), (109,25), (25,26), (33,26), (34,26), (35,26), (36,26), (37, 26), (40, 26), (43, 26), (104, 26), (111, 26), (33, 28), (34, 28), (35, 28), (36, 28), (37, 28), (40,28), (43,28), (104,28), (77,30), (81,30), (88,30), (77, 32), (81, 32), (88, 32), (77, 34), (88, 34) satisfy only the Axiom (L-CP).

This theorem gives 135 examples of "implication and negation" pairs that are solutions of Problem 1.7.1 (ii).

Let us call the Problem in its present form (i.e., searching of *some* implications and *some* negations) a "weak problem". Then, the "strong problem" (this classification is not discussed in [124]) is be related to search of implications and the negations generated by them, which satisfy only (L-CP). The answer to this problem is given by

Theorem 9.12: The pairs (21, 3), (29, 7), (111, 26) satisfy only Axiom (L-CP).

101 implications and 29 negations participate in some pairs.

In [57], it is shown that IF negations that are different from the classical negation ($\neg_1$ from Table 9.11) do not satisfy LEM, but some of the non-classical negations satisfy the Modified LEM. Using the idea of Modified LEM, we can change equalities (CP), $(L - CP)$ and $(R - CP)$ with the equalities

$$I(N(N(x)), N(N(y))) = I(N(y), N(x)), \qquad (CP')$$

$$I(N(x), N(N(y))) = I(N(y), N(N(x))), \qquad (L - CP')$$

$$I(N(N(x)), N(y)) = I(N(N(y)), N(x)). \qquad (R - CP')$$

Here, following [91], we give examples of pairs of implications and negations that satisfy an extended form of Problem 1.7.1 (ii), where the assertions are related to (CP'), $(L - CP')$ and $(R - CP')$.

Note that there are 1322 pairs (m, n) satisfying the three equalities.

Theorem 9.13: The pairs $(1, 19)$, $(6, 19)$, $(7, 19)$, $(9, 19)$, $(1, 20)$, $(6, 20)$, $(7, 20)$, $(9, 20)$, $(109, 20)$, $(1, 25)$, $(6, 25)$, $(7, 25)$, $(9, 25)$ satisfy two equalities and more precisely, they satisfy $(L - CP')$ and $(R - CP')$.

We did not find any pairs of implications and negations that solve the extended form of Problem 1.7.1 (iii).

Also, we did not find any pairs of implications and negations that satisfy the first equality only, i.e., we cannot give examples for the case of the extended form of Problem 1.7.1 (i).

Another result of our research is

Theorem 9.14: The pairs $(21, 3)$, $(25, 3)$, $(33, 3)$, $(34, 3)$, $(35, 3)$, $(36, 3)$, $(37, 3)$, $(40, 3)$, $(43, 3)$, $(104, 3)$, $(12, 7)$, $(46, 7)$, $(49, 7)$, $(50, 7)$, $(51, 7)$, $(53, 7)$, $(54, 7)$, $(58, 7)$, $(59, 7)$, $(60, 7)$, $(91, 7)$, $(92, 7)$, $(93, 7)$, $(94, 7)$, $(95, 7)$, $(96, 7)$, $(97, 7)$, $(98, 7)$, $(99, 7)$, $(107, 7)$, $(111, 7)$, $(119, 7)$, $(120, 7)$, $(121, 7)$, $(122, 7)$, $(125, 7)$, $(127, 7)$, $(134, 7)$, $(135, 7)$, $(136, 7)$, $(137, 7)$, $(12, 9)$, $(15, 9)$, $(24, 9)$, $(46, 9)$, $(47, 9)$, $(48, 9)$, $(49, 9)$, $(50, 9)$, $(51, 9)$, $(52, 9)$, $(53, 9)$, $(54, 9)$, $(55, 9)$, $(56, 9)$, $(57, 9)$, $(58, 9)$, $(59, 9)$, $(60, 9)$, $(62, 9)$, $(63, 9)$, $(65, 9)$, $(68, 9)$, $(69, 9)$, $(70, 9)$, $(72, 9)$, $(73, 9)$, $(78, 9)$, $(91, 9)$, $(92, 9)$, $(93, 9)$, $(94, 9)$, $(95, 9)$, $(96, 9)$, $(98, 9)$, $(99, 9)$, $(107, 9)$, $(119, 9)$, $(120, 9)$, $(121, 9)$, $(122, 9)$, $(135, 9)$, $(136, 9)$, $(137, 9)$, $(12, 16)$, $(15, 16)$, $(46, 16)$, $(47, 16)$, $(48, 16)$, $(49, 16)$, $(50, 16)$, $(51, 16)$, $(52, 16)$, $(53, 16)$, $(54, 16)$, $(55, 16)$, $(56, 16)$, $(57, 16)$, $(58, 16)$, $(59, 16)$, $(60, 16)$, $(69, 16)$, $(72, 16)$, $(73, 16)$, $(91, 16)$, $(92, 16)$, $(93, 16)$, $(94, 16)$, $(95, 16)$, $(96, 16)$, $(97, 16)$, $(98, 16)$, $(99, 16)$, $(119, 16)$, $(120, 16)$, $(121, 16)$, $(122, 16)$, $(134, 16)$, $(135, 16)$, $(136, 16)$, $(137, 16)$, $(30, 17)$, $(31, 17)$, $(32, 17)$, $(33, 17)$, $(34, 17)$, $(35, 17)$, $(36, 17)$, $(37, 17)$, $(38, 17)$, $(39, 17)$, $(40, 17)$, $(41, 17)$, $(42, 17)$, $(43, 17)$, $(44, 17)$, $(45, 17)$, $(62, 17)$, $(63, 17)$, $(65, 17)$, $(68, 17)$, $(70, 17)$, $(82, 17)$, $(83, 17)$, $(84, 17)$, $(85, 17)$, $(86, 17)$, $(87, 17)$, $(88, 17)$, $(89, 17)$, $(90, 17)$, $(114, 17)$, $(115, 17)$, $(116, 17)$, $(117, 17)$, $(129, 17)$, $(130, 17)$, $(131, 17)$, $(132, 17)$, $(2, 19)$, $(3, 19)$, $(8, 19)$, $(10, 19)$, $(11, 19)$, $(15, 19)$, $(16, 19)$, $(20, 19)$, $(24, 19)$, $(30, 19)$, $(36, 19)$, $(38, 19)$, $(69, 19)$, $(73, 19)$, $(76, 19)$, $(104, 19)$, $(105, 19)$, $(114, 19)$, $(133, 19)$, $(2, 20)$, $(3, 20)$, $(8, 20)$, $(11, 20)$, $(14, 20)$, $(15, 20)$, $(16, 20)$, $(20, 20)$, $(24, 20)$, $(30, 20)$, $(36, 20)$, $(38, 20)$, $(69, 20)$, $(76, 20)$, $(77, 20)$, $(104, 20)$, $(105, 20)$, $(114, 20)$, $(133, 20)$, $(2, 25)$, $(3, 25)$, $(8, 25)$, $(10, 25)$, $(11, 25)$, $(14, 25)$, $(15, 25)$, $(16, 25)$, $(20, 25)$, $(24, 25)$, $(30, 25)$, $(36, 25)$, $(38, 25)$, $(69, 25)$, $(73, 25)$, $(76, 25)$, $(104, 25)$, $(105, 25)$, $(114, 25)$, $(133, 25)$, $(25, 26)$, $(33, 26)$, $(34, 26)$, $(35, 26)$, $(36, 26)$, $(37, 26)$, $(40, 26)$, $(43, 26)$, $(104, 26)$, $(111, 26)$, $(25, 28)$, $(33, 28)$, $(34, 28)$, $(35, 28)$, $(36, 28)$, $(37, 28)$, $(40, 28)$, $(43, 28)$, $(104, 28)$, $(55, 30)$, $(81, 30)$, $(55, 32)$, $(81, 32)$, satisfy only equality $(R - CP')$.

The most interesting is the following:

Theorem 9.15: The pairs $(12,3)$, $(17,3)$, $(49,3)$, $(50,3)$, $(51,3)$, $(52,3)$, $(55,3)$, $(58,3)$, $(107,3)$, $(15,7)$, $(24,7)$, $(25,7)$, $(26,7)$, $(28,7)$, $(30,7)$, $(31,7)$, $(32,7)$, $(33,7)$, $(34,7)$, $(35,7)$, $(36,7)$, $(37,7)$, $(38,7)$, $(39,7)$, $(40,7)$, $(41,7)$, $(43,7)$, $(44,7)$, $(45,7)$, $(81,7)$, $(82,7)$, $(83,7)$, $(84,7)$, $(85,7)$, $(86,7)$, $(87,7)$, $(88,7)$, $(89,7)$, $(90,7)$, $(104,7)$, $(114,7)$, $(115,7)$, $(116,7)$, $(117,7)$, $(129,7)$, $(130,7)$, $(131,7)$, $(132,7)$, $(2,9)$, $(3,9)$, $(11,9)$, $(16,9)$, $(30,9)$, $(31,9)$, $(32,9)$, $(33,9)$, $(34,9)$, $(35,9)$, $(36,9)$, $(38,9)$, $(39,9)$, $(41,9)$, $(42,9)$, $(43,9)$, $(44,9)$, $(45,9)$, $(82,9)$, $(85,9)$, $(86,9)$, $(89,9)$, $(90,9)$, $(104,9)$, $(115,9)$, $(116,9)$, $(117,9)$, $(129,9)$, $(130,9)$, $(131,9)$, $(132,9)$, $(2,16)$, $(3,16)$, $(8,16)$, $(11,16)$, $(16,16)$, $(30,16)$, $(31,16)$, $(32,16)$, $(33,16)$, $(34,16)$, $(35,16)$, $(36,16)$, $(37,16)$, $(38,16)$, $(39,16)$, $(40,16)$, $(41,16)$, $(42,16)$, $(43,16)$, $(44,16)$, $(45,16)$, $(82,16)$, $(83,16)$, $(84,16)$, $(85,16)$, $(86,16)$, $(87,16)$, $(88,16)$, $(89,16)$, $(90,16)$, $(114,16)$, $(115,16)$, $(116,16)$, $(117,16)$, $(129,16)$, $(130,16)$, $(131,16)$, $(132,16)$, $(12,17)$, $(14,17)$, $(15,17)$, $(18,17)$, $(19,17)$, $(46,17)$, $(47,17)$, $(48,17)$, $(49,17)$, $(50,17)$, $(51,17)$, $(52,17)$, $(53,17)$, $(54,17)$, $(55,17)$, $(56,17)$, $(57,17)$, $(58,17)$, $(59,17)$, $(60,17)$, $(91,17)$, $(92,17)$, $(93,17)$, $(94,17)$, $(95,17)$, $(96,17)$, $(97,17)$, $(98,17)$, $(99,17)$, $(119,17)$, $(120,17)$, $(121,17)$, $(122,17)$, $(134,17)$, $(135,17)$, $(136,17)$, $(137,17)$, $(4,19)$, $(5,19)$, $(12,19)$, $(13,19)$, $(17,19)$, $(18,19)$, $(19,19)$, $(21,19)$, $(26,19)$, $(27,19)$, $(28,19)$, $(46,19)$, $(47,19)$, $(48,19)$, $(49,19)$, $(50,19)$, $(51,19)$, $(53,19)$, $(54,19)$, $(56,19)$, $(57,19)$, $(58,19)$, $(59,19)$, $(60,19)$, $(61,19)$, $(64,19)$, $(66,19)$, $(67,19)$, $(71,19)$, $(75,19)$, $(91,19)$, $(94,19)$, $(95,19)$, $(98,19)$, $(99,19)$, $(100,19)$, $(106,19)$, $(107,19)$, $(108,19)$, $(110,19)$, $(112,19)$, $(119,19)$, $(120,19)$, $(121,19)$, $(122,19)$, $(123,19)$, $(124,19)$, $(125,19)$, $(126,19)$, $(127,19)$, $(128,19)$, $(134,19)$, $(135,19)$, $(136,19)$, $(137,19)$, $(138,19)$, $(4,20)$, $(5,20)$, $(12,20)$, $(13,20)$, $(17,20)$, $(18,20)$, $(19,20)$, $(21,20)$, $(25,20)$, $(26,20)$, $(27,20)$, $(28,20)$, $(29,20)$, $(46,20)$, $(47,20)$, $(48,20)$, $(49,20)$, $(50,20)$, $(51,20)$, $(52,20)$, $(53,20)$, $(54,20)$, $(55,20)$, $(56,20)$, $(57,20)$, $(58,20)$, $(59,20)$, $(60,20)$, $(61,20)$, $(64,20)$, $(66,20)$, $(67,20)$, $(71,20)$, $(75,20)$, $(78,20)$, $(79,20)$, $(80,20)$, $(81,20)$, $(91,20)$, $(92,20)$, $(93,20)$, $(94,20)$, $(95,20)$, $(96,20)$, $(97,20)$, $(98,20)$, $(99,20)$, $(100,20)$, $(106,20)$, $(107,20)$, $(108,20)$, $(110,20)$, $(111,20)$, $(112,20)$, $(119,20)$, $(120,20)$, $(121,20)$, $(122,20)$, $(123,20)$, $(124,20)$, $(125,20)$, $(126,20)$, $(127,20)$, $(128,20)$, $(134,20)$, $(135,20)$, $(136,20)$, $(137,20)$, $(138,20)$, $(4,25)$, $(5,25)$, $(12,25)$, $(13,25)$, $(17,25)$, $(18,25)$, $(19,25)$, $(21,25)$, $(26,25)$, $(27,25)$, $(28,25)$, $(46,25)$, $(47,25)$, $(48,25)$, $(49,25)$, $(50,25)$, $(51,25)$, $(53,25)$, $(54,25)$, $(56,25)$, $(57,25)$, $(58,25)$, $(59,25)$, $(60,25)$, $(61,25)$, $(64,25)$, $(66,25)$, $(67,25)$, $(71,25)$, $(75,25)$, $(91,25)$, $(94,25)$, $(95,25)$, $(98,25)$, $(99,25)$, $(100,25)$, $(106,25)$, $(107,25)$, $(108,25)$, $(110,25)$, $(112,25)$, $(119,25)$, $(120,25)$, $(121,25)$, $(122,25)$, $(123,25)$, $(124,25)$, $(125,25)$, $(126,25)$, $(127,25)$, $(128,25)$, $(134,25)$, $(135,25)$, $(136,25)$, $(137,25)$, $(138,25)$, $(12,26)$, $(49,26)$, $(50,26)$, $(51,26)$, $(52,26)$, $(55,26)$, $(58,26)$, $(107,26)$, $(110,26)$, $(112,26)$, $(12,28)$, $(49,28)$, $(50,28)$, $(51,28)$, $(52,28)$, $(55,28)$, $(58,28)$, $(107,28)$, $(23,30)$, $(31,30)$, $(32,30)$, $(34,30)$, $(37,30)$, $(40,30)$, $(42,30)$, $(45,30)$, $(83,30)$, $(84,30)$, $(90,30)$,

$(23, 32)$, $(31, 32)$, $(32,32)$, $(34,32)$, $(37,32)$, $(40, 32)$, $(42, 32)$, $(45, 32)$, $(83, 32)$, $(84, 32)$, $(90, 32)$ satisfy only equality $(L - CP')$.

Theorem 9.15 gives 375 examples of pairs of implications and negations that are solutions to the extended form of Problem 1.7.1 (ii).

Let us call the extended form of the Problem in its present form (i.e., searching of *some* implications and *some* negations) a "weak problem". Then, the "strong problem" (this classification is not discussed in [124]) is related to search of implications and the negations generated by them (see Table 9.12), which satisfy only (L-CP'). The answer to this problem is given by

Theorem 9.16: The pairs $(17, 3)$, $(25, 7)$, $(28, 7)$, $(43, 9)$, $(44, 9)$, $(45, 9)$, $(90, 16)$, $(99, 17)$, $(108, 20)$, $(107, 25)$, $(110, 26)$, and $(112, 26)$ satisfy only property $(L - CP')$.

9.2.3   IF Negations and the Operators over IFSs

Here we give the relations between the IF negations and the operators over IFSs, following papers [96, 261, 262, 263, 264] of Chris Hinde and the author.

Theorem 9.17: For every IFS A, the following properties are valid:

(1) $\neg_2 \square A = \square \neg_2 A$,

(2) $\neg_2 \Diamond A \subset \Diamond \neg_2 A$,

(3) $\neg_3 \square A \supset \square \neg_3 A$,

(4) $\neg_3 \Diamond A \subset \Diamond \neg_3 A$,

(5) $\neg_4 \square A \supset \square \neg_4 A$,

(6) $\neg_4 \Diamond A = \Diamond \neg_4 A$,

(7) $\neg_5 \Diamond A = \Diamond \neg_5 A$,

(8) $\neg_6 \Diamond A = \Diamond \neg_6 A$,

(9) $\neg_7 \square A \supset \square \neg_7 A$,

(10) $\neg_7 \Diamond A \subset \Diamond \neg_7 A$,

(11) $\neg_8 \square A = \square \neg_8 A$,

(12) $\neg_8 \Diamond A \subset \Diamond \neg_8 A$,

(13) $\neg_9 \square A \supset \square \neg_9 A$,

(14) $\neg_9 \Diamond A \subset \Diamond \neg_9 A$,

(15) $\neg_{10} \square A \supset \square \neg_{10} A$,

(16) $\neg_{11} \square A = \square \neg_{11} A$,

(17) $\neg_{11} \Diamond A = \Diamond \neg_{11} A$,

(18) $\neg_{13} \square A = \square \neg_{12} A$,

(19) $\neg_{15} \square A \supset \square \neg_{15} A$,

(20) $\neg_{15}\Diamond A \subset \Diamond\neg_{15}A$,

(21) $\neg_{16}\Box A \supset \Box\neg_{16}A$,
(22) $\neg_{16}\Diamond A \subset \Diamond\neg_{16}A$,

(23) $\neg_{17}\Box A \supset \Box\neg_{17}A$,
(24) $\neg_{17}\Diamond A \subset \Diamond\neg_{17}A$,

(25) $\neg_{18}\Box A \supset \Box\neg_{18}A$,
(26) $\neg_{18}\Diamond A \subset \Diamond\neg_{18}A$,

(27) $\neg_{19}\Box A \supset \Box\neg_{19}A$,
(28) $\neg_{19}\Diamond A \subset \Diamond\neg_{19}A$,

(29) $\neg_{20}\Box A \supset \Box\neg_{20}A$,
(30) $\neg_{20}\Diamond A = \Diamond\neg_{20}A$,

(31) $\neg_{21}\Box A \supset \Box\neg_{21}A$,
(32) $\neg_{21}\Diamond A \subset \Diamond\neg_{21}A$,

(33) $\neg_{22}\Box A \supset \Box\neg_{22}A$,
(34) $\neg_{22}\Diamond A \subset \Diamond\neg_{22}A$.

In Section **9.1.3**, the cases in which some intuitionistic fuzzy (non-classical) negations do not satisfy De Morgan's laws are shown. Now, by analogy with this result, we study the De Morgans' form of modal logic operators (see, e.g. [227]):

$$\Box A = \neg\Diamond\neg A,$$

$$\Diamond A = \neg\Box\neg A$$

and formulate the following assertions.

Theorem 9.18: For every IFS A, the following properties are valid:

(1) $\neg_1\Box\neg_1 A = \Diamond A$,

(2) $\neg_1\Diamond\neg_1 A = \Box A$,

(3) $\neg_3\Box\neg_3 A = \Diamond A$,

(4) $\neg_4\Box\neg_4 A = \Diamond A$,

(5) $\neg_4\Diamond\neg_4 A \supset \Box A$,

(6) $\neg_7\Diamond\neg_7 A \subset \Box A$,

(7) $\neg_8\Diamond\neg_8 A = \Box A$,

(8) $\neg_9\Diamond\neg_9 A \subset \Box A$.

Theorem 9.19: For every IFS A, and for every $\alpha, \beta \in [0,1]$ such that $\alpha + \beta \leq 1$, the following properties are valid:

(1) $\neg_1 F_{\alpha,\beta}(A) = F_{\beta,\alpha}(\neg_1 A)$,

(2) $\neg_2 F_{\alpha,\beta}(A) \subset F_{\alpha,\beta}(\neg_2 A)$,

(3) $\neg_4 F_{\alpha,\beta}(A) \supset F_{\alpha,\beta}(\neg_4 A)$,

(4) $\neg_5 F_{\alpha,\beta}(A) \supset F_{\alpha,\beta}(\neg_5 A)$,

(5) $\neg_8 F_{\alpha,\beta}(A) \subset F_{\alpha,\beta}(\neg_8 A)$.

(6) $\neg_{11} F_{\alpha,\beta}(A) \supset F_{\alpha,\beta}(\neg_{11} A)$.

Proof: Let $\alpha, \beta \in [0,1]$ be given such that $\alpha + \beta \leq 1$, and let A be an IFS. Then, we directly obtain,

$$\neg_1 F_{\alpha,\beta}(A)$$

$$= \neg_1 \{\langle x, \mu_A(x) + \alpha.\pi_A(x), \nu_A(x) + \beta.\pi_A(x)\rangle | x \in E\}$$

$$= \{\langle x, \nu_A(x) + \beta.\pi_A(x), \mu_A(x) + \alpha.\pi_A(x)\rangle | x \in E\}$$

$$= F_{\beta,\alpha}(\{\langle x, \nu_A(x), \mu_A(x)\rangle | x \in E\})$$

$$= F_{\beta,\alpha}(\neg_1 A).$$

Therefore, equality (1) is valid.

The rest of the assertions can be proved in an another way. Let us prove, for example (5).

Let $\alpha, \beta \in [0,1]$ be given so that $\alpha + \beta \leq 1$, and let A be an IFS. Then:

$$\neg_8 F_{\alpha,\beta}(A)$$

$$= \neg_8 \{\langle x, \mu_A(x) + \alpha.\pi_A(x), \nu_A(x) + \beta.\pi_A(x)\rangle | x \in E\}$$

$$= \{\langle x, 1 - \mu_A(x) - \alpha.\pi_A(x), \mu_A(x) + \alpha.\pi_A(x)\rangle | x \in E\}$$

and

$$F_{\alpha,\beta}(\neg_8 A) = F_{\alpha,\beta}(\{\langle x, 1 - \mu_A(x), \mu_A(x)\rangle | x \in E\})$$

$$= \{\langle x, 1 - \mu_A(x), \mu_A(x)\rangle | x \in E\}.$$

Now, we easily see that

$$1 - \mu_A(x) - (1 - \mu_A(x) - \alpha.\pi_A(x)) = \alpha.\pi_A(x)) \geq 0$$

and

$$\mu_A(x) + \alpha.\pi_A(x) - \mu_A(x) \geq 0.$$

Therefore, inclusion (5) is valid.

Theorem 9.20: For every IFS A, and for every $\alpha, \beta \in [0,1]$ the following properties are valid:

(1) $\neg_1 G_{\alpha,\beta}(A) = G_{\beta,\alpha}(\neg_1 A)$,

(2) $\neg_7 G_{\alpha,\beta}(A) \subset G_{\beta,\alpha}(\neg_7 A)$,

(3) $\neg_{15} G_{\alpha,\beta}(A) \subset G_{\beta,\alpha}(\neg_{15} A)$,

(4) $\neg_{19} G_{\alpha,\beta}(A) \subset G_{\beta,\alpha}(\neg_{19} A)$,

(5) $\neg_{20} G_{\alpha,\beta}(A) = G_{\beta,\alpha}(\neg_{20} A)$,

(6) $\neg_{25} G_{\alpha,\beta}(A) \supset G_{\beta,\alpha}(\neg_{25} A)$.

There are other, more complex relations, e.g., if $0 \leq \alpha \leq \beta \leq 1$, then for the IFS A, the inclusions:

$$\neg_8 G_{\alpha,\beta}(A) \supset G_{\beta,\alpha}(\neg_8 A),$$

$$\neg_9 G_{\alpha,\beta}(A) \supset G_{\beta,\alpha}(\neg_9 A),$$

are valid.

Theorem 9.21: For every IFS A, and for every $\alpha, \beta \in [0,1]$ the following properties are valid:

(1) $\neg_1 H_{\alpha,\beta}(A) = J_{\beta,\alpha}(\neg_1 A)$,

(2) $\neg_i H_{\alpha,\beta}(A) \supset H_{\alpha,\beta}(\neg_i A)$, for $2 \leq i \leq 25$.

Proof: Let $\alpha, \beta \in [0,1]$ be given and let A be an IFS. Then,

$$\neg_1 H_{\alpha,\beta}(A) = \neg_1 \{\langle x, \alpha.\mu_A(x), \nu_A(x) + \beta.\pi_A(x)\rangle | x \in E\}$$

$$= \{\langle x, \nu_A(x) + \beta.\pi_A(x), \alpha.\mu_A(x)\rangle | x \in E\}$$

$$= J_{\beta,\alpha}(\{\langle x, \nu_A(x), \mu_A(x)\rangle | x \in E\})$$

$$= J_{\beta,\alpha}(\neg_1 A).$$

Therefore, equality (1) is valid.

The rest of the assertions can be proved in a similar way.

Theorem 9.22: For every IFS A, and for every $\alpha, \beta \in [0,1]$, the following properties are valid:

(1) $\neg_1 J_{\alpha,\beta}(A) = H_{\beta,\alpha}(\neg_1 A)$,

(2) $\neg_i J_{\alpha,\beta}(A) \subset J_{\alpha,\beta}(\neg_i A)$, for $2 \leq i \leq 25$.

Theorem 9.23: For every IFS A, and for every $\alpha, \beta \in [0, 1]$, the following properties are valid:

(1) $\neg_1 H^*_{\alpha,\beta}(A) = J^*_{\beta,\alpha}(\neg_1 A)$,

(2) $\neg_i H^*_{\alpha,\beta}(A) \supset H^*_{\alpha,\beta}(\neg_i A)$, for $2 \leq i \leq 25$.

Theorem 9.24: For every IFS A, and for every $\alpha, \beta \in [0, 1]$, the following properties are valid:

(1) $\neg_1 J^*_{\alpha,\beta}(A) = H^*_{\beta,\alpha}(\neg_1 A)$,

(2) $\neg_i J^*_{\alpha,\beta}(A) \subset J^*_{\alpha,\beta}(\neg_i A)$, for $2 \leq i \leq 25$.

Theorem 9.25: For every IFS A, and for every $\alpha, \beta \in [0, 1]$ so that $\alpha + \beta \leq 1$, the following properties are valid:

(1) $\neg_1 P_{\alpha,\beta}(A) = Q_{\beta,\alpha}(\neg_1 A)$,

(2) $\neg_2 P_{\alpha,\beta}(A) \subset P_{\alpha,\beta}(\neg_2 A)$,

(3) $\neg_4 P_{\alpha,\beta}(A) \subset P_{\alpha,\beta}(\neg_4 A)$,

(4) $\neg_5 P_{\alpha,\beta}(A) \subset P_{\alpha,\beta}(\neg_5 A)$,

(5) $\neg_6 P_{\alpha,\beta}(A) \subset P_{\alpha,\beta}(\neg_6 A)$,

(6) $\neg_7 P_{\alpha,\beta}(A) \subset P_{\alpha,\beta}(\neg_7 A)$,

(7) $\neg_8 P_{\alpha,\beta}(A) \subset P_{\alpha,\beta}(\neg_8 A)$,

(8) $\neg_9 P_{\alpha,\beta}(A) \supset P_{\alpha,\beta}(\neg_9 A)$,

(9) $\neg_{10} P_{\alpha,\beta}(A) \subset P_{\alpha,\beta}(\neg_{10} A)$,

(10) $\neg_{11} P_{\alpha,\beta}(A) \subset P_{\alpha,\beta}(\neg_{11})$,

(11) $\neg_{13} P_{\alpha,\beta}(A) \subset P_{\alpha,\beta}(\neg_{13} A)$,

(12) $\neg_{14} P_{\alpha,\beta}(A) \subset P_{\alpha,\beta}(\neg_{14} A)$,

(13) $\neg_{15} P_{\alpha,\beta}(A) \subset P_{\alpha,\beta}(\neg_{15} A)$,

(14) $\neg_{16} P_{\alpha,\beta}(A) \subset P_{\alpha,\beta}(\neg_{16} A)$,

(15) $\neg_{17} P_{\alpha,\beta}(A) \subset P_{\alpha,\beta}(\neg_{17} A)$,

(16) $\neg_{20} P_{\alpha,\beta}(A) \subset P_{\alpha,\beta}(\neg_{20} A)$,

(17) $\neg_{25} P_{\alpha,\beta}(A) \supset P_{\alpha,\beta}(\neg_{25} A)$.

Proof: Let $\alpha, \beta \in [0, 1]$ be given so that $\alpha + \beta \leq 1$, and let A be an IFS. Then we directly obtain that

$$\neg_1 P_{\alpha,\beta}(A) = \neg_1 \{\langle x, \max(\mu_A(x), \alpha), \min(\nu_A(x), \beta)\rangle | x \in E\}$$

$$= \{\langle x, \min(\nu_A(x), \beta), \max(\mu_A(x), \alpha)\rangle | x \in E\}$$

$$= Q_{\beta,\alpha}(\{\langle x, \nu_A(x), \mu_A(x)\rangle | x \in E\})$$

$$= Q_{\beta,\alpha}(\neg_1 A).$$

Therefore, equality (1) is valid.

The rest of the assertions can be proved in another way. Let us prove, for example (4).

Let $\alpha, \beta \in [0, 1]$ be given so that $\alpha + \beta \leq 1$, and let A be an IFS. Then

$$\neg_5 P_{\alpha,\beta}(A) = \neg_5\{\langle x, \max(\mu_A(x), \alpha), \min(\nu_A(x), \beta)\rangle | x \in E\}$$

$$= \{\langle \overline{sg}(1 - \min(\nu_A(x), \beta)), sg(1 - \min(\nu_A(x), \beta))\rangle | x \in E\}$$

and

$$P_{\alpha,\beta}(\neg_5 A) = P_{\alpha,\beta}(\{\langle \overline{sg}(1 - \nu_A(x)), sg(1 - \nu_A(x))\rangle | x \in E\}$$

$$= \{\langle \max(\overline{sg}(1 - \nu_A(x)), \alpha), \min(sg(1 - \nu_A(x)), \beta)\rangle | x \in E\}.$$

Now, let

$$X \equiv \max(\overline{sg}(1 - \nu_A(x)), \alpha) - \overline{sg}(1 - \min(\nu_A(x), \beta)).$$

If $\nu_A(x) = 1$, then

$$X = \max(\overline{sg}(0), \alpha) - \overline{sg}(1 - \min(1, \beta))$$

$$= \max(1, \alpha) - \overline{sg}(1 - \beta) = 1 - \overline{sg}(1 - \beta) \geq 0.$$

If $\nu_A(x) < 1$, then

$$X = \max(0, \alpha) - \overline{sg}(1 - \min(\nu_A(x), \beta)) = \alpha \geq 0.$$

Let

$$Y \equiv sg(1 - \min(\nu_A(x), \beta)) - \min(sg(1 - \nu_A(x)), \beta).$$

If $\nu_A(x) = 1$, then

$$Y \equiv sg(1 - \min(1, \beta)) - \min(sg(0), \beta)$$

$$= sg(1 - \beta) - \min(0, \beta) = sg(1 - \beta) \geq 0.$$

If $\nu_A(x) < 1$, then $1 - \min(\nu_A(x), \beta) > 0$ and

$$Y = 1 - \min(1, \beta) = 1 - \beta \geq 0.$$

Therefore, inclusion (4) is valid.

Theorem 9.26: For every IFS A, and for every $\alpha, \beta \in [0, 1]$ such that $\alpha + \beta \leq 1$, the following properties are valid:
(1) $\neg_1 Q_{\alpha,\beta}(A) = P_{\beta,\alpha}(\neg_1 A)$,
(2) $\neg_2 Q_{\alpha,\beta}(A) \supset Q_{\alpha,\beta}(\neg_2 A)$,
(3) $\neg_4 Q_{\alpha,\beta}(A) \supset Q_{\alpha,\beta}(\neg_4 A)$,
(4) $\neg_6 Q_{\alpha,\beta}(A) \supset Q_{\alpha,\beta}(\neg_6 A)$,
(5) $\neg_7 Q_{\alpha,\beta}(A) \supset Q_{\alpha,\beta}(\neg_7 A)$,
(6) $\neg_8 Q_{\alpha,\beta}(A) \subset Q_{\alpha,\beta}(\neg_8 A)$,

(7) $\neg_9 Q_{\alpha,\beta}(A) \supset Q_{\alpha,\beta}(\neg_9 A)$,

(8) $\neg_{10} Q_{\alpha,\beta}(A) \supset Q_{\alpha,\beta}(\neg_{10} A)$,

(9) $\neg_{11} Q_{\alpha,\beta}(A) \subset Q_{\alpha,\beta}(\neg_{11})$,

(10) $\neg_{13} Q_{\alpha,\beta}(A) \supset Q_{\alpha,\beta}(\neg_{13} A)$,

(11) $\neg_{14} Q_{\alpha,\beta}(A) \supset Q_{\alpha,\beta}(\neg_{14} A)$,

(12) $\neg_{15} Q_{\alpha,\beta}(A) \supset Q_{\alpha,\beta}(\neg_{15} A)$,

(13) $\neg_{16} Q_{\alpha,\beta}(A) \supset Q_{\alpha,\beta}(\neg_{16} A)$,

(14) $\neg_{17} Q_{\alpha,\beta}(A) \supset Q_{\alpha,\beta}(\neg_{17} A)$,

(15) $\neg_{20} Q_{\alpha,\beta}(A) \supset Q_{\alpha,\beta}(\neg_{20} A)$,

(16) $\neg_{24} Q_{\alpha,\beta}(A) \supset Q_{\alpha,\beta}(\neg_{24} A)$.

The validity of these assertions is checked analogously as above.

9.3   Definitions and Properties of Some IF Subtractions

In Section **2.4**, some types of "subtraction" operations were defined. Here, on the basis of the negations and implications, we give the definitions of new types of "subtraction" operations over IFSs. Some of these definitions are introduced in [71, 72, 83, 414, 415].

As a basis of the new versions of operation "subtraction", we use the well-known formula from set theory:

$$A - B = A \cap \neg B$$

where A and B are given sets. In the IFS-case, if the IFSs A and B are given, we define the following versions of operation "subtraction":

$$A -_i' B = A \cap \neg_i B,$$

$$A -_i'' B = \neg_i \neg_i A \cap \neg_i B,$$

where $i = 1, 2, ..., 34$.

Of course, for every two IFSs A and B, it is valid that

$$A -_1' B = A -_1'' B,$$

because the first negation will satisfy the LEM, but in the other cases this equality is not valid.

All subtractions are given in Table 9.15.

Table 9.15 List of intuitionistic fuzzy subtractions

$-'_1$	$\{\langle x, \min(\mu_A(x), \nu_B(x)), \max(\nu_A(x), \mu_B(x))\rangle | x \in E\}$
$-'_2$	$\{\langle x, \min(\mu_A(x), \overline{sg}(\mu_B(x))), \max(\nu_A(x), sg(\mu_B(x)))\rangle | x \in E\}$
$-'_3$	$\{\langle x, \min(\mu_A(x), \nu_B(x)), \max(\nu_A(x), \mu_B(x).\nu_B(x) + \mu_B(x)^2)\rangle | x \in E\}$
$-'_4$	$\{\langle x, \min(\mu_A(x), \nu_B(x)), \max(\nu_A(x), 1 - \nu_B(x))\rangle | x \in E\}$
$-'_5$	$\{\langle x, \min(\mu_A(x), \overline{sg}(1 - \nu_B(x))), \max(\nu_A(x), sg(1 - \nu_B(x)))\rangle | x \in E\}$
$-'_6$	$\{\langle x, \min(\mu_A(x), \overline{sg}(1 - \nu_B(x))), \max(\nu_A(x), sg(\mu_B(x)))\rangle | x \in E\}$
$-'_7$	$\{\langle x, \min(\mu_A(x), \overline{sg}(1 - \nu_B(x))), \max(\nu_A(x), \mu_B(x))\rangle | x \in E\}$
$-'_8$	$\{\langle x, \min(\mu_A(x), 1 - \mu_B(x)), \max(\nu_A(x), \mu_B(x))\rangle | x \in E\}$
$-'_9$	$\{\langle x, \min(\mu_A(x), \overline{sg}(\mu_B(x))), \max(\nu_A(x), \mu_B(x))\rangle | x \in E\}$
$-'_{10}$	$\{\langle x, \min(\mu_A(x), \overline{sg}(1 - \nu_B(x))), \max(\nu_A(x), 1 - \nu_B(x))\rangle | x \in E\}$
$-'_{11}$	$\{\langle x, \min(\mu_A(x), sg(\nu_B(x))), \max(\nu_A(x), \overline{sg}(\nu_B(x)))\rangle | x \in E\}$
$-'_{12}$	$\{\langle x, \min(\mu_A(x), \nu_B(x).(\mu_B(x) + \nu_B(x))),$ $\max(\nu_A(x), \mu_B(x).(\nu_B(x)^2 + \mu_B(x) + \mu_B(x).\nu_B(x)))\rangle | x \in E\}$
$-'_{13}$	$\{\langle x, \min(\mu_A(x), sg(1 - \mu_B(x))), \max(\nu_A(x), \overline{sg}(1 - \mu_B(x)))\rangle | x \in E\}$
$-'_{14}$	$\{\langle x, \min(\mu_A(x), sg(\nu_B(x))), \max(\nu_A(x), \overline{sg}(1 - \mu_B(x)))\rangle | x \in E\}$
$-'_{15}$	$\{\langle x, \min(\mu_A(x), \overline{sg}(1 - \nu_B(x))), \max(\nu_A(x), \overline{sg}(1 - \mu_B(x)))\rangle | x \in E\}$
$-'_{16}$	$\{\langle x, \min(\mu_A(x), \overline{sg}(\mu_B(x))), \max(\nu_A(x), \overline{sg}(1 - \mu_B(x)))\rangle | x \in E\}$
$-'_{17}$	$\{\langle x, \min(\mu_A(x), \overline{sg}(1 - \nu_B(x))), \max(\nu_A(x), \overline{sg}(\nu_B(x)))\rangle | x \in E\}$
$-'_{18}$	$\{\langle x, \min(\mu_A(x), \nu_B(x), sg(\mu_B(x))),$ $\max(\nu_A(x), \min(\mu_B(x), sg(\nu_B(x))))\rangle | x \in E\}$
$-'_{19}$	$\{\langle x, \min(\mu_A(x), \nu_B(x), sg(\mu_B(x))), \nu_A(x)\rangle | x \in E\}$
$-'_{20}$	$\{\langle x, \min(\mu_A(x), \nu_B(x)), \nu_A(x)\rangle | x \in E\}$
$-'_{21}$	$\{\langle x, \min(\mu_A(x), 1 - \mu_B(x), sg(\mu_B(x))),$ $\max(\nu_A(x), \min(\mu_B(x), sg(1 - \mu_B(x))))\rangle | x \in E\}$
$-'_{22}$	$\{\langle x, \min(\mu_A(x), 1 - \mu_B(x), sg(\mu_B(x))), \nu_A(x)\rangle | x \in E\}$
$-'_{23}$	$\{\langle x, \min(\mu_A(x), 1 - \mu_B(x)), \nu_A(x)\rangle | x \in E\}$
$'_{24}$	$\{\langle x, \min(\mu_A(x), \nu_B(x), sg(1 - \nu_B(x))),$ $\max(\nu_A(x), \min(1 - \nu_B(x), sg(\nu_B(x))))\rangle | x \in E\}$
$-'_{25}$	$\{\langle x, \min(\mu_A(x), \nu_B(x), sg(1 - \nu_B(x))), \nu_A(x)\rangle | x \in E\}$
$-'_{26}$	$\{\langle x, \min(\mu_A(x), \nu_B(x)),$ $\max(\nu_A(x), \mu_B(x).\nu_B(x) + \overline{sg}(1 - \mu_B(x)))\rangle | x \in E\}$
$-'_{27}$	$\{\langle x, \min(\mu_A(x), 1 - \mu_B(x)),$ $\max(\nu_A(x), \mu_B(x).(1 - \mu_B(x)) + \overline{sg}(1 - \mu_B(x)))\rangle | x \in E\}$
$-'_{28}$	$\{\langle x, \min(\mu_A(x), \nu_B(x)),$ $\max(\nu_A(x), (1 - \nu_B(x)).\nu_B(x) + \overline{sg}(\nu_B(x)))\rangle | x \in E\}$
$-'_{29}$	$\{\langle x, \min(\mu_A(x), \max(0, \mu_B(x).\nu_B(x) + \overline{sg}(1 - \nu_B(x)))),$ $\max(\nu_A(x), \mu_B(x).(\mu_B(x).\nu_B(x)$ $+ \overline{sg}(1 - \nu_B(x))) + \overline{sg}(1 - \mu_B(x)))\rangle | x \in E\}$

Table 9.15 (*continued*)

$-'_{30}$	$\{\langle x, \min(\mu_A(x), \mu_B(x).\nu_B(x)),$ $\quad \max(\nu_A(x), \mu_B(x).(\mu_B(x).\nu_B(x)$ $\quad +\overline{sg}(1-\nu_B(x))) + \overline{sg}(1-\mu_B(x)))\rangle | x \in E\}$
$-'_{31}$	$\{\langle x, \min(\mu_A(x), (1-\mu_B(x)).\mu_B(x) + \overline{sg}(\mu_B(x))),$ $\quad \max(\nu_A(x), \mu_B(x).((1-\mu_B(x)).\mu_B(x)$ $\quad +\overline{sg}(\mu_B(x))) + \overline{sg}(1-\mu_B(x)))\rangle | x \in E\}$
$-'_{32}$	$\{\langle x, \min(\mu_A(x), (1-\mu_B(x)).\mu_B(x)),$ $\quad \max(\nu_A(x), \mu_B(x).((1-\mu_B(x)).\mu_B(x)$ $\quad +\overline{sg}(\mu_B(x))) + \overline{sg}(1-\mu_B(x)))\rangle | x \in E\}$
$-'_{33}$	$\{\langle x, \min(\mu_A(x), \nu_B(x).(1-\nu_B(x)) + \overline{sg}(1-\nu_B(x))),$ $\quad \max(\nu_A(x), (1-\nu_B(x)).(\nu_B(x).(1-\nu_B(x))$ $\quad +\overline{sg}(1-\nu_B(x))) + \overline{sg}(\nu_B(x)))\rangle | x \in E\}$
$-'_{34}$	$\{\langle x, \min(\mu_A(x), \nu_B(x).(1-\nu_B(x))),$ $\quad \max(\nu_A(x), (1-\nu_B(x)).(\nu_B(x).(1-\nu_B(x))$ $\quad +\overline{sg}(1-\nu_B(x))) + \overline{sg}(\nu_B(x)))\rangle | x \in E\}$
$-''_1$	$\{\langle x, \min(\mu_A(x), \nu_B(x)), \max(\nu_A(x), \mu_B(x))\rangle | x \in E\}$
$-''_2$	$\{\langle x, \min(\text{sg}(\mu_A(x)), \overline{sg}(\mu_B(x))),$ $\quad \max(\overline{sg}(\mu_A(x)), \text{sg}(\mu_B(x)))\rangle | x \in E\}$
$-''_3$	$\{\langle x, \min(\mu_A(x).\nu_A(x) + \mu_A(x)^2, \nu_B(x)),$ $\quad \max(\nu_A(x).(\mu_A(x).\nu_A(x) + \mu_A(x)^2) + \nu_A(x)^2,$ $\quad \mu_B(x).\nu_B(x) + \mu_B(x)^2)\rangle | x \in E\}$
$-''_4$	$\{\langle x, \min(1-\nu_A(x), \nu_B(x)), \max(\nu_A(x), 1-\mu_B(x))\rangle | x \in E\}$
$-''_5$	$\{\langle x, \min(\text{sg}(1-\nu_A(x)), \overline{sg}(1-\nu_B(x))),$ $\quad \max(\overline{sg}(1-\nu_A(x)), \text{sg}(1-\nu_B(x)))\rangle | x \in E\}$
$-''_6$	$\{\langle x, \min(\text{sg}(\mu_A(x)), \overline{sg}(1-\nu_B(x))),$ $\quad \max(\overline{sg}(1-\nu_A(x)), \text{sg}(\mu_B(x)))\rangle | x \in E\}$
$-''_7$	$\{\langle x, \min(\overline{sg}(1-\mu_A(x)), \overline{sg}(1-\nu_B(x))),$ $\quad \max(\overline{sg}(1-\nu_A(x)), \mu_B(x))\rangle | x \in E\}$
$-''_8$	$\{\langle x, \min(\mu_A(x), 1-\mu_B(x)), \max(1-\mu_A(x), \mu_B(x))\rangle | x \in E\}$
$-''_9$	$\{\langle x, \min(\text{sg}(\mu_A(x)), \overline{sg}(\mu_B(x))), \max(\overline{sg}(\mu_A(x)), \mu_B(x))\rangle | x \in E\}$
$-''_{10}$	$\{\langle x, \min(\overline{sg}(\nu_A(x)), \overline{sg}(1-\nu_B(x))), \max(\nu_A(x), 1-\nu_B(x))\rangle | x \in E\}$
$-''_{11}$	$\{\langle x, \min(\overline{sg}(\nu_A(x)), \text{sg}(\nu_B(x))),$ $\quad \max(\text{sg}(\nu_A(x)), \overline{sg}(\nu_B(x)))\rangle | x \in E\}$
$-''_{12}$	$\{\langle x, \min(\mu_A(x).(\nu_A(x)^2 + \mu_A(x) + \mu_A(x).\nu_A(x)).(\mu_A(x).(\nu_A(x)^2$ $\quad +\mu_A(x) + \mu_A(x).\nu_A(x)) + \nu_A(x).(\mu_A(x) + \nu_A(x))),$ $\quad \nu_B(x).(\mu_B(x) + \nu_B(x))),$ $\quad \max(\nu_A(x).(\mu_A(x) + \nu_A(x)).(\mu_A(x)^2.(\nu_A(x)^2 + \mu_A(x)$ $\quad +\mu_A(x).\nu_A(x))^2 + \nu_A(x).(\mu_A(x) + \nu_A(x))) + \mu_A(x).\nu_A(x)$ $\quad .(\nu_A(x)^2 + \mu_A(x) + \mu_A(x).\nu_A(x)).(\mu_A(x) + \nu_A(x)),$ $\quad \mu_B(x).(\nu_B(x)^2 + \mu_B(x) + \mu_B(x).\nu_B(x)))\rangle | x \in E\}$

Table 9.15 (*continued*)

$-''_{13}$	$\{\langle x, \min(\overline{sg}(1-\mu_A(x)), sg(1-\mu_B(x))),$ $\max(sg(1-\mu_A(x)), \overline{sg}(1-\mu_B(x)))\rangle \vert x \in E\}$
$-''_{14}$	$\{\langle x, \min(\overline{sg}(1-\mu_A(x)), sg(\nu_B(x))),$ $\max(sg(\nu_A(x)), \overline{sg}(1-\mu_B(x)))\rangle \vert x \in E\}$
$-''_{15}$	$\{\langle x, \min(\overline{sg}(1-\mu_A(x)), \overline{sg}(1-\nu_B(x))),$ $\max(\overline{sg}(1-\nu_A(x)), \overline{sg}(1-\mu_B(x)))\rangle \vert x \in E\}$
$-''_{16}$	$\{\langle x, \min(sg(\mu_A(x)), \overline{sg}(\mu_B(x))),$ $\max(\overline{sg}(\mu_A(x)), \overline{sg}(1-\mu_B(x)))\rangle \vert x \in E\}$
$-''_{17}$	$\{\langle x, \min(\overline{sg}(\nu_A(x)), \overline{sg}(1-\nu_B(x))),$ $\max(sg(\nu_A(x)), \overline{sg}(\nu_B(x)))\rangle \vert x \in E\}$
$-''_{18}$	$\{\langle x, \min(\mu_A(x), sg(\nu_A(x)), \nu_B(x), sg(\mu_B(x))),$ $\max(\min(\nu_A(x), sg(\mu_A(x))), \min(\mu_B(x), sg(\nu_B(x))))\rangle \vert x \in E\}$
$-''_{19}$	$\{\langle x, 0, 0\rangle \vert x \in E\}$
$-''_{20}$	$\{\langle x, 0, 0\rangle \vert x \in E\}$
$-''_{21}$	$\{\langle x, \mu_A(x).sg(1-\mu_A(x)),$ $\max((1-\mu_A(x)).sg(\mu_A(x)), \min(\mu_B(x), sg(1-\mu_B(x))))\rangle \vert x \in E\}$
$-''_{22}$	$\{\langle x, \min(\mu_A(x).sg(\mu_A(x)), 1-\mu_B(x), sg(\mu_B(x))), 0\rangle \vert x \in E\}$
$-''_{23}$	$\{\langle x, \min(\mu_A(x), 1-\mu_B(x)), 0\rangle \vert x \in E\}$
$-''_{24}$	$\{\langle x, \min(1-\nu_A(x), sg(\nu_A(x)), \nu_B(x), sg(1-\nu_B(x))),$ $\max(\min(1-\min(1-\nu_A(x), sg(\nu_A(x))), sg(\min(1-\nu_A(x), sg(\nu_A(x))))),$ $\min(1-\nu_B(x), sg(\nu_B(x))))\rangle \vert x \in E\}$
$-''_{25}$	$\{\langle x, 0, 0\rangle \vert x \in E\}$
$-''_{26}$	$\{\langle x, \min(\mu_A(x).\nu_A(x) + \overline{sg}(1-\mu_A(x)), \nu_B(x)),$ $\max(\nu_A(x).(\mu_A(x).\nu_A(x) + \overline{sg}(1-\mu_A(x))) + \overline{sg}(1-\nu_A(x)),$ $\mu_B(x).\nu_B(x) + \overline{sg}(1-\mu_B(x)))\rangle \vert x \in E\}$
$-''_{27}$	$\{\langle x, \min(\mu_A(x), 1-\mu_B(x)),$ $\max((1-\mu_A(x)).\mu_A(x) + \overline{sg}(\mu_A(x)),$ $\mu_B(x).(1-\mu_B(x)) + \overline{sg}(1-\mu_B(x)))\rangle \vert x \in E\}$
$-''_{28}$	$\{\langle x, \min((1-\nu_A(x)).\nu_A(x) + \overline{sg}(\nu_A(x)), \nu_B(x)),$ $\max((1-(1-\nu_A(x)).\nu_A(x)) - \overline{sg}(\nu_A(x))).((1-\nu_A(x)).\nu_A(x)$ $+\overline{sg}((1-\nu_A(x)).\nu_A(x) + \overline{sg}(\nu_A(x))),$ $(1-\nu_B(x)).\nu_B(x) + \overline{sg}(\nu_B(x)))\rangle \vert x \in E\}$
$-''_{29}$	$\{\langle x, \min((\mu_A(x).(\mu_A(x).\nu_A(x) + \overline{sg}(1-\nu_A(x))) + \overline{sg}(1-\mu_A(x)))$ $.(\mu_A(x).\nu_A(x) + \overline{sg}(1-\nu_A(x))) + \overline{sg}(1-\mu_A(x).(\mu_A(x).\nu_A(x)$ $+\overline{sg}(1-\nu_A(x))) - \overline{sg}(1-\mu_A(x)), \mu_B(x).\nu_B(x) + \overline{sg}(1-\nu_B(x))),$ $\max((\mu_A(x).\nu_A(x) + \overline{sg}(1-\nu_A(x))).((\mu_A(x).(\mu_A(x).\nu_A(x)$ $+\overline{sg}(1-\nu_A(x))) + \overline{sg}(1-\mu_A(x))).(\mu_A(x).\nu_A(x) + \overline{sg}(1-\nu_A(x)))$ $+\overline{sg}(1-\mu_A(x).(\mu_A(x).\nu_A(x) + \overline{sg}(1-\nu_A(x))) - \overline{sg}(1-\mu_A(x))))$ $+\overline{sg}(1-\mu_A(x).\nu_A(x) - \overline{sg}(1-\nu_A(x))),$ $\mu_B(x).(\mu_B(x).\nu_B(x) + \overline{sg}(1-\nu_B(x))) + \overline{sg}(1-\mu_B(x)))\rangle \vert x \in E\}$

Table 9.15 (continued)

$-''_{30}$	$\{\langle x, \min(((\mu_A(x).(\mu_A(x).\nu_A(x) + \overline{sg}(1 - \nu_A(x))) + \overline{sg}(1 - \mu_A(x)))$ $.\mu_A(x).\nu_A(x)), \mu_B(x).\nu_B(x)),$ $\max(\mu_A(x).\nu_A(x).((\mu_A(x).(\mu_A(x).\nu_A(x) + \overline{sg}(1 - \nu_A(x)))$ $+\overline{sg}(1 - \mu_A(x))).\mu_A(x).\nu_A(x) + \overline{sg}(1 - \mu_A(x).(\mu_A(x).\nu_A(x)$ $+\overline{sg}(1 - \nu_A(x))) - \overline{sg}(1 - \mu_A(x)))) + \overline{sg}(1 - (\mu_A(x).\nu_A(x))),$ $\mu_B(x).(\mu_B(x).\nu_B(x) + \overline{sg}(1 - \nu_B(x))) + \overline{sg}(1 - \mu_B(x)))\rangle | x \in E\}$
$-''_{31}$	$\{\langle x, \min((1 - (1 - \mu_A(x)).\mu_A(x) - \overline{sg}(\mu_A(x))).((1 - \mu_A(x)).\mu_A(x)$ $+\overline{sg}(\mu_A(x))) + \overline{sg}(((1 - \mu_A(x)).\mu_A(x) + \overline{sg}(\mu_A(x)))), (1 - \mu_B(x))$ $.\mu_B(x) + \overline{sg}(\mu_B(x))),$ $\max(((1 - \mu_A(x)).\mu_A(x) + \overline{sg}(\mu_A(x))).((1 - (1 - \mu_A(x)).\mu_A(x))$ $-\overline{sg}(\mu_A(x)).((1 - \mu_A(x)).\mu_A(x) + \overline{sg}(\mu_A(x))) + \overline{sg}((1 - \mu_A(x))$ $.\mu_A(x) + \overline{sg}(\mu_A(x)))) + \overline{sg}(1 - (1 - \mu_A(x)).\mu_A(x) - \overline{sg}(\mu_A(x))),$ $\mu_B(x).((1 - \mu_B(x)).\mu_B(x) + \overline{sg}(\mu_B(x))) + \overline{sg}(1 - \mu_B(x)))\rangle | x \in E\}$
$-''_{32}$	$\{\langle x, \min((1 - (1 - \mu_A(x)).\mu_A(x)).(1 - \mu_A(x)).\mu_A(x), (1 - \mu_B(x)).\mu_B(x)),$ $\max(((1 - \mu_A(x)).\mu_A(x).((1 - (1 - \mu_A(x)).\mu_A(x)).(1 - \mu_A(x)).\mu_A(x)$ $+\overline{sg}((1 - \mu_A(x)).\mu_A(x))) + \overline{sg}(1 - (1 - \mu_A(x)).\mu_A(x)),$ $\mu_B(x).((1 - \mu_B(x)).\mu_B(x) + \overline{sg}(\mu_B(x))) + \overline{sg}(1 - \mu_B(x))))\rangle | x \in E\}$
$-''_{33}$	$\{\langle x, \min(((1 - \nu_A(x)).(\nu_A(x).(1 - \nu_A(x)) + \overline{sg}(1 - \nu_A(x))) + \overline{sg}(\nu_A(x)))$ $.(1 - (1 - \nu_A(x)).(\nu_A(x).(1 - \nu_A(x)) + \overline{sg}(1 - \nu_A(x))) - \overline{sg}(\nu_A(x)))$ $+\overline{sg}(1 - (1 - \nu_A(x)).(\nu_A(x).(1 - \nu_A(x))$ $+\overline{sg}(1 - \nu_A(x))) - \overline{sg}(\nu_A(x))), \nu_B(x).(1 - \nu_B(x)) + \overline{sg}(1 - \nu_B(x))),$ $\max((1 - (1 - \nu_A(x)).(\nu_A(x).(1 - \nu_A(x)) + \overline{sg}(1 - \nu_A(x))) - \overline{sg}(\nu_A(x)))$ $.(((1 - \nu_A(x)).(\nu_A(x).(1 - \nu_A(x)) + \overline{sg}(1 - \nu_A(x))) + \overline{sg}(\nu_A(x)))$ $.(1 - (1 - \nu_A(x)).(\nu_A(x).(1 - \nu_A(x)) + \overline{sg}(1 - \nu_A(x))) - \overline{sg}(\nu_A(x)))$ $+\overline{sg}(1 - (1 - \nu_A(x)).(\nu_A(x).(1 - \nu_A(x)) + \overline{sg}(1 - \nu_A(x))) - \overline{sg}(\nu_A(x))))$ $+\overline{sg}((1 - \nu_A(x)).(\nu_A(x).(1 - \nu_A(x)) + \overline{sg}(1 - \nu_A(x))) + \overline{sg}(\nu_A(x))),$ $(1 - \nu_B(x)).(\nu_B(x).(1 - \nu_B(x)) + \overline{sg}(1 - \nu_B(x))) + \overline{sg}(\nu_B(x)))\rangle | x \in E\}$
$-''_{34}$	$\{\langle x, \min(((1 - \nu_A(x)).(\nu_A(x).(1 - \nu_A(x)) + \overline{sg}(1 - \nu_A(x))) + \overline{sg}(\nu_A(x)))$ $.(1 - (1 - \nu_A(x)).(\nu_A(x).(1 - \nu_A(x)) + \overline{sg}(1 - \nu_A(x))) - \overline{sg}(\nu_A(x))),$ $\nu_B(x).(1 - \nu_B(x))),$ $\max(((1 - (1 - \nu_A(x)).(\nu_A(x).(1 - \nu_A(x))$ $+\overline{sg}(1 - \nu_A(x))) - \overline{sg}(\nu_A(x))).(((1 - \nu_A(x)).(\nu_A(x).(1 - \nu_A(x))$ $+\overline{sg}(1 - \nu_A(x))) + \overline{sg}(\nu_A(x))).(1 - (1 - \nu_A(x)).(\nu_A(x).(1 - \nu_A(x))$ $+\overline{sg}(1 - \nu_A(x))) - \overline{sg}(\nu_A(x))) + \overline{sg}(1 - (1 - \nu_A(x))$ $.(\nu_A(x).(1 - \nu_A(x)) + \overline{sg}(1 - \nu_A(x))) - \overline{sg}(\nu_A(x))))) + \overline{sg}((1 - \nu_A(x)).$ $.(\nu_A(x).(1 - \nu_A(x)) + \overline{sg}(1 - \nu_A(x))) + \overline{sg}(\nu_A(x))), (1 - \nu_B(x)).(\nu_B(x)$ $.(1 - \nu_B(x)) + \overline{sg}(1 - \nu_B(x))) + \overline{sg}(\nu_B(x)))\rangle | x \in E\}$

Some of the most important properties of the subtractions are:

(a) $A - E^* = O^*$,

(b) $A - O^* = A$,

(c) $E^* - A = \neg A$,

(d) $O^* - A = O^*$,

(e) $(A - B) \cap C = (A \cap C) - B = A \cap (C - B)$,

(f) $(A \cap B) - C = (A - C) \cap (B - C)$,
(g) $(A \cup B) - C = (A - C) \cup (B - C)$,
(h) $(A - B) - C = (A - C) - B$,
(i) $(A - C) \cap B = A \cap (B - C)$.

On Table 9.16 the subtractions that satisfy these properties are shown.

Table 9.16

	(a)	(b)	(c)	(d)	(e)	(f)	(g)	(h)	(i)
$-'_1$	+	+	+	+	+	+	+	+	+
$-'_2$	+	+	+	+	+	+	+	+	+
$-'_3$	+	+	+	+	+	+	+	+	+
$-'_4$	+	+	+	+	+	+	+	+	+
$-'_5$	+	+	+	+	+	+	+	+	+
$-'_6$	+	+	+	+	+	+	+	+	+
$-'_7$	+	+	+	+	+	+	+	+	+
$-'_8$	+	+	+	+	+	+	+	+	+
$-'_9$	+	+	+	+	+	+	+	+	+
$-'_{10}$	+	+	+	+	+	+	+	+	+
$-'_{11}$	+	+	+	+	+	+	+	+	+
$-'_{12}$	+	+	+	+	+	+	+	+	+
$-'_{13}$	+	+	+	+	+	+	+	+	+
$-'_{14}$	+	+	+	+	+	+	+	+	+
$-'_{15}$	+	+	+	+	+	+	+	+	+
$-'_{16}$	+	+	+	+	+	+	+	+	+
$-'_{17}$	+	+	+	+	+	+	+	+	+
$-'_{18}$	-	-	+	+	+	+	+	+	+
$-'_{19}$	-	-	+	+	+	+	+	+	+
$-'_{20}$	-	+	+	+	+	+	+	+	+
$-'_{21}$	-	-	+	+	+	+	+	+	+
$-'_{22}$	-	-	+	+	+	+	+	+	+
$-'_{23}$	-	+	+	+	+	+	+	+	+
$-'_{24}$	-	-	+	+	+	+	+	+	+
$-'_{25}$	-	-	+	+	+	+	+	+	+
$-'_{26}$	+	+	+	+	+	+	+	+	+
$-'_{27}$	+	+	+	+	+	+	+	+	+
$-'_{28}$	+	+	+	+	+	+	+	+	+
$-'_{29}$	+	+	+	+	+	+	+	+	+
$-'_{30}$	+	-	+	+	+	+	+	+	+
$-'_{31}$	+	+	+	+	+	+	+	+	+
$-'_{32}$	+	-	+	+	+	+	+	+	+
$-'_{33}$	+	+	+	+	+	+	+	+	+
$-'_{34}$	+	-	+	+	+	+	+	+	+

Table 9.16 (*continued*)

	(a)	(b)	(c)	(d)	(e)	(f)	(g)	(h)	(i)
$-''_1$	+	+	+	+	+	+	+	+	+
$-''_2$	+	-	+	+	-	+	+	+	-
$-''_3$	+	-	+	+	-	-	-	-	-
$-''_4$	+	-	+	+	-	+	+	+	-
$-''_5$	+	-	+	+	-	+	+	+	-
$-''_6$	+	-	+	+	-	+	+	+	-
$-''_7$	+	-	+	+	-	+	+	-	-
$-''_8$	+	-	+	+	-	+	+	+	-
$-''_9$	+	-	+	+	-	+	+	-	-
$-''_{10}$	+	-	+	+	-	+	+	+	-
$-''_{11}$	+	-	+	+	-	+	+	+	-
$-''_{12}$	+	-	+	+	-	-	-	-	-
$-''_{13}$	+	-	+	+	-	+	+	+	-
$-''_{14}$	+	-	+	+	-	+	+	+	-
$-''_{15}$	+	-	+	+	-	+	+	+	-
$-''_{16}$	+	-	+	+	-	+	+	-	-
$-''_{17}$	+	-	+	+	-	+	+	-	-
$-''_{18}$	-	-	-	-	-	-	-	-	-
$-''_{19}$	-	-	-	-	-	+	+	+	-
$-''_{20}$	-	-	-	-	-	+	+	+	-
$-''_{21}$	-	-	-	-	-	-	-	-	-
$-''_{22}$	-	-	-	-	-	-	-	+	-
$-''_{23}$	-	-	+	-	-	+	+	+	-
$-''_{24}$	-	-	-	-	-	-	-	-	-
$-''_{25}$	-	-	-	-	-	+	+	+	-
$-''_{26}$	+	-	+	+	-	-	-	-	-
$-''_{27}$	+	-	+	+	-	-	-	-	-
$-''_{28}$	+	-	+	+	-	-	-	-	-
$-''_{29}$	+	-	+	+	-	-	-	-	-
$-''_{30}$	+	-	-	-	-	-	-	-	-
$-''_{31}$	+	-	+	+	-	-	-	-	-
$-''_{32}$	+	-	-	-	-	-	-	-	-
$-''_{33}$	+	-	+	+	-	-	-	-	-
$-''_{34}$	+	-	-	+	-	-	-	-	-

As we have seen in Section **2.4**, the validity of the following equalities

(a) $O^* - U^* = O^*$,   (b) $O^* - E^* = O^*$,   (c) $U^* - O^* = U^*$,

(d) $U^* - E^* = O^*$,   (e) $E^* - O^* = E^*$,   (f) $E^* - U^* = O^*$

is interesting to check. In Table 9.17, the subtractions that satisfy these equalities are given.

Table 9.17

	(a)	(b)	(c)	(d)	(e)	(f)		(a)	(b)	(c)	(d)	(e)	(f)
$-'_1$	+	+	+	+	+	-	$-''_1$	+	+	+	+	+	-
$-'_2$	+	+	+	+	+	-	$-''_2$	+	+	-	+	+	-
$-'_3$	+	+	+	+	+	-	$-''_3$	+	+	+	+	+	-
$-'_4$	+	+	+	+	+	+	$-''_4$	+	+	-	+	+	+
$-'_5$	+	+	+	+	+	+	$-''_5$	+	+	-	+	+	+
$-'_6$	+	+	+	+	+	-	$-''_6$	+	+	+	+	+	-
$-'_7$	+	+	+	+	+	-	$-''_7$	+	+	+	+	+	-
$-'_8$	+	+	+	+	+	-	$-''_8$	+	+	-	+	+	-
$-'_9$	+	+	+	+	+	-	$-''_9$	+	+	-	+	+	-
$-'_{10}$	+	+	+	+	+	+	$-''_{10}$	+	+	-	+	+	+
$-'_{11}$	+	+	+	+	+	+	$-''_{11}$	+	+	-	+	+	+
$-'_{12}$	+	+	+	+	+	-	$-''_{12}$	+	+	+	+	+	-
$-'_{13}$	+	+	+	+	+	-	$-''_{13}$	+	+	-	+	+	-
$-'_{14}$	+	+	+	+	+	-	$-''_{14}$	+	+	+	+	+	-
$-'_{15}$	+	+	+	+	+	-	$-''_{15}$	+	+	+	+	+	-
$-'_{16}$	+	+	+	+	+	-	$-''_{16}$	+	+	-	+	+	-
$-'_{17}$	+	+	+	+	+	+	$-''_{17}$	+	+	-	+	+	+
$-'_{18}$	+	+	+	-	-	-	$-''_{18}$	-	-	+	-	-	-
$-'_{19}$	+	+	+	-	-	-	$-''_{19}$	-	-	+	-	-	-
$-'_{20}$	+	+	+	-	+	-	$-''_{20}$	-	-	+	-	-	-
$-'_{21}$	+	+	+	-	-	-	$-''_{21}$	-	-	+	-	-	-
$-'_{22}$	+	+	+	-	-	-	$-''_{22}$	-	-	+	-	-	-
$-'_{23}$	+	+	+	-	+	-	$-''_{23}$	-	-	+	-	+	-
$-'_{24}$	+	+	+	-	-	-	$-''_{24}$	-	-	+	-	-	-
$-'_{25}$	+	+	+	-	-	-	$-''_{25}$	-	-	+	-	-	-
$-'_{26}$	+	+	+	+	+	-	$-''_{26}$	+	+	+	+	+	-
$-'_{27}$	+	+	+	+	+	-	$-''_{27}$	+	+	-	+	+	-
$-'_{28}$	+	+	+	+	+	+	$-''_{28}$	+	+	-	+	+	+
$-'_{29}$	+	+	+	+	+	-	$-''_{29}$	+	+	+	+	+	-
$-'_{30}$	+	+	+	+	-	-	$-''_{30}$	-	+	+	+	-	-
$-'_{31}$	+	+	+	+	+	-	$-''_{31}$	+	+	-	+	+	-
$-'_{32}$	+	+	+	+	-	-	$-''_{32}$	-	+	+	+	-	-
$-'_{33}$	+	+	+	+	+	+	$-''_{33}$	+	+	-	+	+	+
$-'_{34}$	+	+	+	+	-	+	$-''_{34}$	+	+	+	+	-	+

9.4 (ε, η)-Negations, Implications, Subtractions, Norms and Distances

Here, following the ideas from [68, 72], we introduce a set of new negations and implications over IFSs. They generalize the classical negation over IFSs, but on the other hand, they have some non-classical properties. Based on this, we construct, firstly, new forms of operation subtraction, secondly, new forms of norms of the elements of a given IFS, and, thirdly, new forms of distances between two elements of a given IFS, or between two elements (in one universe) of two different IFSs.

We construct a set of IF-negations that has the form

$$\mathcal{N} = \{\neg^{\varepsilon,\eta} \mid 0 \leq \varepsilon < 1 \ \& \ 0 \leq \eta < 1\},$$

where for each IFS A,

$$\neg^{\varepsilon,\eta} A = \{\langle x, \min(1, \nu_A(x) + \varepsilon), \max(0, \mu_A(x) - \eta)\rangle | x \in E\}. \tag{9.1}$$

Below, we study some basic properties of an arbitrary element of $\mathcal{N}$.

For ε and η there are two cases.

• $\eta < \varepsilon$, but this case is impossible, because, for example, if $\mu_A(x) = 0.6, \nu_A(x) = 0.3, \varepsilon = 0.8, \eta = 0.4$, then

$$\min(1, \nu_A(x) + \varepsilon) + \max(0, \mu_A(x) - \eta) = \min(1, 1.1) + \max(0, 0.2) = 1.2 > 1.$$

• $\eta \geq \varepsilon$

Let everywhere below $0 \leq \varepsilon \leq \eta < 1$ be fixed.

First, we see that set $\neg^{\varepsilon,\eta} A$ is an IFS, because for each $x \in E$ if $\mu_A(x) \leq \eta$ then,

$$\min(1, \nu_A(x) + \varepsilon) + \max(0, \mu_A(x) - \eta) = \min(1, \nu_A(x) + \varepsilon) \leq 1;$$

if $\mu_A(x) \geq \eta$ then,

$$\min(1, \nu_A(x) + \varepsilon) + \max(0, \mu_A(x) - \eta) = \min(1, \nu_A(x) + \varepsilon) + \mu_A(x) - \eta$$

$$\leq \nu_A(x) + \varepsilon + \mu_A(x) - \eta \leq \nu_A(x) + \mu_A(x) \leq 1.$$

In Fig. 9.3, x and $\neg_1 x$ are shown, while in Fig. 9.4 and Fig. 9.5, y and $\neg^{\varepsilon} y$ and z and $\neg^{\varepsilon} z$ are shown.

Second, we construct a new implication, generated by the new negation as

$$A \rightarrow^{\varepsilon,\eta} B = \{\langle x, \max(\mu_B(x), \min(1, \nu_A(x) + \varepsilon)),$$

$$\min(\nu_B(x), \max(0, \mu_A(x) - \eta))\rangle | x \in E\}$$

$$= \{\langle x, \min(1, \max(\mu_B(x), \nu_A(x) + \varepsilon)),$$

$$\max(0, \min(\nu_B(x), \mu_A(x) - \eta))\rangle | x \in E\}. \tag{9.2}$$

In [68], it is shown that for the George Klir and Bo Yuan's axioms implication $\rightarrow^{\varepsilon,\eta}$ and negation $\neg^{\varepsilon,\eta}$:

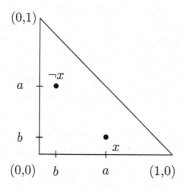

Fig. 9.3

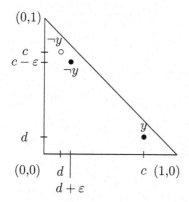

Fig. 9.4

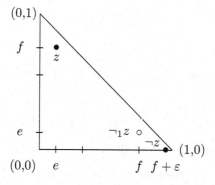

Fig. 9.5

(a) satisfy Axioms 1, 2, 3, 6 and 9;
(b) satisfy Axioms 4 and 5 as IFTs, but not as tautologies;
(c) satisfy Axiom 8 in the form
Axiom 8': $(\forall x, y)(I(x, y) \leq I(N(y), N(x)))$.

These properties can be transformed for the case of IFSs.

Theorem 9.27: For every three IFSs A, B and C, the IFSs from (a)-(q) from Section **9.2** are IFTSs, but they are not always equal to E^*.

Theorem 9.28: For each IFS A:

(a) $A \cup \neg^{\varepsilon,\eta} A$ is an IFTS, but not always equal to E^*;

(b) $\neg^{\varepsilon,\eta}\neg^{\varepsilon,\eta} A \cup \neg^{\varepsilon,\eta} A$ is an IFTS, but not always equal to E^*.

Usually, in set theory the De Morgan's Laws have the forms:

$$\neg A \cap \neg B = \neg(A \cup B), \tag{9.3}$$

$$\neg A \cup \neg B = \neg(A \cap B). \tag{9.4}$$

or

$$\neg(\neg A \cap \neg B) = A \cup B, \tag{9.5}$$

$$\neg(\neg A \cup \neg B) = A \cap B. \tag{9.6}$$

but, as we discussed in Section **9.1**, they can also have the forms:

$$\neg(\neg A \cap \neg B) = \neg\neg A \cup \neg\neg B, \tag{9.7}$$

$$\neg(\neg A \cup \neg B) = \neg\neg A \cap \neg\neg B. \tag{9.8}$$

Theorem 9.29: For every two IFSs A and B:

(a) the IFSs from (9.3) and (9.4) are IFTSs, but not always equal to E^*;

(b) the IFSs from (9.5) – (9.8) are not always IFTSs or not always equal to E^*.

For every IFS A:

$$\neg^{\varepsilon,\eta} \square A \supset \square \neg^{\varepsilon,\eta} A,$$

$$\neg^{\varepsilon,\eta} \Diamond A \subset \Diamond \neg^{\varepsilon,\eta} A.$$

Let us prove, for example, the second inclusion. The rest of the assertions can be proved analogously. Let $0 \leq \varepsilon \leq \eta \leq 1$ for some ε and η. Then,

$$\neg^{\varepsilon,\eta} \Diamond A = \neg^{\varepsilon,\eta}\{\langle x, 1 - \nu_A(x), \nu_A(x)\rangle | x \in E\}$$

$$= \{\langle x, \min(1, \nu_A(x) + \varepsilon), \max(0, 1 - \nu_A(x) - \eta)\rangle | x \in E\}.$$

$$\Diamond\neg^{\varepsilon,\eta} A = \Diamond\{\langle x, \min(1, \nu_A(x) + \varepsilon), \max(0, \mu_A(x) - \eta)\rangle | x \in E\}$$

$$= \{\langle x, 1 - \max(0, \mu_A(x) - \eta), \max(0, \mu_A(x) - \eta)\rangle | x \in E\}.$$

Let
$$X \equiv 1 - \max(0, \mu_A(x) - \eta) - \min(1, \nu_A(x) + \varepsilon).$$

If $\nu_A(x) + \varepsilon \geq 1$, then

$$\mu_A(x) - \eta \leq 1 - \nu_A(x) - \eta \leq \varepsilon - \eta \leq 0$$

and
$$X = 1 - 1 - 0 = 0.$$

If $\nu_A(x) + \varepsilon \leq 1$, then there are two subcases. If $\mu_A(x) - \eta \leq 0$, then

$$X = 1 - (\nu_A(x) + \varepsilon) - 0 \geq 0$$

and if $\mu_A(x) - \eta \geq 0$, then

$$X = 1 - (\nu_A(x) + \varepsilon) - \mu_A(x) + \eta = 1 - \mu_A(x) - \mu_A(x) + \eta - \varepsilon \geq 0.$$

Therefore, the first component of the second term is higher than the first component of the first term, while the inequality

$$\max(0, 1 - \nu_A(x) - \eta) - \max(0, \mu_A(x) - \eta) \geq 0$$

is obvious. Therefore, inclusion is valid.

For every IFS A and for every $\alpha, \beta \in [0, 1]$, the following properties are valid:

(1) $\neg^\varepsilon G_{\alpha,\beta}(A) \supset G_{\beta,\alpha}(\neg^\varepsilon A)$, where $0 \leq \varepsilon \leq 1$,

(2) $\neg^{\varepsilon,\eta} G_{\alpha,\beta}(A) \supset G_{\beta,\alpha}(\neg^{\varepsilon,\eta} A)$, where $0 \leq \varepsilon \leq \eta \leq 1$.

Really, let $\alpha, \beta \in [0, 1]$ be given such that $\alpha + \beta \leq 1$, let A be an IFS and let ε, η be given, such that, $0 \leq \varepsilon \leq \eta \leq 1$. Then,

$$\neg^{\varepsilon,\eta} G_{\alpha,\beta}(A) = \neg^{\varepsilon,\eta} \{\langle x, \alpha.\mu_A(x), \beta.\nu_A(x)\rangle | x \in E\}$$

$$= \{\langle x, \min(1, \beta.\nu_A(x) + \varepsilon), \max(0, \alpha.\mu_A(x) - \eta)\rangle | x \in E\}$$

and
$$G_{\beta,\alpha}(\neg^{\varepsilon,\eta} A)$$

$$= G_{\beta,\alpha}(\{\langle x, \min(1, \nu_A(x) + \varepsilon), \max(0, \mu_A(x) - \eta)\rangle | x \in E\})$$

$$= \{\langle x, \beta.\min(1, \nu_A(x) + \varepsilon), \alpha.\max(0, \mu_A(x) - \eta)\rangle | x \in E\}.$$

Now, we have

$$\min(1, \beta.\nu_A(x) + \varepsilon) - \beta.\min(1, \nu_A(x) + \varepsilon)$$

$$= \min(1, \beta.\nu_A(x) + \varepsilon) - \min(\beta, \beta.\nu_A(x) + \beta.\varepsilon) \geq 0$$

and
$$\alpha.\max(0, \mu_A(x) - \eta) - \max(0, \alpha.\mu_A(x) - \eta)$$

$$= \max(0, \alpha.\mu_A(x) - \alpha.\eta) - \max(0, \alpha.\mu_A(x) - \eta) \geq 0.$$

Therefore, inclusion (2) is valid.

For every IFS A and for every $\alpha, \beta \in [0,1]$, such that $\alpha + \beta \leq 1$ the following properties are valid:

(1) $\neg^\varepsilon H_{\alpha,\beta}(A) \supset H_{\beta,\alpha}(\neg^\varepsilon A)$, where $0 \leq \varepsilon \leq 1$,

(2) $\neg^{\varepsilon,\eta} H_{\alpha,\beta}(A) \supset H_{\beta,\alpha}(\neg^{\varepsilon,\eta} A)$, where $0 \leq \varepsilon \leq \eta \leq 1$,

(3) $\neg^\varepsilon J_{\alpha,\beta}(A) \subset J_{\beta,\alpha}(\neg^\varepsilon A)$, where $0 \leq \varepsilon \leq 1$,

(4) $\neg^{\varepsilon,\eta} J_{\alpha,\beta}(A) \subset J_{\beta,\alpha}(\neg^{\varepsilon,\eta} A)$, where $0 \leq \varepsilon \leq \eta \leq 1$.

Let us prove (4). Let $\alpha, \beta \in [0,1]$ be given and let A be an IFS. Let ε, η be given such that $0 \leq \varepsilon \leq \eta \leq 1$. Then

$$\neg^{\varepsilon,\eta} J_{\alpha,\beta}(A) = \neg^{\varepsilon,\eta} \{\langle x, \mu_A(x) + \alpha.\pi_A(x), \beta.\nu_A(x)\rangle | x \in E\}$$

$$= \{\langle x, \min(1, \beta.\nu_A(x) + \varepsilon), \max(0, \mu_A(x) + \alpha.\pi_A(x) - \eta)\rangle | x \in E\}$$

$$\subset \{\langle x, \min(1, \nu_A(x) + \varepsilon) + \alpha.(), \max(0, \mu_A(x) - \eta)\rangle | x \in E\}$$

$$= J_{\alpha,\beta}(A)(\{\langle x, x, \min(1, \nu_A(x) + \varepsilon), \max(0, \mu_A(x) - \varepsilon)\rangle | x \in E\})$$

$$= J_{\beta,\alpha}(\neg^{\varepsilon,\eta} A).$$

For every IFS A and for every $\alpha, \beta \in [0,1]$, such that $\alpha + \beta \leq 1$, the following properties are valid:

(1) $\neg^\varepsilon H^*_{\alpha,\beta}(A) \supset H^*_{\beta,\alpha}(\neg^\varepsilon A)$, where $0 \leq \varepsilon \leq 1$,

(2) $\neg^{\varepsilon,\eta} H^*_{\alpha,\beta}(A) \supset H^*_{\beta,\alpha}(\neg^{\varepsilon,\eta} A)$, where $0 \leq \varepsilon \leq \eta \leq 1$,

(3) $\neg^\varepsilon J^*_{\alpha,\beta}(A) \subset J^*_{\beta,\alpha}(\neg^\varepsilon A)$, where $0 \leq \varepsilon \leq 1$,

(4) $\neg^{\varepsilon,\eta} J^*_{\alpha,\beta}(A) \subset J^*_{\beta,\alpha}(\neg^{\varepsilon,\eta} A)$, where $0 \leq \varepsilon \leq \eta \leq 1$,

(5) $\neg^\varepsilon P_{\alpha,\beta}(A) \subset P_{\alpha,\beta}(\neg^\varepsilon A)$,

(6) $\neg P_{\alpha,\beta}(A) \subset P_{\alpha,\beta}(\neg^{\varepsilon,\eta} A)$,

(7) $\neg^\varepsilon Q_{\alpha,\beta}(A) \supset Q_{\alpha,\beta}(\neg^\varepsilon A)$,

(8) $\neg Q_{\alpha,\beta}(A) \supset Q_{\alpha,\beta}(\neg^{\varepsilon,\eta} A)$.

Third, in Section **9.3**, a series of new versions of operation "subtraction" has been introduced on the basis of formula:

$$A -' B = A \cap \neg B. \tag{9.9}$$

On the other hand, as we discussed above, the LEM is not always valid in IFS theory. Hence, we introduce a new series of "subtraction" operations, that have the form,

$$A -'' B = \neg\neg A \cap \neg B. \tag{9.10}$$

Using (9.9), we obtained the following two operations of subtraction:

$$A -'^{\varepsilon,\eta} B = A \cap \neg^{\varepsilon,\eta} B$$

$$= \{\langle x, \min(\mu_A(x), 1, \nu_B(x) + \varepsilon), \max(\nu_A(x), 0, \mu_B(x) - \eta)\rangle | x \in E\}$$
$$= \{\langle x, \min(\mu_A(x), \nu_B(x) + \varepsilon), \max(\nu_A(x), \mu_B(x) - \eta)\rangle | x \in E\}.$$

Also, using (9.10) and having in mind that

$$\neg^{\varepsilon,\eta} \neg^{\varepsilon,\eta} A = \neg^{\varepsilon,\eta} \{\langle x, \min(1, \nu_A(x) + \varepsilon), \max(0, \mu_A(x) - \eta)\rangle x, |x \in E\}$$

$$= \{\langle x, \min(1, \max(0, \mu_A(x) - \eta) + \varepsilon), \max(0, \min(1, \nu_A(x) + \varepsilon) - \eta)\rangle x, |x \in E\}$$
$$= \{\langle x, \min(1, \max(\varepsilon, \mu_A(x) - \eta + \varepsilon)), \max(0, \min(1 - \eta, \nu_A(x) + \varepsilon - \eta)))\rangle x, |x \in E\}$$
$$= \{\langle x, \max(\varepsilon, \mu_A(x) - \eta + \varepsilon), \max(0, \min(1 - \eta, \nu_A(x) + \varepsilon - \eta)))\rangle x, |x \in E\},$$

we obtain the following form of the operation $-''^{\varepsilon,\eta}$:

$$A -''^{\varepsilon,\eta} B = \neg^{\varepsilon,\eta} \neg^{\varepsilon,\eta} A \cap \neg^{\varepsilon,\eta} B$$

$$= \{\langle x, \max(\varepsilon, \mu_A(x) - \eta + \varepsilon), \max(0, \min(1 - \eta, \nu_A(x) + \varepsilon - \eta)))\rangle x, |x \in E\}$$
$$\cap \{\langle x, \min(1, \nu_B(x) + \varepsilon), \max(0, \mu_B(x) - \eta)\rangle x, |x \in E\}$$
$$= \{\langle x, \min(\max(\varepsilon, \mu_A(x) - \eta + \varepsilon), 1, \nu_B(x) + \varepsilon),$$
$$\max(0, \min(1 - \eta, \nu_A(x) + \varepsilon - \eta), \mu_B(x) - \eta)\rangle x, |x \in E\}$$
$$= \{\langle x, \min(\max(\varepsilon, \mu_A(x) - \eta + \varepsilon), \nu_B(x) + \varepsilon),$$
$$\max(0, \min(1 - \eta, \nu_A(x) + \varepsilon - \eta), \mu_B(x) - \eta)\rangle x, |x \in E\}.$$

Using (9.9), we obtain the following form of the operation $-'^{\varepsilon,\eta}$:

$$A -'^{\varepsilon,\eta} B = A \cap \neg^{\varepsilon,\eta} B$$

$$= \{\langle x, \min(\mu_A(x), 1, \nu_B(x) + \varepsilon), \max(\nu_A(x), 0, \mu_B(x) - \eta)\rangle | x \in E\}$$
$$= \{\langle x, \min(\mu_A(x), \nu_B(x) + \varepsilon), \max(\nu_A(x), \mu_B(x) - \eta)\rangle | x \in E\}.$$

First, we must check that as a result of the operation we obtain an IFS or not. For two given IFSs A and B and for each $x \in E$, we obtain
(a) if $\nu_A(x) \le \mu_B(x) - \eta$, then

$$0 \le \min(\mu_A(x), \nu_B(x) + \varepsilon) + \max(\nu_A(x), \mu_B(x) - \eta)$$

$$= \min(\mu_A(x), \nu_B(x) + \varepsilon) + \mu_B(x) - \eta$$
$$= \min(\mu_A(x) + \mu_B(x) - \eta, \nu_B(x) + \varepsilon + \mu_B(x) - \eta) \le 1,$$

because $\nu_B(x) + \mu_B(x) + \varepsilon - \eta \le 1$;
(b) if $\nu_A(x) > \mu_B(x) - \eta$, then

$$0 \le \min(\mu_A(x), \nu_B(x) + \varepsilon) + \max(\nu_A(x), \mu_B(x) - \eta)$$

$$= \min(\mu_A(x), \nu_B(x) + \varepsilon) + \nu_A(x)$$

$$= \min(\mu_A(x) + \nu_A(x), \nu_B(x) + \varepsilon + \nu_A(x)) \le 1$$

because $\mu_A(x) + \nu_B(x) \le 1$.

Define the ε-uncertain IFS, and the η-*empty IFS* (see [39]) by:

$$\varepsilon^* = \{\langle x, \varepsilon, 0\rangle | x \in E\}, \tag{9.11}$$

$$\eta^* = \{\langle x, 0, 1 - \eta\rangle | x \in E\}, \tag{9.12}$$

and also two other special sets:

$$(\varepsilon, \eta)^* = \{\langle x, \varepsilon, 1 - \eta\rangle | x \in E\}, \tag{9.13}$$

$$(\varepsilon, \eta)^{**} = \{\langle x, 1 - \eta + \varepsilon, 0\rangle | x \in E\} \tag{9.14}$$

(see, also (2.13), (2.14), (2.15)).

For every two IFSs A and B:

(a) $A -^{\prime\varepsilon,\eta} E^* = O^*$,

(b) $A -^{\prime\varepsilon,\eta} O^* = A$,

(c) $E^* -^{\prime\varepsilon,\eta} A = \neg^{\varepsilon,\eta} A$,

(d) $O^* -^{\prime\varepsilon,\eta} A = O^*$,

(e) $(A -^{\prime\varepsilon,\eta} B) \cap C = (A \cap C) -^{\prime\varepsilon,\eta} B = A \cap (C -^{\prime\varepsilon,\eta} B)$,

(f) $(A \cap B) -^{\prime\varepsilon,\eta} C = (A -^{\prime\varepsilon,\eta} C) \cap (B -^{\prime\varepsilon,\eta} C)$,

(g) $(A \cup B) -^{\prime\varepsilon,\eta} C = (A -^{\prime\varepsilon,\eta} C) \cup (B -^{\prime\varepsilon,\eta} C)$,

(h) $(A -^{\prime\varepsilon,\eta} B) -^{\prime\varepsilon,\eta} C = (A -^{\prime\varepsilon,\eta} C) -^{\prime\varepsilon,\eta} B$.

Obviously

$$O^* -^{\prime\varepsilon,\eta} U^* = O^*,$$

$$O^* -^{\prime\varepsilon,\eta} E^* = O^*,$$

$$O^* -^{\prime\varepsilon,\eta} O^* = O^*,$$

$$U^* -^{\prime\varepsilon,\eta} U^* = O^*,$$

$$U^* -^{\prime\varepsilon,\eta} O^* = U^*,$$

$$U^* -^{\prime\varepsilon,\eta} E^* = \eta^*,$$

$$E^* -^{\prime\varepsilon,\eta} O^* = E^*,$$

$$E^* -^{\prime\varepsilon,\eta} U^* = \varepsilon^*,$$

$$E^* -^{\prime\varepsilon,\eta} E^* = (\varepsilon, \eta)^*.$$

Now, using (9.10) and having in mind that

$$\neg^{\varepsilon,\eta} \neg^{\varepsilon,\eta} A = \neg^{\varepsilon,\eta} \{\langle x, \min(1, \nu_A(x) + \varepsilon), \max(0, \mu_A(x) - \eta)\rangle x, |x \in E\}$$

$$= \{\langle x, \min(1, \max(0, \mu_A(x) - \eta) + \varepsilon),$$

$$\max(0, \min(1, \nu_A(x) + \varepsilon) - \eta)\rangle x, |x \in E\}$$

$$= \{\langle x, \min(1, \max(\varepsilon, \mu_A(x) - \eta + \varepsilon)),$$
$$\max(0, \min(1 - \eta, \nu_A(x) + \varepsilon - \eta))\rangle x, |x \in E\}$$
$$= \{\langle x, \max(\varepsilon, \mu_A(x) - \eta + \varepsilon),$$
$$\max(0, \min(1 - \eta, \nu_A(x) + \varepsilon - \eta))\rangle x, |x \in E\}$$

we obtain the following form of the operation $-''^{\varepsilon,\eta}$:

$$A -''^{\varepsilon,\eta} B = \neg^{\varepsilon,\eta}\neg^{\varepsilon,\eta} A \cap \neg^{\varepsilon,\eta} B$$

$$= \{\langle x, \max(\varepsilon, \mu_A(x) - \eta + \varepsilon), \max(0, \min(1 - \eta, \nu_A(x) + \varepsilon - \eta))\rangle x, |x \in E\}$$
$$\cap\{\langle x, \min(1, \nu_B(x) + \varepsilon), \max(0, \mu_B(x) - \eta)\rangle x, |x \in E\}$$
$$= \{\langle x, \min(\max(\varepsilon, \mu_A(x) - \eta + \varepsilon), 1, \nu_B(x) + \varepsilon),$$
$$\max(0, \min(1 - \eta, \nu_A(x) + \varepsilon - \eta), \mu_B(x) - \eta))\rangle x, |x \in E\}$$
$$= \{\langle x, \min(\max(\varepsilon, \mu_A(x) - \eta + \varepsilon), \nu_B(x) + \varepsilon),$$
$$\max(0, \min(1 - \eta, \nu_A(x) + \varepsilon - \eta), \mu_B(x) - \eta))\rangle x, |x \in E\}. \tag{9.15}$$

For every IFS A:
(a) $A -''^{\varepsilon,\eta} E^* = (\varepsilon, \eta)^*$,
(b) $A -''^{\varepsilon,\eta} O^* = \neg^{\varepsilon,\eta}\neg^{\varepsilon,\eta} A$,
(c) $O^* -''^{\varepsilon,\eta} A = (\varepsilon, \eta)^*$,
(d) $(A \cap B) -'^{\varepsilon,\eta} C = (A -'^{\varepsilon,\eta} C) \cap (B -'^{\varepsilon,\eta} C)$,
(e) $(A \cup B) -'^{\varepsilon,\eta} C = (A -'^{\varepsilon,\eta} C) \cup (B -'^{\varepsilon,\eta} C)$,
(f) $(A -''^{\varepsilon,\eta} B) \cap \neg^{\varepsilon,\eta}\neg^{\varepsilon,\eta} C = (C -''^{\varepsilon,\eta} B) \cap \neg^{\varepsilon,\eta}\neg^{\varepsilon,\eta} A$.

Obviously,
$$O^* -'^{\varepsilon,\eta} U^* = (\varepsilon, \eta)^*,$$
$$O^* -'^{\varepsilon,\eta} E^* = (\varepsilon, \eta)^*,$$
$$O^* -'^{\varepsilon,\eta} O^* = (\varepsilon, \eta)^*,$$
$$U^* -'^{\varepsilon,\eta} U^* = \varepsilon^*,$$
$$U^* -'^{\varepsilon,\eta} E^* = (\varepsilon, \eta)^*,$$
$$U^* -'^{\varepsilon,\eta} O^* = \eta^*,$$
$$E^* -'^{\varepsilon,\eta} O^* = (\varepsilon, \eta)^{**},$$
$$E^* -'^{\varepsilon,\eta} U^* = \varepsilon^*,$$
$$E^* -'^{\varepsilon,\eta} E^* = (\varepsilon, \eta)^*.$$

Fourth, following [77], using the subtraction of sets A and $\neg^{\varepsilon,\eta} A$, we introduce the following two (ε, η)-norms for element $x \in E$ as

$$||x||'_{\varepsilon,\eta} = \langle \min(\mu_A(x), \nu_A(x) + \varepsilon), \max(\nu_A(x), \mu_A(x) - \eta)\rangle, \tag{9.16}$$

$$||x||_{\varepsilon,\eta}'' = \langle \min(\max(\varepsilon, \mu_A(x) - \eta + \varepsilon), \nu_A(x) + \varepsilon),$$

$$\max(0, \min(1 - \eta, \nu_A(x) + \varepsilon - \eta), \mu_A(x) - \eta)\rangle. \tag{9.17}$$

Theorem 9.30: The two (ε, η)-norms are intuitionistic fuzzy pairs.
Proof: Let

$$X \equiv \min(\mu_A(x), \nu_A(x) + \varepsilon) + \max(\nu_A(x), \mu_A(x) - \eta).$$

If $\nu_A(x) \geq \mu_A(x) - \eta$, then

$$X = \min(\mu_A(x), \nu_A(x) + \varepsilon) + \nu_A(x) \leq \mu_A(x) + \nu_A(x) \leq 1.$$

If $\nu_A(x) > \mu_A(x) - \eta$, then

$$X = \min(\mu_A(x), \nu_A(x) + \varepsilon) + \mu_A(x) - \eta$$

$$\leq \nu_A(x) + \varepsilon + \mu_A(x) - \eta \leq 1 + \varepsilon - \eta < 1.$$

Therefore, the first norm is an intuitionistic fuzzy pair.
Let

$$Y \equiv \min(\max(\varepsilon, \mu_A(x) - \eta + \varepsilon), \nu_A(x) + \varepsilon)$$

$$+ \max(0, \min(1 - \eta, \nu_A(x) + \varepsilon - \eta), \mu_A(x) - \eta).$$

If $\max(0, \min(1 - \eta, \nu_A(x) + \varepsilon - \eta), \mu_A(x) - \eta) = 0$, then

$$Y = \min(\max(\varepsilon, \mu_A(x) - \eta + \varepsilon), \nu_A(x) + \varepsilon) + 0 \leq \max(\varepsilon, \mu_A(x) - \eta + \varepsilon) \leq 1.$$

If $\max(0, \min(1 - \eta, \nu_A(x) + \varepsilon - \eta), \mu_A(x) - \eta) = \min(1 - \eta, \nu_A(x) + \varepsilon - \eta)$,
then

$$\min(1 - \eta, \nu_A(x) + \varepsilon - \eta) \geq \mu_A(x) - \eta,$$

i.e., $\nu_A(x) + \varepsilon \geq \mu_A(x)$. Hence,

$$\mu_A(x) - \eta \leq \nu_A(x) + \varepsilon - \eta \leq \nu_A(x)$$

and therefore,

$$Y = \min(\max(\varepsilon, \mu_A(x) - \eta + \varepsilon), \nu_A(x) + \varepsilon) + \min(1 - \eta, \nu_A(x) + \varepsilon - \eta)$$

$$= \min(\varepsilon + \max(0, \mu_A(x) - \eta), \nu_A(x) + \varepsilon) + \min(1 - \eta, \nu_A(x) + \varepsilon - \eta)$$

$$= \varepsilon + \min(\max(0, \mu_A(x) - \eta), \nu_A(x)) + \min(1 - \eta, \nu_A(x) + \varepsilon - \eta)$$

$$= \varepsilon + \max(0, \mu_A(x) - \eta) + \min(1, \nu_A(x) + \varepsilon) - \eta$$

$$\leq \max(0, \mu_A(x) - \eta) + \min(1, \nu_A(x) + \varepsilon).$$

If $\mu_A(x) \geq \eta$, then

$$Y \le \mu_A(x) - \eta + \min(1, \nu_A(x) + \varepsilon) \le \mu_A(x) - \eta + \nu_A(x) + \varepsilon \le 1 - \eta + \varepsilon \le 1;$$

if $\mu_A(x) < \eta$, then $Y \le \min(1, \nu_A(x) + \varepsilon) \le 1$.

If $\max(0, \min(1 - \eta, \nu_A(x) + \varepsilon - \eta), \mu_A(x) - \eta) = \mu_A(x) - \eta$, then

$$Y = \min(\max(\varepsilon, \mu_A(x) - \eta + \varepsilon), \nu_A(x) + \varepsilon) + \mu_A(x) - \eta$$

$$\le \nu_A(x) + \varepsilon + \mu_A(x) - \eta \le 1 + \varepsilon - \eta \le 1.$$

Therefore, the second norm is also an intuitionistic fuzzy pair.

All norms and distances, defined over IFSs so far (see, e.g. [128, 166, 242, 243, 245, 249, 440, 465, 468, 471, 472, 476, 480, 513, 565, 585, 597]), have been real numbers, that may in some cases be normalized to the $[0,1]$ interval. As we have seen, the two norms (9.16) and (9.17) have the form of intuitionistic fuzzy pairs and they are the first of this form.

Let $e^*, o^*, u^* \in E$ so that $\mu_A(e^*) = 1$, $\nu_A(e^*) = 0$, $\mu_A(o^*) = 0$, $\nu_A(o^*) = 1$, $\mu_A(u^*) = 0$, $\nu_A(u^*) = 0$. Then, the norms of these three elements are:

x	$\|x\|'_{\varepsilon,\eta}$	$\|x\|''_{\varepsilon,\eta}$
e^*	$\langle \varepsilon, 1 - \eta \rangle$	$\langle \varepsilon, 1 - \eta \rangle\rangle$
o^*	$\langle 0, 1 \rangle$	$\langle \varepsilon, 1 - \eta \rangle$
u^*	$\langle 0, 0 \rangle$	$\langle \varepsilon, 0 \rangle$

Fifth, we mention that there are two ways of introducing distances between elements of a fixed universe E, as shown in Fig. 9.6.

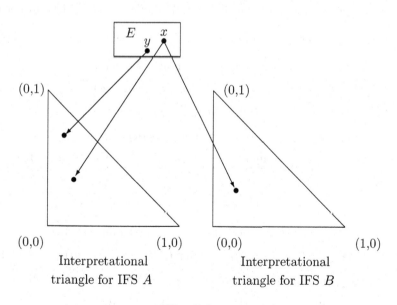

Interpretational   Interpretational
triangle for IFS A triangle for IFS B

Fig. 9.6

The distances and norms so far discussed have been real numbers. Now, we introduce distances, having forms of intuitionistic fuzzy pairs.

First, we introduce the following five (ε, η)-distances between the values of one element $x \in E$ about two IFSs A and B, similar to the first (ε, η)-norm:

$$d'_{\varepsilon,\eta;str\_opt}(A, B)(x) = \langle \min(\mu_A(x), \nu_B(x) + \varepsilon) + \min(\mu_B(x), \nu_A(x) + \varepsilon)$$

$$- \min(\mu_A(x), \nu_B(x) + \varepsilon) . \min(\mu_B(x), \nu_A(x) + \varepsilon),$$

$$\max(\nu_A(x), \mu_B(x) - \eta) . \max(\nu_B(x), \mu_A(x) - \eta) \rangle, \tag{9.18}$$

$$d'_{\varepsilon,\eta;opt}(A, B)(x) = \langle \max(\min(\mu_A(x), \nu_B(x) + \varepsilon), \min(\mu_B(x), \nu_A(x) + \varepsilon)),$$

$$\min(\max(\nu_A(x), \mu_B(x) - \eta), \max(\nu_B(x), \mu_A(x) - \eta)) \rangle, \tag{9.19}$$

$$d'_{\varepsilon,\eta;aver}(A, B)(x) = \langle \frac{\min(\mu_A(x), \nu_B(x) + \varepsilon) + \min(\mu_B(x), \nu_A(x) + \varepsilon)}{2},$$

$$\frac{\max(\nu_A(x), \mu_B(x) - \eta) + \max(\nu_B(x), \mu_A(x) - \eta)}{2} \rangle, \tag{9.20}$$

$$d'_{\varepsilon,\eta;pes}(A, B)(x) = \langle \min(\min(\mu_A(x), \nu_B(x) + \varepsilon), \min(\mu_B(x), \nu_A(x) + \varepsilon)),$$

$$\max(\max(\nu_A(x), \mu_B(x) - \eta), \max(\nu_B(x), \mu_A(x) - \eta)) \rangle, \tag{9.21}$$

$$d'_{\varepsilon,\eta;str\_pes}(A, B)(x) = \langle \min(\mu_A(x), \nu_B(x) + \varepsilon) . \min(\mu_B(x), \nu_A(x) + \varepsilon),$$

$$\max(\nu_A(x), \mu_B(x) - \eta) + \max(\nu_B(x), \mu_A(x) - \eta)$$

$$- \max(\nu_A(x), \mu_B(x) - \eta) . \max(\nu_B(x), \mu_A(x) - \eta) \rangle. \tag{9.22}$$

Second, we introduce the following five (ε, η)-distances between the values of two elements $x, y \in E$ about the IFS A, similar to the first (ε, η)-norm:

$$d'_{\varepsilon,\eta;str\_opt}(A)(x, y) = \langle \min(\mu_A(x), \nu_A(y) + \varepsilon) + \min(\mu_A(y), \nu_A(x) + \varepsilon)$$

$$- \min(\mu_A(x), \nu_A(y) + \varepsilon) . \min(\mu_A(y), \nu_A(x) + \varepsilon),$$

$$\max(\nu_A(x), \mu_A(y) - \eta) . \max(\nu_A(y), \mu_A(x) - \eta) \rangle, \tag{9.23}$$

$$d'_{\varepsilon,\eta;opt}(A)(x, y) = \langle \max(\min(\mu_A(x), \nu_A(y) + \varepsilon), \min(\mu_A(y), \nu_A(x) + \varepsilon)),$$

$$\min(\max(\nu_A(x), \mu_A(y) - \eta), \max(\nu_A(y), \mu_A(x) - \eta)) \rangle, \tag{9.24}$$

$$d'_{\varepsilon,\eta;aver}(A)(x, y) = \langle \frac{\min(\mu_A(x), \nu_A(y) + \varepsilon) + \min(\mu_A(y), \nu_A(x) + \varepsilon)}{2},$$

$$\frac{\max(\nu_A(x), \mu_A(y) - \eta) + \max(\nu_A(y), \mu_A(x) - \eta)}{2} \rangle, \tag{9.25}$$

$$d'_{\varepsilon,\eta;pes}(A)(x, y) = \langle \min(\min(\mu_A(x), \nu_A(y) + \varepsilon), \min(\mu_A(y), \nu_A(x) + \varepsilon)),$$

$$\max(\max(\nu_A(x), \mu_A(y) - \eta), \max(\nu_A(y), \mu_A(x) - \eta)) \rangle, \tag{9.26}$$

$$d'_{\varepsilon,\eta;str\_pes}(A)(x, y) = \langle \min(\mu_A(x), \nu_A(y) + \varepsilon) . \min(\mu_A(y), \nu_A(x) + \varepsilon),$$

$$\max(\nu_A(x), \mu_A(y) - \eta) + \max(\nu_A(y), \mu_A(x) - \eta)$$

$$- \max(\nu_A(x), \mu_A(y) - \eta) . \max(\nu_A(y), \mu_A(x) - \eta)\rangle. \tag{9.27}$$

Theorem 9.31: The ten (ε, η)-distances (9.18) – (9.27) are intuitionistic fuzzy pairs.

The proof is similar to the above.

As an illustration, we calculate the distances between the pairs (e^*, o^*), (e^*, u^*) and (u^*, o^*).

	(e^*, o^*)	(e^*, u^*)	(o^*, u^*)
$d'_{\varepsilon,\eta;str\_opt}(A)$	$\langle 1, 0 \rangle$	$\langle 0, 1 \rangle$	$\langle 0, 0 \rangle$
$d'_{\varepsilon,\eta;aver}(A)$	$\langle 1, 0 \rangle$	$\langle 0, 1 \rangle$	$\langle 0, 0 \rangle$
$d'_{\varepsilon,\eta;opt}(A)$	$\langle \frac{1}{2}, \frac{1}{2} \rangle$	$\langle 0, 1 \rangle$	$\langle 0, \frac{1}{2} \rangle$
$d'_{\varepsilon,\eta;pes}(A)$	$\langle 0, 1 \rangle$	$\langle 0, 1 \rangle$	$\langle 0, 1 \rangle$
$d'_{\varepsilon,\eta;str\_pes}(A)$	$\langle 0, 1 \rangle$	$\langle 0, 1 \rangle$	$\langle 0, 1 \rangle$

Third, we introduce the following five (ε, η)-distances between the values of two elements $x, y \in E$ about the IFS A, similar to the second (ε, η)-norm:

$$d''_{\varepsilon,\eta;str\_opt}(A, B)(x) = \langle \min(\max(\varepsilon, \mu_A(x) - \eta + \varepsilon), \nu_B(x) + \varepsilon)$$

$$+ \min(\max(\varepsilon, \mu_B(x) - \eta + \varepsilon), \nu_A(x) + \varepsilon) - \min(\max(\varepsilon, \mu_A(x) - \eta + \varepsilon), \nu_B(x) + \varepsilon)$$

$$. \min(\max(\varepsilon, \mu_B(x) - \eta + \varepsilon), \nu_A(x) + \varepsilon), \max(0, \min(1 - \eta, \nu_A(x) + \varepsilon - \eta), \mu_B(x) - \eta)$$

$$. \max(0, \min(1 - \eta, \nu_B(x) + \varepsilon - \eta), \mu_A(x) - \eta)\rangle, \tag{9.28}$$

$$d''_{\varepsilon,\eta;opt}(A, B)(x) = \langle \max(\min(\max(\varepsilon, \mu_A(x) - \eta + \varepsilon), \nu_B(x) + \varepsilon),$$

$$\min(\max(\varepsilon, \mu_B(x) - \eta + \varepsilon), \nu_A(x) + \varepsilon)),$$

$$\min(\max(0, \min(1 - \eta, \nu_A(x) + \varepsilon - \eta), \mu_B(x) - \eta),$$

$$\max(0, \min(1 - \eta, \nu_B(x) + \varepsilon - \eta), \mu_A(x) - \eta)))\rangle, \tag{9.29}$$

$$d''_{\varepsilon,\eta;aver}(A, B)(x) = \langle \frac{1}{2} . (\min(\max(\varepsilon, \mu_A(x) - \eta + \varepsilon), \nu_B(x) + \varepsilon)$$

$$+ \min(\max(\varepsilon, \mu_B(x) - \eta + \varepsilon), \nu_A(x) + \varepsilon)),$$

$$\frac{1}{2} . (\max(0, \min(1 - \eta, \nu_A(x) + \varepsilon - \eta), \mu_B(x) - \eta)$$

$$+ \max(0, \min(1 - \eta, \nu_B(x) + \varepsilon - \eta), \mu_A(x) - \eta)))\rangle, \tag{9.30}$$

$$d''_{\varepsilon,\eta;pes}(A, B)(x) = \langle \min(\max(\varepsilon, \mu_A(x) - \eta + \varepsilon), \nu_B(x) + \varepsilon),$$

$$\max(\varepsilon, \mu_B(x) - \eta + \varepsilon), \nu_A(x) + \varepsilon),$$

$$\max(0, \min(1 - \eta, \nu_A(x) + \varepsilon - \eta), \mu_B(x) - \eta,$$

$$\min(1 - \eta, \nu_B(x) + \varepsilon - \eta), \mu_A(x) - \eta)\rangle, \tag{9.31}$$

$$d''_{\varepsilon,\eta;str\_pes}(A, B)(x) = \langle \min(\max(\varepsilon, \mu_A(x) - \eta + \varepsilon), \nu_B(x) + \varepsilon).$$

$$\min(\max(\varepsilon, \mu_B(x) - \eta + \varepsilon), \nu_A(x) + \varepsilon),$$

$$\max(0, \min(1 - \eta, \nu_A(x) + \varepsilon - \eta), \mu_B(x) - \eta)$$
$$+ \max(0, \min(1 - \eta, \nu_B(x) + \varepsilon - \eta), \mu_A(x) - \eta)$$
$$- \max(0, \min(1 - \eta, \nu_A(x) + \varepsilon - \eta), \mu_B(x) - \eta)$$
$$. \max(0, \min(1 - \eta, \nu_B(x) + \varepsilon - \eta), \mu_A(x) - \eta)\rangle. \tag{9.32}$$

Fourth, we introduce the following five (ε, η)-distances between the values of two elements $x, y \in E$ about the IFS A, similar to the second (ε, η)-norm:

$$d''_{\varepsilon,\eta;str\_opt}(A)(x,y) = \langle \min(\max(\varepsilon, \mu_A(x) - \eta + \varepsilon), \nu_A(y) + \varepsilon)$$
$$+ \min(\max(\varepsilon, \mu_A(y) - \eta + \varepsilon), \nu_A(x) + \varepsilon)$$
$$- \min(\max(\varepsilon, \mu_A(x) - \eta + \varepsilon), \nu_A(y) + \varepsilon)$$
$$. \min(\max(\varepsilon, \mu_A(y) - \eta + \varepsilon), \nu_A(x) + \varepsilon),$$
$$\max(0, \min(1 - \eta, \nu_A(x) + \varepsilon - \eta), \mu_A(y) - \eta)$$
$$. \max(0, \min(1 - \eta, \nu_A(y) + \varepsilon - \eta), \mu_A(x) - \eta)\rangle, \tag{9.33}$$

$$d''_{\varepsilon,\eta;opt}(A)(x,y) = \langle \max(\min(\max(\varepsilon, \mu_A(x) - \eta + \varepsilon), \nu_A(y) + \varepsilon),$$
$$\min(\max(\varepsilon, \mu_A(y) - \eta + \varepsilon), \nu_A(x) + \varepsilon)),$$
$$\min(\max(0, \min(1 - \eta, \nu_A(x) + \varepsilon - \eta), \mu_A(y) - \eta),$$
$$\max(0, \min(1 - \eta, \nu_A(y) + \varepsilon - \eta), \mu_A(x) - \eta)))\rangle, \tag{9.34}$$

$$d''_{\varepsilon,\eta;aver}(A)(x,y) = \langle \frac{1}{2}.(\min(\max(\varepsilon, \mu_A(x) - \eta + \varepsilon), \nu_A(y) + \varepsilon)$$
$$+ \min(\max(\varepsilon, \mu_A(y) - \eta + \varepsilon), \nu_A(x) + \varepsilon)),$$
$$\frac{1}{2}.(\max(0, \min(1 - \eta, \nu_A(x) + \varepsilon - \eta), \mu_A(y) - \eta)$$
$$+ \max(0, \min(1 - \eta, \nu_A(y) + \varepsilon - \eta), \mu_A(x) - \eta)))\rangle, \tag{9.35}$$

$$d''_{\varepsilon,\eta;pes}(A)(x,y) = \langle \min(\max(\varepsilon, \mu_A(x) - \eta + \varepsilon, \nu_A(y) + \varepsilon),$$
$$\max(\varepsilon, \mu_A(y) - \eta + \varepsilon), \nu_A(x) + \varepsilon),$$
$$\max(0, \min(1 - \eta, \nu_A(x) + \varepsilon - \eta), \mu_A(y) - \eta,$$
$$\min(1 - \eta, \nu_A(y) + \varepsilon - \eta), \mu_A(x) - \eta)\rangle, \tag{9.36}$$

$$d''_{\varepsilon,\eta;str\_pes}(A)(x,y) = \langle \min(\max(\varepsilon, \mu_A(x) - \eta + \varepsilon), \nu_A(y) + \varepsilon)$$
$$. \min(\max(\varepsilon, \mu_A(y) - \eta + \varepsilon), \nu_A(x) + \varepsilon),$$
$$\max(0, \min(1 - \eta, \nu_A(x) + \varepsilon - \eta), \mu_A(y) - \eta)$$
$$+ \max(0, \min(1 - \eta, \nu_A(y) + \varepsilon - \eta), \mu_A(x) - \eta)$$
$$- \max(0, \min(1 - \eta, \nu_A(x) + \varepsilon - \eta), \mu_A(y) - \eta)$$
$$. \max(0, \min(1 - \eta, \nu_A(y) + \varepsilon - \eta), \mu_A(x) - \eta)\rangle. \tag{9.37}$$

Theorem 9.32: The last ten (ε, η)-distances $(9.28) - (9.37)$ are intuitionistic fuzzy pairs.

The proof is similar to the above.

Now, we discuss new distances between two given IFSs A and B. So far, they have been real numbers or also been intuitionistic fuzzy pairs. Here, for the first time, we introduce a whole IFS, representing the distances between the origins of each element $x \in E$ with respect to the two sets. This set-form of distances can have different forms, but we describe the following ten of them:

$$D'(A, B)_{\varepsilon,\eta;type} = \{\langle x, \mu_{d'(A,B)_{\varepsilon,\eta;type}}(x), \nu_{d'(A,B)_{\varepsilon,\eta;type}}(x)\rangle | x \in E\}, \quad (9.38)$$

$$D''(A, B)_{\varepsilon,\eta;type} = \{\langle x, \mu_{d''(A,B)_{\varepsilon,\eta;type}}(x), \nu_{d''(A,B)_{\varepsilon,\eta;type}}(x)\rangle | x \in E\}, \quad (9.39)$$

where

$$\langle \mu_{d'(A,B)_{\varepsilon,\eta;type}}(x), \nu_{d'(A,B)_{\varepsilon,\eta;type}}(x)\rangle = d'_{\varepsilon,\eta;type}(A, B)(x),$$

$$\langle \mu_{d''(A,B)_{\varepsilon,\eta;type}}(x), \nu_{d''(A,B)_{\varepsilon,\eta;type}}(x)\rangle = d''_{\varepsilon,\eta;type}(A, B)(x)$$

for "type" $\in \{$"str_opt", "opt", "aver", "pes", "str_pes"$\}$.

The so constructed sets are IFSs and the proof of this fact is similar to the above one.

9.5   Concluding Remarks

Here, we will shortly discuss the author's plans for the future research.

The list of the introduced implications and negations is already longer, but here we have not included those operations, that were defined after 2010. Now, there are some new implications and negations for which all the above research must be repeated. Soon, at least two other types of implications will be introduced. Doing this, we will finish the classification of the implications and negations, that was mentioned here by the way. As a result of the classification, we will determine the operations that are the most suitable for application. Of course, they will satisfy all axioms of some axiomatic system, e.g. of intuitionistic logic. The non-classical implications and negations can generate new types of conjunctions and disjunctions and they also will be an object of a future research.

Up to now, we have not discussed the so called T- and S-norms. One of the reasons for this is that the new forms of De Morgan's Laws from Section **9.1.3** show the necessity to have a new theory developed of these norms in near future.

On Two New Extensions of Intuitionistic Fuzzy Sets

10.1 On Some Intuitionistic Fuzzy Extensions

An year after defining IFSs, in 1984, Stefka Stoeva and the author extended the concept of IFS to "Intuitionistic L-fuzzy set", where L is some lattice, and they studied some of its basic properties [103]. All these results are included in [39] and hence, they not be discussed here.

The idea for IFSs of second type (2-IFSs) and more general, of p-IFSs, was studied by the author in 1992-1993 and in the last years it was developed by Parvathi Rangasamy and Peter Vassilev (see [395, 557]). The most important result of the three authors is the following (see [398]). Let $p \in [1, +\infty)$ and let

$$A = \{\langle x, \mu_A(x), \nu_A(x)\rangle | x \in E\}$$

be a p-IFS. Then for each $q \in [1, +\infty)$ the set

$$\{\langle x, (\mu_A(x))^{\frac{p}{q}}, (\nu_A(x))^{\frac{p}{q}}\rangle | x \in E\}$$

is a q-IFS.

In [93], George Gargov and the author introduced another IFS-extension, called "IFS with interval values" (IVIFS). The author's results on this concept obtained before 1997 are also included in [39], but during the last years some new results were published. The author plans to publish them in a separate book and so, they are not discussed here.

Another direction of extension is related to introducing a new type of IFSs with the form

$$\{\langle x, \rho_1(x), \rho_2(x), ..., \rho_s(x)\rangle | x \in E\},$$

where for every $x \in E$:

$$0 \leq \rho_1(x) + \rho_s(x) + ... + \rho_s(x) \leq 1.$$

Of course, such an object will be an essential extension of the ordinary IFS, but the question for its sense and of the sense of its degrees $\rho_1(x), \rho_2(x), ...,$

K.T. Atanassov: On Intuitionistic Fuzzy Sets Theory, STUDFUZZ 283, pp. 259–272.
springerlink.com   © Springer-Verlag Berlin Heidelberg 2012

$\rho_s(x)$ will arise. In the IFS-case we have two explicitly defined degrees (μ and ν) that generate the third degree (π). Therefore, in the case of an IFS with s explicit degrees $(\rho_1, \rho_2, ..., \rho_s)$, we have a $(s+1)$−st degree, so that for every $x \in E$:

$$\rho_{s+1}(x) = 1 - \rho_1(x) - \rho_2(x) - ... - \rho_s(x)$$

and it stays for the uncertainty.

We assume that in all cases ρ_{s+1} is the degree of uncertainty.

In case of $s = 4$, the new object coincides with the IVIFS, if

$$\rho_1(x) = \inf M_A(x),$$

$$\rho_2(x) = \sup M_A(x) - \inf M_A(x),$$

$$\rho_3(x) = \inf N_A(x),$$

$$\rho_4(x) = \sup N_A(x) - \inf N_A(x),$$

where the IVIFS has the form

$$\{\langle x, M_A(x), N_A(x)\rangle | x \in E\},$$

and $M_A(x), N_A(x) \subset [0, 1]$, so that $\sup M_A(x) + \sup N_A(x) \le 1$.

In [265, 266], Chris Hinde, R. Patching and S. McCoy discussed the possibility to construct an IFS extension with $s = 3$. In his version of IFS, the first two degrees are the standard ones, while the third one denotes the degree of contradiction.

The future rsearch will show which of these extensions will be used for real applications.

In this Chapter, we extend and illustrate the results on Temporal IFSs (TIFSs) that were mentioned in [39] and the Multidimentional IFSs (MIFSs) introduced in the last years.

10.2 Temporal IFSs

Let E be a universe, and T be a non-empty set. We call the elements of T "time-moments". Based on the definition of IFS, we define another type of an IFS (see, e.g., [22, 46, 47]).

We define a *Temporal IFS (TIFS)* as the following:

$$A(T) = \{\langle x, \mu_A(x,t), \nu_A(x,t)\rangle | \langle x,t\rangle \in E \times T\}, \tag{10.1}$$

where

(a) $A \subset E$ is a fixed set,
(b) $\mu_A(x,t) + \nu_A(x,t) \le 1$ for every $\langle x,t\rangle \in E \times T$,
(c) $\mu_A(x,t)$ and $\nu_A(x,t)$ are the degrees of membership and non-membership, respectively, of the element $x \in E$ at the time-moment $t \in T$.

For brevity, we write A instead of $A(T)$ when this does not cause confusion.

Obviously, every ordinary IFS can be regarded as a TIFS for which T is a singleton set.

All operations and operators on the IFSs can be defined for the TIFSs. Suppose that we have two TIFSs:

$$A(T') = \{\langle x, \mu_A(x,t), \nu_A(x,t)\rangle | \langle x,t\rangle \in E \times T'\},$$

and

$$B(T'') = \{\langle x, \mu_B(x,t), \nu_B(x,t)\rangle | \langle x,t\rangle \in E \times T''\},$$

where T' and T'' have finite number of distinct time-elements or they are time-intervals. Then we can define the above operations ($\cap$, $\cup$, etc.) and the topological (C and I) and modal ($\square$ and $\diamondsuit$) operators. For example,

$$A(T') \cup B(T'') = \{\langle x, \mu_{A(T')\cup B(T'')}(x,t), \nu_{A(T')\cup B(T'')}(x,t)\rangle | \langle x,t\rangle \in E \times T'\},$$

where

$$\langle x, \mu_{A(T')\cup B(T'')}(x,t), \nu_{A(T')\cup B(T'')}(x,t)\rangle$$

$$= \begin{cases} \langle x, \mu_A(x,t'), \nu_A(x,t')\rangle, & \text{if } t = t' \in T' - T'' \\[2mm] \langle x, \mu_B(x,t''), \nu_A(x,t'')\rangle, & \text{if } t = t'' \in T'' - T' \\[2mm] \langle x, \max(\mu_A(x,t'), \mu_B(x,t'')), \min(\nu_A(x,t'), \nu_A(x,t''))\rangle, \\ \qquad\qquad\qquad\qquad \text{if } t = t' = t'' \in T' \cap T'' \\[2mm] \langle x, 0, 1\rangle, & \text{otherwise} \end{cases}$$

$$(10.2)$$

The specific operators over TIFSs are:

$$C^*(A(T)) = \{\langle x, \sup_{t \in T} \mu_{A(T)}(x,t), \inf_{t \in T} \nu_{A(T)}(x,t)\rangle | x \in E\}, \qquad (10.3)$$

$$I^*(A(T)) = \{\langle x, \inf_{t \in T} \mu_{A(T)}(x,t), \sup_{t \in T} \nu_{A(T)}(x,t)\rangle | x \in E\}. \qquad (10.4)$$

We have the following important equalities for every TIFS $A(T)$:

(a) $C^*(C^*(A(T))) = C^*(A(T))$,

(b) $C^*(I^*(A(T))) = I^*(A(T))$,

(c) $I^*(C^*(A(T))) = C^*(A(T))$,

(d) $I^*(I^*(A(T))) = I^*(A(T))$,

(e) $C(C^*(A(T))) = C^*(C(A(T)))$,

(f) $I(I^*(A(T))) = I^*(I(A(T)))$,

(g) $\overline{\mathcal{C}^*(\overline{A(T)})} = \mathcal{I}^*(A(T))$,

(h) $\mathcal{C}^*(\square A(T)) = \square\mathcal{C}^*(A(T))$,

(i) $\mathcal{C}^*(\Diamond A(T)) = \Diamond\mathcal{C}^* = (A(T))$,

(j) $\mathcal{I}^*(\square A(T)) = \square\mathcal{I}^*(A(T))$,

(k) $\mathcal{I}^*(\Diamond A(T)) = \Diamond\mathcal{I}^* = (A(T))$.

For every two TIFSs $A(T')$ and $B(T'')$:

(a) $\mathcal{C}^*(A(T') \cap B(T'')) \subset \mathcal{C}^*(A(T')) \cap \mathcal{C}^*(B(T''))$,

(b) $\mathcal{C}^*(A(T') \cup B(T'')) = \mathcal{C}^*(A(T')) \cup \mathcal{C}^*(B(T''))$,

(c) $\mathcal{I}^*(A(T') \cap B(T'')) = \mathcal{I}^*(A(T')) \cap \mathcal{I}^*(B(T''))$,

(d) $\mathcal{I}^*(A(T') \cup B(T'')) \supset \mathcal{I}^*(A(T')) \cup \mathcal{I}^*(B(T''))$.

So far, the research on TIFSs has been oriented mainly to the properties of the elements of the universe E. Only operators $\mathcal{C}^*$ and $\mathcal{I}^*$ have been related to a fixed time-scale T. Now, we introduce modal type of operators, related to the time-components.

Consider the TIFS $A(T)$. For every $x \in E$, we can construct the set

$$T(A, x) = \{\langle t, \mu_A(x, t), \nu_A(x, t)\rangle | t \in T\}.$$

It is directly seen that $T(A, x)$ is an IFS, but now over universe T.

In [39, 47], the following example of the necessity of fuzzy sets, IFSs and TIFSs is given, that we use below. Imagine someone has just arrived for a trip in a foreign country. How shall we answer to his question about the weather today if the day had started with a beautiful sunrise, and some hours later it was already raining cats and dogs. The ordinary logic cannot help us. And we cannot answer *"yes"* only due to the sun in the morning, and we cannot answer *"no"* just for the rain. To provide a possible accurate answer, we must have stayed the whole day, a chronometer in hand, and measured that sun has been observed in $S\%$ of the daytime, in $Q\%$ it has been cloudy and in the rest $(100 - S - Q)\%$ it could have been seen through the clouds, not brightly shining. And whereas the fuzzy set theory gives us the means to determine that *"Yes, it has been sunny in $S\%$ of the day"*, the apparatus of the IFSs helps us for the more comprehensive estimation: *"It has been sunny in $S\%$ of the day and cloudy in $Q\%$"*. Therefore, the possibility for a more precise appraisal, the IFSs theory gives us, can be the means to satisfy our desire of plausibility and correctness. Just in the way we demand more exhaustive answers than simply *"yes"* or *"no"*, the propositional calculus provides us, and we find them in the fuzzy sets, by analogy, we use the intuitionistic fuzzy sets as a higher level to accuracy. However, no matter the evaluation technique we use, our appraisal will only render the results of the whole day, at the expense of the observations of a part of the day.

If we decide to use the example about the status of the Sun at the past day, the apparatus of the TIFS will give us possibility to trace the changes of the Sun status for the whole observed time-period (see, Fig. 10.1).

If the universe E is a set of different towns, A is the set of towns of a fixed country, then the ordinary IFS A can be interpreted as a set of the estimations of the degrees of sunshine and of the degree of cloudiness (e.g., for a fixed time-moment) for all towns of the country. Set $T(A, x)$ can be interpreted as the set of the above estimations, but now, only about the fixed town $x \in A$ and for the different time-moments from time-scale T. Finally, set $A(T)$ corresponds to the set of the estimations of the degrees of sunniness and of the degrees of cloudiness for all towns of the given country and for all time-moments from time-scale T.

Now, operators C^* and $\mathcal{I}^*$ defined over set $A(T)$ determine the highest and the lowest degrees of the above discussed estimations for each town $x \in A$, while they, defined over set $T(A, x)$, determine the highest and the lowest degrees of these estimations for the fixed town $x \in A$.

On the other hand, operators C and I, defined oves set $A(T)$, determine the highest and the lowest degrees of the same estimations but about all towns from A in general, i.e., for the fixed country.

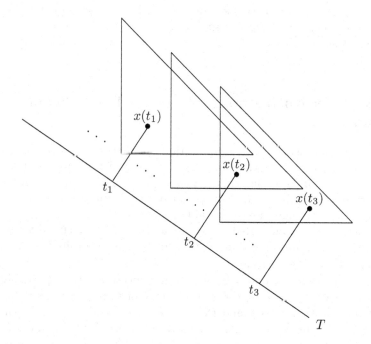

Fig. 10.1

We can also construct the set

$$T(A, E) = \{T(A, x) | x \in E\}$$

and it can be easily seen that for all couples $\langle x, t \rangle \in E \times T$ the μ- and ν-degrees in sets $T(A, E)$ and $A(T)$ coincide.

Now, we return to the extended modal operators. They can be used for obtaining more detailed estimations. For our example, this detailization can be expressed in the following three forms.

Let apart from us (as observers of the Sun), there are one or more other observers (let us call them "experts"). Let them be able to detailize the degree of uncertainty, that we show, because we cannot decide whether it is sunny or cloudy (the Sun shines through the cloud, but not clearly enough). The expert opinion can also be in IF-form. Let them determine the uncertainty for us time-period as sunny in $\alpha\%$ and as cloudy in $\beta\%$ (he also can be not totally sure about the Sun status). Then, we can use operator $F_{\alpha,\beta}$ and decrease our degree of uncertainty, increasing both other degrees.

Using the expert estimations we can change our estimations in either a pessimistic (by operators $H_{\alpha,\beta}$ or $H_{\alpha,\beta}^*$) or an optimistic (by operators $J_{\alpha,\beta}$ or $J_{\alpha,\beta}^*$) direction.

If we have more than one expert, we can calculate their estimations in different ways (see Section **1.6**), including possibility to use ordinary and temporal intuitionistic fuzzy graphs (see [38]).

This shows that TIFSs permit more detail estimations of real processes flowing in time.

10.3   Example: An IF-Interpretation of RGB-Colour System

Here we give intuitionistic fuzzy interpretation of "Rod-Green-Blue (RGB) colour system". Let us use the IF-geometrical interpretation in an equilateral triangle (see Section **3.3**). If the length of the triangle median is 1, then for each point that lies in this triangle the sum of the distances between the point and the three triangle sides is equal to 1.

Let the three basic colours "Red", "Green" and "Blue" correspond to the three vertices of the triangle (see Fig. 10.2). They have the coordinates $\langle 0, 0 \rangle$, $\langle \frac{2\sqrt{3}}{3}, 0 \rangle$, and $\langle \frac{\sqrt{3}}{3}, 1 \rangle$, respectively.

Let for a fixed point x, the lengths of the three distances are respectively r, g and b. Then we can interpret, e.g., the degree of membership as r, the degree of non-membership as g and the degree of uncertainty as b. Then, for example, a point in the triangle, that corresponds to "Violet", will have coordinates $\langle \frac{1}{4}, \frac{\sqrt{3}}{4} \rangle$. The point with coordinates $\langle \frac{1}{2}, \frac{\sqrt{3}}{6} \rangle$ will correspond to colour "Grey".

Therefore, all colours from the colour pallette that do not contain "Black" and "White" colours, can obtain IF-interpretation. Now, we discuss the case, when we like to add the colours "Black", "White" and all their derivatives. We see directly, that only colours "Black", "White" and all in the greyscale can be interpreted as points of a section. Now, when we like to add the rest of the colours, we can use the interpretation shown in Fig. 10.3. Now, the basic colours "Red", "Green", "Blue", "Black" and "White" have the coordinates $\langle 1,0,0\rangle$, $\langle 0,1,0\rangle$, $\langle 0,0,1\rangle$, $\langle 0,0,0\rangle$ and $\langle \frac{2}{3}, \frac{2}{3}, \frac{2}{3}\rangle$, respectively.

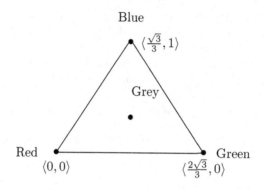

Fig. 10.2

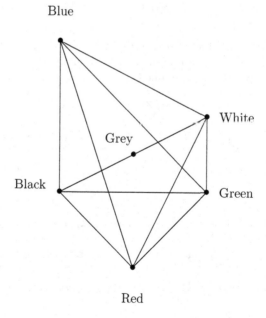

Fig. 10.3

Now, we see that the used set is not IFSs but a set from, TIFS type or an IF three-dimensional set. So, these sets can obtain real applications.

10.4   IF Multidimensional Sets

In four papers [104, 105, 106, 107] Eulalia Szmidt, Janusz Kacprzyk, Parvathi Rangasamy and the author introduced a new extension as of the IFSs, as well as of the TIFSs. In the new set, we change the temporal component t with an n-dimensional vector $\langle z_1, z_2, ..., z_n \rangle$. Let sets $Z_1, Z_2, ..., Z_n$ be fixed and let for each i $(1 \leq i \leq n) : z_i \in Z_i$.

Let the set E be fixed. The Intuitionistic Fuzzy Multi-Dimensional Set (IFMDS) A in $E, Z_1, Z_2, ..., Z_n$ is an object of the form

$$A(Z_1, Z_2, ..., Z_n) = \{\langle x, \mu_A(x, z_1, z_2, ..., z_n), \nu_A(x, z_1, z_2, ..., z_n)\rangle |$$

$$\langle x, z_1, z_2, ..., z_n \rangle \in E \times Z_1 \times Z_2 \times ... \times Z_n\}, \tag{10.5}$$

where:

(a) $\mu_A(x, z_1, z_2, ..., z_n) + \nu_A(x, z_1, z_2, ..., z_n) \leq 1$ for every
$\langle x, z_1, z_2, ..., z_n \rangle \in E \times Z_1 \times Z_2 \times ... \times Z_n$,

(b) $\mu_A(x, z_1, z_2, ..., z_n)$ and $\nu_A(x, z_1, z_2, ..., z_n)$ are the degrees of membership and non-membership, respectively, of the element $x \in E$ at the time-moment $t \in T$.

For every two IFSs A and B over sets $E, Z_1, Z_2, ..., Z_n$ all above relations, operations and operators can be defined by analogy.

Now, we extend the above defined operators $\mathcal{C}^*$ and $\mathcal{I}^*$ to the forms:

$$\mathcal{C}_i^*(A(Z_1, Z_2, ..., Z_n))$$

$$= \{\langle x, \sup_{t_i \in Z_i} \mu_{A(Z_1, Z_2, ..., Z_n)}(x, z_1, ..., z_{i-1}, t_i, z_{i+1}, ..., z_n),$$

$$\inf_{t_i \in Z_i} \nu_{A(Z_1, Z_2, ..., Z_n)}(x, z_1, ..., z_{i-1}, t_i, z_{i+1}, ..., z_n)\rangle | x \in E\}, \tag{10.6}$$

$$\mathcal{I}_i^*(A(Z_1, Z_2, ..., Z_n))$$

$$= \{\langle x, \inf_{t_i \in Z_i} \mu_{A(Z_1, Z_2, ..., Z_n)}(x, z_1, ..., z_{i-1}, t_i, z_{i+1}, ..., z_n),$$

$$\sup_{t_i \in Z_i} \nu_{A(Z_1, Z_2, ..., Z_n)}(x, z_1, ..., z_{i-1}, t_i, z_{i+1}, ..., z_n)\rangle | x \in E\}. \tag{10.7}$$

Theorem 10.1: For every IFMDS $A(Z_1, Z_2, ..., Z_n)$, and for every i $(1 \leq i \leq n)$, we have

(a) $\mathcal{C}_i^*(\mathcal{C}_i^*(A(Z_1, Z_2, ..., Z_n))) = \mathcal{C}_i^*(A(Z_1, Z_2, ..., Z_n))$,

(b) $\mathcal{C}_i^*(\mathcal{I}_i^*(A(Z_1, Z_2, ..., Z_n))) = \mathcal{I}_i^*(A(Z_1, Z_2, ..., Z_n))$,

(c) $\mathcal{I}_i^*(\mathcal{C}_i^*(A(Z_1, Z_2, ..., Z_n))) = \mathcal{C}_i^*(A(Z_1, Z_2, ..., Z_n))$,

(d) $\mathcal{I}_i^*(\mathcal{I}_i^*(A(Z_1, Z_2, ..., Z_n))) = \mathcal{I}_i^*(A(Z_1, Z_2, ..., Z_n))$,

(e) $\mathcal{C}(\mathcal{C}_i^*(A(Z_1, Z_2, ..., Z_n))) = \mathcal{C}_i^*(\mathcal{C}(A(Z_1, Z_2, ..., Z_n)))$,

(f) $\mathcal{I}(\mathcal{I}_i^*(A(Z_1, Z_2, ..., Z_n))) = \mathcal{I}_i^*(\mathcal{I}(A(Z_1, Z_2, ..., Z_n)))$,

(g) $\overline{\mathcal{C}_{\mathcal{I}}^*(\overline{A(Z_1, Z_2, ..., Z_n)})} = \mathcal{I}_i^*(A(Z_1, Z_2, ..., Z_n))$,

(h) $\overline{\mathcal{I}_{\mathcal{I}}^*(\overline{A(Z_1, Z_2, ..., Z_n)})} = \mathcal{C}_i^*(A(Z_1, Z_2, ..., Z_n))$,

(i) $\mathcal{C}_i^*(\square A(Z_1, Z_2, ..., Z_n)) = \square\mathcal{C}_i^*(A(Z_1, Z_2, ..., Z_n))$,

(j) $\mathcal{C}_i^*(\lozenge A(Z_1, Z_2, ..., Z_n)) = \lozenge\mathcal{C}_i^*(A(Z_1, Z_2, ..., Z_n))$,

(k) $\mathcal{I}_i^*(\square A(Z_1, Z_2, ..., Z_n)) = \square\mathcal{I}_i^*(A(Z_1, Z_2, ..., Z_n))$,

(l) $\mathcal{I}_i^*(\lozenge A(Z_1, Z_2, ..., Z_n)) = \lozenge\mathcal{I}_i^*(A(Z_1, Z_2, ..., Z_n))$.

Theorem 10.2: For every IFMDS $A(Z_1, Z_2, ..., Z_n)$, and for every two numbers i, j $(1 \le i, j \le n)$, we have

(a) $\mathcal{C}_i^*(\mathcal{C}_j^*(A(Z_1, Z_2, ..., Z_n))) = \mathcal{C}_j^*(\mathcal{C}_i^*(A(Z_1, Z_2, ..., Z_n)))$,

(b) $\mathcal{C}_i^*(\mathcal{I}_j^*(A(Z_1, Z_2, ..., Z_n))) = \mathcal{I}_j^*(\mathcal{I}_i^*(A(Z_1, Z_2, ..., Z_n)))$,

(c) $\mathcal{I}_i^*(\mathcal{C}_j^*(A(Z_1, Z_2, ..., Z_n))) = \mathcal{C}_j^*(\mathcal{C}_i^*(A(Z_1, Z_2, ..., Z_n)))$,

(d) $\mathcal{I}_i^*(\mathcal{I}_j^*(A(Z_1, Z_2, ..., Z_n))) = \mathcal{I}_j^*(\mathcal{I}_i^*(A(Z_1, Z_2, ..., Z_n)))$.

Theorem 10.3: For every two IFMDSs $A(Z_1, Z_2, ..., Z_n)$ and $B(Z_1, Z_2, ..., Z_n)$,

(a) $\mathcal{C}^*(A(Z_1, Z_2, ..., Z_n) \cap B(Z_1, Z_2, ..., Z_n)) \subset \mathcal{C}^*(A(Z_1, Z_2, ..., Z_n))$
$\cap \mathcal{C}^*(B(Z_1, Z_2, ..., Z_n))$,

(b) $\mathcal{C}^*(A(Z_1, Z_2, ..., Z_n) \cup B(Z_1, Z_2, ..., Z_n)) = \mathcal{C}^*(A(Z_1, Z_2, ..., Z_n))$
$\cup \mathcal{C}^*(B(Z_1, Z_2, ..., Z_n))$,

(c) $\mathcal{I}^*(A(Z_1, Z_2, ..., Z_n) \cap B(Z_1, Z_2, ..., Z_n)) = \mathcal{I}^*(A(Z_1, Z_2, ..., Z_n))$
$\cap \mathcal{I}^*(B(Z_1, Z_2, ..., Z_n))$,

(d) $\mathcal{I}^*(A(Z_1, Z_2, ..., Z_n) \cup B(Z_1, Z_2, ..., Z_n)) \supset \mathcal{I}^*(A(Z_1, Z_2, ..., Z_n))$
$\cup \mathcal{I}^*(B(Z_1, Z_2, ..., Z_n))$.

(e) $\mathcal{C}^*(A(Z_1, Z_2, ..., Z_n)@B(Z_1, Z_2, ..., Z_n)) \subset \mathcal{C}^*(A(Z_1, Z_2, ..., Z_n))$
$@\mathcal{C}^*(B(Z_1, Z_2, ..., Z_n))$,

(f) $\mathcal{I}^*(A(Z_1, Z_2, ..., Z_n)@B(Z_1, Z_2, ..., Z_n)) \supset \mathcal{I}^*(A(Z_1, Z_2, ..., Z_n))$
$@\mathcal{I}^*(B(Z_1, Z_2, ..., Z_n))$.

In Section **4.5**, some convex hull algorithms are discussed. Let, for a given IFS A, the region obtained by some of these algorithms be marked by $R(A)$. Therefore, $R(A)$ is the minimal convex region containing IFS A.

First, we introduce four operators acting in the IF-interpretation triangle. Let the point $\langle \mu, \nu \rangle$ be given. Then,

$$\gamma_{L,\nu}(\mu) = \mu',$$

$$\gamma_{R,\nu}(\mu) = \mu'',$$

$$\gamma_{T,\mu}(\nu) = \nu',$$

$$\gamma_{B,\mu}(\nu) = \nu'',$$

where μ' and μ'' are the left and the right intersection points of $R(A)$ and the line

$$l : y = \nu;$$

ν' and ν'' are the upper and the down intersection points of $R(A)$ and the line

$$l : x = \mu.$$

For an ordinary IFS A, the new operators have the forms:

$$\Gamma_L(A) = \{\langle x, \gamma_{L,\nu}(\mu), \nu \rangle | x \in E\},$$

$$\Gamma_R(A) = \{\langle x, \gamma_{R,\nu}(\mu), \nu \rangle | x \in E\},$$

$$\Gamma_T(A) = \{\langle x, \mu, \gamma_{T,\mu}(\nu) \rangle | x \in E\},$$

$$\Gamma_B(A) = \{\langle x, \mu, \gamma_{B,\mu}(\nu) \rangle | x \in E\}.$$

Now, for the fixed set A, using operators C_i^* and I_i^* sequentially for each i ($1 \le i \le n$), we construct the following points of the interpretation triangle:

$$\langle \mu_i^C, \nu_i^C \rangle = \langle \sup_{t_i \in Z_i} \mu_{A(Z_1, Z_2, ..., Z_n)}(x, z_1, ..., z_{i-1}, t_i, z_{i+1}, ..., z_n),$$

$$\inf_{t_i \in Z_i} \nu_{A(Z_1, Z_2, ..., Z_n)}(x, z_1, ..., z_{i-1}, t_i, z_{i+1}, ..., z_n) \rangle$$

$$\langle \mu_i^I, \nu_i^I \rangle = \langle \inf_{t_i \in Z_i} \mu_{A(Z_1, Z_2, ..., Z_n)}(x, z_1, ..., z_{i-1}, t_i, z_{i+1}, ..., z_n),$$

$$\sup_{t_i \in Z_i} \nu_{A(Z_1, Z_2, ..., Z_n)}(x, z_1, ..., z_{i-1}, t_i, z_{i+1}, ..., z_n) \rangle$$

Therefore, the sets

$$\overline{\mathcal{C}}(A) = \{\langle i^C, \mu_i^C, \nu_i^C \rangle | 1 \le i \le n\}$$

and

$$\overline{\mathcal{I}}(A) = \{\langle i^I, \mu_i^I, \nu_i^I \rangle | 1 \le i \le n\}$$

are ordinary IFSs.

We can construct the sets

$$\overline{C}_L(A) = \Gamma_L(\overline{C}(A)),$$

$$\overline{C}_R(A) = \Gamma_R(\overline{C}(A)),$$

$$\overline{C}_U(A) = \Gamma_T(\overline{C}(A)),$$

$$\overline{C}_D(A) = \Gamma_B(\overline{C}(A)),$$

$$\overline{I}_L(A) = \Gamma_L(\overline{I}(A)),$$

$$\overline{I}_R(A) = \Gamma_R(\overline{I}(A)),$$

$$\overline{I}_U(A) = \Gamma_T(\overline{I}(A)),$$

$$\overline{I}_D(A) = \Gamma_B(\overline{I}(A)).$$

Theorem 10.4: For each IFMDS A,

$$W(\overline{C}(A)@\overline{I}(A)) = W(\overline{C}(A)@W(\overline{I}(A)).$$

Proof: Let IFMDS A be given. Then we obtain sequentially

$$W(\overline{C}(A)@\overline{I}(A))$$

$$= W(\{\langle i, \frac{\mu_i^C + \mu_i^I}{2}, \frac{\nu_i^C + \nu_i^I}{2}\rangle | 1 \le i \le n\})$$

$$= \{\langle i, \frac{\sum\limits_{j=1}^{n}(\mu_i^C + \mu_i^I)}{2n}, \frac{\sum\limits_{j=1}^{n}(\nu_i^C + \nu_i^I)}{2n}\rangle | 1 \le i \le n\}$$

$$= \{\langle i, \frac{\sum\limits_{j=1}^{n}\mu_j^C + \sum\limits_{j=1}^{n}\mu_j^I}{2n}, \frac{\sum\limits_{j=1}^{n}\nu_j^C + \sum\limits_{j=1}^{n}\nu_j^I}{2n}\rangle | 1 \le i \le n\}$$

$$= \{\langle i, \frac{\sum\limits_{j=1}^{n}\mu_j^C}{n}, \frac{\sum\limits_{j=1}^{n}\nu_j^C}{n}\rangle | 1 \le i \le n\}$$

$$@\{\langle i, \frac{\sum\limits_{j=1}^{n}\mu_j^I}{n}, \frac{\sum\limits_{j=1}^{n}\nu_j^I}{n}\rangle | 1 \le i \le n\}$$

$$= W(\overline{C}(A)@W(\overline{I}(A)).$$

Theorem 10.5: For any IFMDS A:

$$(a) \quad \left. \begin{array}{c} \Gamma_L(\overline{C}(A)) \\ \Gamma_T(\overline{C}(A)) \end{array} \right\} \subset \overline{C}(A) \subset \left\{ \begin{array}{c} \Gamma_R(\overline{C}(A)) \\ \Gamma_B(\overline{C}(A)) \end{array} \right. ,$$

$$(b) \quad \left. \begin{array}{c} \Gamma_L(\overline{I}(A)) \\ \Gamma_T(\overline{I}(A)) \end{array} \right\} \subset \overline{I}(A) \subset \left\{ \begin{array}{c} \Gamma_R(\overline{I}(A)) \\ \Gamma_B(\overline{I}(A)) \end{array} \right. .$$

The proofs of these assertions are analogous as above.

Let the IFMDS A be defined over $E \times Z_1' \times Z_2' \times ... \times Z_n'$ and IFMDS B be defined over $E \times Z_1'' \times Z_2'' \times ... \times Z_n''$. The operations $\cap, \cup$ and @ have the forms:

$$A(Z_1', Z_2', ..., Z_n') \cap B(Z_1'', Z_2'', ..., Z_n'')$$

$$= \{ \langle x, \mu_{A(Z_1', Z_2', ..., Z_n') \cap B(Z_1'', Z_2'', ..., Z_n'')}(x, z_1, z_2, ..., z_n),$$

$$\nu_{A(Z_1', Z_2', ..., Z_n') \cap B(Z_1'', Z_2'', ..., Z_n'')}(x, z_1, z_2, ..., z_n) \rangle$$

$$| \langle x, z_1, z_2, ..., z_n \rangle \in E \times Z_1 \times Z_2 \times ... \times Z_n \},$$

where

$$\langle x, \mu_{A(Z_1', Z_2', ..., Z_n') \cap B(Z_1'', Z_2'', ..., Z_n'')}(x, z_1, z_2, ..., z_n),$$

$$\nu_{A(Z_1', Z_2', ..., Z_n') \cap B(Z_1'', Z_2'', ..., Z_n'')}(x, z_1, z_2, ..., z_n) \rangle$$

$$= \begin{cases} \langle x, \min(\mu_{A(Z_1', Z_2', ..., Z_n')}(x, z_1, z_2, ..., z_n), \mu_{B(Z_1'', Z_2'', ..., Z_n'')}(x, z_1, z_2, ..., z_n)), \\ \\ \quad \max(\nu_{A(Z_1', Z_2', ..., Z_n')}(x, z_1, z_2, ..., z_n), \nu_{B(Z_1'', Z_2'', ..., Z_n'')}(x, z_1, z_2, ..., z_n)) \rangle, \\ \\ \quad \text{if } (\forall i : 1 \leq i \leq n)(z_i = z_i' = z_i'' \in Z_i = Z_i' \cap Z_i'') \\ \\ \langle x, 0, 1 \rangle, \quad \text{otherwise} \end{cases}$$

$$A(Z_1', Z_2', ..., Z_n') \cup B(Z_1'', Z_2'', ..., Z_n'')$$

$$= \{ \langle x, \mu_{A(Z_1', Z_2', ..., Z_n') \cup B(Z_1'', Z_2'', ..., Z_n'')}(x, z_1, z_2, ..., z_n),$$

$$\nu_{A(Z_1', Z_2', ..., Z_n') \cup B(Z_1'', Z_2'', ..., Z_n'')}(x, z_1, z_2, ..., z_n) \rangle$$

$$| \langle x, z_1, z_2, ..., z_n \rangle \in E \times Z_1 \times Z_2 \times ... \times Z_n \},$$

where

$$\langle x, \mu_{A(Z_1', Z_2', ..., Z_n') \cup B(Z_1'', Z_2'', ..., Z_n'')}(x, z_1, z_2, ..., z_n),$$

$$\nu_{A(Z_1', Z_2', ..., Z_n') \cup B(Z_1'', Z_2'', ..., Z_n'')}(x, z_1, z_2, ..., z_n) \rangle$$

$$= \begin{cases} \langle x, \mu_{A(Z_1',Z_2',...,Z_n')}(x,z_1,z_2,...,z_n), \nu_{A(Z_1',Z_2',...,Z_n')}(x,z_1,z_2,...,z_n)\rangle, \\ \quad \text{if } (\forall i:1\le i\le n)(z_i=z_i'\in Z_i=Z_i') \\ \qquad \&(\exists i:1\le i\le n)(z_i=z_i'\in Z_i'-Z_i'') \\[2mm] \langle x, \mu_{B(Z_1'',Z_2'',...,Z_n'')}(x,z_1,z_2,...,z_n), \nu_{B(Z_1'',Z_2'',...,Z_n'')}(x,z_1,z_2,...,z_n)\rangle, \\ \quad \text{if } (\forall i:1\le i\le n)(z_i=z_i''\in Z_i=Z_i'') \\ \qquad \&(\exists i:1\le i\le n)(z_i=z_i''\in Z_i''-Z_i') \\[2mm] \langle x, \max(\mu_{A(Z_1',Z_2',...,Z_n')}(x,z_1,z_2,...,z_n), \mu_{B(Z_1'',Z_2'',...,Z_n'')}(x,z_1,z_2,...,z_n)), \\ \quad \min(\nu_{A(Z_1',Z_2',...,Z_n')}(x,z_1,z_2,...,z_n), \nu_{B(Z_1'',Z_2'',...,Z_n'')}(x,z_1,z_2,...,z_n))\rangle, \\ \quad \text{if } (\forall i:1\le i\le n)(z_i=z_i'=z_i''\in Z_i=Z_i'\cap Z_i'') \\[2mm] \langle x,0,1\rangle, \quad \text{otherwise} \end{cases}$$

$$A(Z_1',Z_2',...,Z_n')@B(Z_1'',Z_2'',...,Z_n'')$$
$$= \{\langle x, \mu_{A(Z_1',Z_2',...,Z_n')@B(Z_1'',Z_2'',...,Z_n'')}(x,z_1,z_2,...,z_n),$$
$$\nu_{A(Z_1',Z_2',...,Z_n')@B(Z_1'',Z_2'',...,Z_n'')}(x,z_1,z_2,...,z_n)\rangle$$
$$|\langle x,z_1,z_2,...,z_n\rangle \in E\times Z_1\times Z_2\times...\times Z_n\},$$

where

$$\langle x, \mu_{A(Z_1',Z_2',...,Z_n')@B(Z_1'',Z_2'',...,Z_n'')}(x,z_1,z_2,...,z_n),$$
$$\nu_{A(Z_1',Z_2',...,Z_n')@B(Z_1'',Z_2'',...,Z_n'')}(x,z_1,z_2,...,z_n)\rangle$$

$$= \begin{cases} \langle x, \dfrac{\mu_{A(Z_1',Z_2',...,Z_n')}(x,z_1,z_2,...,z_n)}{2}, \dfrac{\nu_{A(Z_1',Z_2',...,Z_n')}(x,z_1,z_2,...,z_n)+1}{2}\rangle, \\ \quad \text{if } (\forall i:1\le i\le n)(z_i=z_i'\in Z_i=Z_i') \\ \qquad \&(\exists i:1\le i\le n)(z_i=z_i'\in Z_i'-Z_i'') \\[2mm] \langle x, \dfrac{\mu_{B(Z_1'',Z_2'',...,Z_n'')}(x,z_1,z_2,...,z_n)}{2}, \dfrac{\nu_{B(Z_1'',Z_2'',...,Z_n'')}(x,z_1,z_2,...,z_n)+1}{2}\rangle, \\ \quad \text{if } (\forall i:1\le i\le n)(z_i=z_i''\in Z_i=Z_i'') \\ \qquad \&(\exists i:1\le i\le n)(z_i=z_i''\in Z_i''-Z_i') \\[2mm] \langle x, \dfrac{\mu_{A(Z_1',Z_2',...,Z_n')}(x,z_1,z_2,...,z_n)+\mu_{B(Z_1'',Z_2'',...,Z_n'')}(x,z_1,z_2,...,z_n)}{2}, \\ \quad \dfrac{\nu_{A(Z_1',Z_2',...,Z_n')}(x,z_1,z_2,...,z_n)+\nu_{B(Z_1'',Z_2'',...,Z_n'')}(x,z_1,z_2,...,z_n)}{2}\rangle, \\ \quad \text{if } (\forall i:1\le i\le n)(z_i=z_i'=z_i''\in Z_i=Z_i'\cap Z_i'') \\[2mm] \langle x,0,1\rangle, \quad \text{otherwise} \end{cases}$$

Let E^* be defined by (2.14). For the case of IFMDSs over $E \times Z_1 \times Z_2 \times \ldots \times Z_n$:

$$\mathcal{P}(E^*) = \{A(Z_1, Z_2, ..., Z_n) \mid A(Z_1, Z_2, ..., Z_n)$$

$$= \{\langle x, \mu_{A(Z_1, Z_2, ..., Z_n)}(x, z_1, z_2, ..., z_n), \nu_{A(Z_1, Z_2, ..., Z_n)}(x, z_1, z_2, ..., z_n)\rangle$$

$$\mid \langle x, z_1, z_2, ..., z_n \rangle \in E \times Z_1 \times Z_2 \times ... \times Z_n\}\}.$$

Theorem 10.6. For a fixed universe E,

(a) $\langle \mathcal{P}(E^*), \cap, E^* \rangle$ is a commutative monoid;
(b) $\langle \mathcal{P}(E^*), \cup, O^* \rangle$ is a commutative monoid;
(c) $\langle \mathcal{P}(E^*), @ \rangle$ is a groupoid;
(d) None of these objects is a (commutative) group.

11

Concluding Remarks

11.1 The Intuitionistic Fuzzy Sets as Constructive Objects

1. Let me start with a recollection of my first acquaintance with Prof. Lotfi Zadeh. It was in 2001 in Villa Real, Portugal, where Prof. Pedro Melo-Pinto organized a school on fuzzy sets. Prof. Zadeh was invited for a 3-hour lecture, which he concluded with presentation of slides with articles by Samuel Kleene, Kurt Gödel and other luminaries of mathematical logic, who have written against the fuzzy sets. The fact that the sublime mathematician and logician Gödel had sometimes made slips in his judgments can be confirmed by the cosmologists, yet I was astonished by his opinion. Of course, nowadays, when we are aware of the enormous number of publications in the field of fuzzy sets, as well as of the various impressive applications of these, it is easy to say that Gödel had mistaken. However, I have been long tormented by the question why these mathematicians had opposed the fuzzy sets while they did not have anything against the three- and multi-valued logics of Jan Lukasiewicz. Thus I reached the conclusion that the reason for the then negative attitude towards fuzzy sets was hidden in the presence of the $[0,1]$ interval as the set of the fuzzy sets' membership function (see, e.g, [301, 592, 593]). Indeed, the values of the membership function do belong to the $[0,1]$ interval, yet it does not mean that this function obtains all possible values in this interval! If an expert or a group of experts evaluates, for instance, the chances of a political party to win the elections, it is slightly ever probable (if not absurd) for them to use estimations like $\frac{1}{e}$ or $\sqrt{2}-1$. For any unbiased man it is clear that the experts would not use anything more complex than decimal fractions with one or two digits after the decimal point, i.e. rational numbers. Rational and even integer numbers are those which we use to measure the sizes of objects, the daily temperature or the speed of the vehicles, which are often described by fuzzy sets. Yes, the contemporary mathematics is the mathematics of multiple integrals, topological spaces, arithmetic functions, yet all these objects are abstractions of objects, existing in reality, which in

K.T. Atanassov: On Intuitionistic Fuzzy Sets Theory, STUDFUZZ 283, pp. 273–283.
springerlink.com   © Springer-Verlag Berlin Heidelberg 2012

the end of the day are measured, i.e. certain mathematical estimations are constructed for them, hence these estimations are constructive! Fuzzy sets, which use far from trivial mathematical apparatus, in general are based on constructive objects. It is well known that these sets contain ordered pairs (in the case of intuitionistic fuzzy sets, ordered triplets; see [39]), whose first component is a constructive object, hence following Per Matrin-Löf [368], it constitutes a finite configuration of symbols. As we have seen, the following (and eventually the third) component is also a constructive object.

It follows that the fuzzy and intuitionistic fuzzy sets are constructive objects. Therefore, anyone who had been in trouble that the fuzzy sets introduced deconstructivism in science, had no occasion for fear. However, were they right to prefer the multivalued logics? Let us fix a certain natural number, say 11, and let us use an 11-valued scale for estimation of the statement that the political party P would win 30% of the election votes. According to the 11-valued scale, this estimation will be given the form $\frac{3}{10}$, which is perfectly okay. But what shall we do, if we have chosen a 6-valued scale, without knowing in advance that the estimation would be 30%? Now, for our estimation we have to choose between the values $\frac{1}{5}$ and $\frac{2}{5}$ and in both cases we will allow mistake of $\frac{1}{10}$ which hardly makes us happy! In other words, the multivalued logics that use rational numbers, despite being obviously constructive objects, are not always appropriate.

On the other hand, it is clear that if we use an n-valued estimation scale, we may easily say that we work with a fuzzy set, because that scale would be a subset of the $[0, 1]$ interval.

2.   Let the universe E be a (recursively) enumerable set, i.e., its members are no more than the set of natural numbers, whose cardinality is marked by $\aleph_0$.

Let
$$A = \{\langle x, \mu_A(x), \nu_A(x)\rangle | x \in E\}$$
be an IFS and let the values of its functions μ_A and ν_A be rational numbers, i.e. for each $x \in E$:
$$\mu_A(x) = \frac{p(x)}{q(x)}$$
and
$$\nu_A(x) = \frac{r(x)}{s(x)},$$
where $p(x), q(x), r(x), s(x)$ are natural numbers, $p(x), r(x) \geq 0, q(x), s(x) > 0$ and
$$0 \leq \frac{p(x)}{q(x)} + \frac{r(x)}{s(x)} \leq 1.$$

We will juxtapose to A the (natural) number

$$range(A) = LCM(\{q(x), s(x)|x \in E\}) = \underset{x \in E}{LCM}\left(LCM(q(x), s(x))\right), \quad (11.1)$$

where LCM is the lowest common multiple of all natural numbers $q(x), s(x)$ for all $x \in E$.

We call the number $range(A)$ a "range" of IFS A.

Now it is clear that the IFS A will be a constructive object (in the present case – Constructive IFS (CIFS)) if

$$range(A) < \infty$$

while, in the opposite case it will not be a constructive object, because some of its parameters will not be rational numbers.

If for every $x \in E$: $\mu_A(x), \nu_A(x) \in \{0,1\}$, then

$$range(A) = 1.$$

We immediately see that if A and B are two CIFSs, then the IFSs defined by (2.4) – (2.8), (2.19) – (2.23), also are CIFSs. For them, the following equalities are valid:

$$range(A \cup B) = range(A \cap B) = range(A|B) = LCM(range(A), range(B)).$$

$$range(A + B) = range(A.B) = range(A).range(B),$$

$$range(A@B) = LCM(range(A), range(B), 2),$$

$$range(A^n) = range(n.A) = (range(A))^n,$$

$$range(A - B)$$

$$= LCM(range(A), range(B), \underset{x \in E}{LCM}(r_B(x)), \underset{x \in E}{LCM}(q_B(x) - p_B(x))),$$

$$range(A : B)$$

$$= LCM(range(A), range(B), \underset{x \in E}{LCM}(p_B(x)), \underset{x \in E}{LCM}(s_B(x) - r_B(x))),$$

where for each $x \in E$,

$$\mu_B(x) = \frac{p_B(x)}{q_B(x)}$$

and

$$\nu_B(x) = \frac{r_B(x)}{s_B(x)}.$$

For the CIFS A, the following equalities

$$range(\sqcup A) = range(\Diamond A) = range(\mathcal{C}(A)) = range(\mathcal{I}(A))$$

$$range(\mathcal{C}_\mu(A)) = range(\mathcal{C}_\nu(A)) = range(\mathcal{I}_\mu(A)) = range(\mathcal{I}_\nu(A))$$

$$range(\mathcal{C}_\mu^*(A)) = range(\mathcal{I}_\nu^*(A)) = range(A),$$

hold, where operators $\square, \Diamond, \mathcal{C}, \mathcal{I}, \mathcal{C}_\mu, \mathcal{C}_\nu, \mathcal{I}_\mu, \mathcal{I}_\nu, \mathcal{C}_\mu^*, \mathcal{I}_\nu^*$, are defined by (4.1), (4.2), (4.7), (4.10), (4.13) – (4.16), (4.18), (4.19), respectively and

$$range(W(A)) = LCM(card(E), range(A))$$

for weight-center operator, defined by (4.21).

Let $\alpha = \frac{a_1}{a_2}$ and $\beta = \frac{b_1}{b_2}$ be rational numbers for which $a_1, b_1 \geq 0$ and $a_2, b_2 > 0$. Then the IFSs $D_\alpha(A), F_{\alpha,\beta}(A)$(where $\alpha + \beta \leq 1$)$, G_{\alpha,\beta}(A)$, $H_{\alpha,\beta}(A)$, $H^*_{\alpha,\beta}(A)$, $J_{\alpha,\beta}(A)$, $J^*_{\alpha,\beta}(A)$, $d_\alpha(A)$, $f_{\alpha,\beta}(A)$ (where $\alpha + \beta \leq 1$)$, g_{\alpha,\beta}(A), h_{\alpha,\beta}(A), h^*_{\alpha,\beta}(A), j_{\alpha,\beta}(A), j^*_{\alpha,\beta}(A), P_{\alpha,\beta}(A), Q_{\alpha,\beta}(A)$, defined by (5.1) – (5.7), (5.11) – (5.17), and (6.9), (6.10), respectively, are CIFSs and for them

$$range(D_\alpha(A)) = range(d_\alpha(A)) = LCM(a_2, range(A)),$$

$$range(F_{\alpha,\beta}(A)) = range(G_{\alpha,\beta}(A)) = range(H_{\alpha,\beta}(A)) = range(H^*_{\alpha,\beta}(A))$$

$$= range(J_{\alpha,\beta}(A)) = range(J^*_{\alpha,\beta}(A)) = range(f_{\alpha,\beta}(A)) = range(g_{\alpha,\beta}(A))$$

$$= range(h_{\alpha,\beta}(A)) = range(h^*_{\alpha,\beta}(A)) = range(j_{\alpha,\beta}(A)) = range(j^*_{\alpha,\beta}(A))$$

$$= range(P_{\alpha,\beta}(A)) = range(Q_{\alpha,\beta}(A)) = LCM(a_2, b_2, range(A)).$$

Let it be valid for $a, b, c, d, e, f \in [0, 1]$ and

$$a + e - e.f \leq 1,$$

$$b + d - b.c \leq 1$$

that $a = \frac{a_1}{a_2}$, $b = \frac{b_1}{b_2}$, $c = \frac{c_1}{c_2}$, $d = \frac{d_1}{d_2}$, $e = \frac{e_1}{e_2}$, $f = \frac{f_1}{f_2}$. Let a, b, c, d, e, f be rational numbers for which $a_1, b_1, c_1, d_1, e_1, f_1 \geq 0$ and $a_2, b_2, c_2, d_2, e_2, f_2 > 0$. Then, the IFSs, defined by $X_{a,b,c,d,e,f}(A)$ and $x_{a,b,c,d,e,f}(A)$, defined by (5.8) and (5.19) are CIFSs and for them

$$range(X_{a,b,c,d,e,f}(A)) = range(x_{a,b,c,d,e,f}(A))$$

$$= LCM(a_2, b_2, c_2, d_2, e_2, f_2, range(A)).$$

When A and B are CIFSs, then the IFSs $F_B(A), G_B(A), H_B(A), H^*_B(A)$, $J_B(A), J^*_B(A), f_B(A), g_B(A), h_B(A), h^*_B(A), j_B(A), j^*_B(A), P_B(A), Q_B(A)$, defined by (5.20) –(5.31) and (6.11), (6.12), are also CIFSs for which

$$range(F_B(A)) = range(G_B(A)) = range(H_B(A)) = range(H^*_B(A))$$

$$= range(J_B(A)) = range(J^*_B(A)) = range(f_B(A)) = range(g_B(A))$$

$$= range(h_B(A)) = range(h^*_B(A)) = range(j_B(A)) = range(j^*_B(A))$$

$$= range(P_B(A)) = range(Q_B(A)) = LCM(range(B), range(A)).$$

Let $\alpha, \beta, \gamma, \delta, \varepsilon, \zeta$ be rational numbers for which $\alpha = \frac{a_1}{a_2}$, $\beta = \frac{b_1}{b_2}$, $\gamma = \frac{c_1}{c_2}$, $\delta = \frac{d_1}{d_2}$, $\varepsilon = \frac{e_1}{e_2}$, $\zeta = \frac{f_1}{f_2}$, $a_1, b_1, c_1, d_1, e_1, f_1 \geq 0$ and $a_2, b_2, c_2, d_2, e_2, f_2 > 0$. Then, for operators defined by (6.13) – (6.23) the following equalities hold:

$$range(\boxplus A) = range(\boxtimes A) = LCM(range(A), 2),$$

$$range(\boxplus_\alpha A) = range(\boxtimes_\alpha A) = LCM(range(A), a_2),$$

$$range(\boxplus_{\alpha,\beta} A) = range(\boxtimes_{\alpha,\beta} A) = range(E_{\alpha,\beta} A)$$
$$= LCM(range(A), a_2, b_2),$$

where $\alpha + \beta \le 1$,

$$range(\boxplus_{\alpha,\beta,\gamma} A) = range(\boxtimes_{\alpha,\beta,\gamma} A) = LCM(range(A), a_2, b_2, c_2),$$

where $\max(\alpha, \beta) + \gamma \le 1$,

$$range(\bullet_{\alpha,\beta,\gamma,\delta} A) = LCM(range(A), a_2, b_2, c_2, d_2),$$

where $\max(\alpha, \beta) + \gamma + \delta \le 1$,

$$range(\bigcirc_{\alpha,\beta,\gamma,\delta,\varepsilon,\zeta} A) = LCM(range(A), a_2, b_2, c_2, d_2, e_2, f_2),$$

where $0 \le \min(\alpha - \zeta, \beta - \varepsilon) + \gamma + \delta \le \max(\alpha - \zeta, \beta - \varepsilon) + \gamma + \delta \le 1$.
For the operations "Cartesian product", defined by (8.1) – (8.6) we obtain

$$range(A \times_1 B) = range(A \times_2 B) = range(A \times_3 B) = range(A).range(B),$$

$$range(A \times_4 B) = range(A \times_5 B) = range(A \times_6 B)$$
$$= LCM(range(A), range(B)).$$

For the CIFS A, the IFSs $\neg_1 A, \neg_2 A, ..., \neg_{34} A$, defined in Section **9.2** are CIFSs for which

$$range(\neg_2 A) = range(\neg_5 A) = range(\neg_6 A) = range(\neg_7 A) = range(\neg_{13} A)$$
$$= range(\neg_{14} A) = range(\neg_{15} A) = range(\neg_{16} A) = range(\neg_{17} A) = 1,$$

$$range(\neg_1 A) = range(\neg_4 A) = range(\neg_7 A) = range(\neg_8 A) = range(\neg_9 A)$$
$$= range(\neg_{10} A) = range(\neg_{11} A) = range(\neg_{18} A) = range(\neg_{19} A)$$
$$= range(\neg_{20} A) = range(\neg_{21} A) = range(\neg_{22} A) = range(\neg_{23} A)$$
$$= range(\neg_{24} A) = range(\neg_{25} A) = range(A),$$

$$range(\neg_3 A) = range(\neg_{26} A) = range(\neg_{27} A) = range(\neg_{28} A) = range(A)^2,$$

$$range(\neg_{12} A) = range(\neg_{29} A) = range(\neg_{30} A) = range(\neg_{31} A)$$
$$= range(\neg_{32} A) = range(\neg_{33} A) = range(\neg_{34} A) = range(A)^3.$$

When A and B are CIFSs, then the IFSs $A \to_1 B, A \to_2 B, ..., A \to_{138} B$, defined in Section **9.2**, are CIFSs for which

$$range(A \to_{15} B) = range(A \to_{20} B) = range(A \to_{23} B) = range(A \to_{24} B)$$

$$= range(A \to_{40} B) = range(A \to_{42} B) = range(A \to_{55} B)$$

$$= range(A \to_{57} B) = range(A \to_{74} B) = range(A \to_{79} B)$$

$$= range(A \to_{88} B) = range(A \to_{97} B) = 1;$$

$$range(A \to_{29} B) = range(A \to_{45} B) = range(A \to_{60} B) = range(A \to_{62} B)$$

$$= range(A \to_{63} B) = range(A \to_{68} B) = range(A \to_{69} B)$$

$$= range(A \to_{70} B) = range(A \to_{73} B) = range(A \to_{78} B)$$

$$= range(A \to_{80} B) = range(A \to_{83} B) = range(A \to_{84} B)$$

$$= range(A \to_{87} B) = range(A \to_{89} B) = range(A \to_{92} B)$$

$$= range(A \to_{93} B) = range(A \to_{96} B) = range(A \to_{98} B) = range(A);$$

$$range(A \to_2 B) = range(A \to_3 B) = range(A \to_{11} B) = range(A \to_{14} B)$$

$$= range(A \to_{16} B) = range(A \to_{19} B) = range(A \to_{26} B)$$

$$= range(A \to_{27} B) = range(A \to_{31} B) = range(A \to_{32} B)$$

$$= range(A \to_{41} B) = range(A \to_{43} B) = range(A \to_{47} B)$$

$$= range(A \to_{48} B) = range(A \to_{54} B) = range(A \to_{56} B)$$

$$= range(A \to_{77} B) = range(A \to_{81} B) = range(A \to_{90} B)$$

$$= range(A \to_{99} B) = range(B);$$

$$range(A \to_1 B) = range(A \to_4 B) = range(A \to_5 B) = range(A \to_8 B)$$

$$= range(A \to_{10} B) = range(A \to_{12} B) = range(A \to_{18} B)$$

$$= range(A \to_{22} B) = range(A \to_{25} B) = range(A \to_{28} B)$$

$$= range(A \to_{30} B) = range(A \to_{33} B) = range(A \to_{34} B)$$

$$= range(A \to_{35} B) = range(A \to_{36} B) = range(A \to_{37} B)$$

$$= range(A \to_{39} B) = range(A \to_{44} B) = range(A \to_{46} B)$$

$$= range(A \to_{49} B) = range(A \to_{51} B) = range(A \to_{52} B)$$

$$= range(A \to_{58} B) = range(A \to_{59} B) = range(A \to_{61} B)$$

$$= range(A \to_{65} B) = range(A \to_{67} B) = range(A \to_{72} B)$$

$$= range(A \rightarrow_{76} B) = range(A \rightarrow_{82} B) = range(A \rightarrow_{86} B)$$
$$= range(A \rightarrow_{91} B) = range(A \rightarrow_{95} B) = range(A \rightarrow_{100} B)$$
$$= range(A \rightarrow_{101} B) = range(A \rightarrow_{102} B) = range(A \rightarrow_{103} B)$$
$$= range(A \rightarrow_{104} B) = range(A \rightarrow_{105} B) = range(A \rightarrow_{106} B)$$
$$= range(A \rightarrow_{107} B) = range(A \rightarrow_{108} B) = range(A \rightarrow_{109} B)$$
$$= range(A \rightarrow_{110} B) = LCM(range(A), range(B)).$$

$$range(A \rightarrow_{6} B) = range(A \rightarrow_{13} B) = range(A \rightarrow_{50} B)$$
$$= range(A \rightarrow_{64} B) = range(A \rightarrow_{112} B) = range(A).range(B);$$
$$range(A \rightarrow_{66} B) = range(A \rightarrow_{85} B) = range(A).range(B)^2;$$

$$range(A \rightarrow_{9} B) = range(A \rightarrow_{17} B) = range(A \rightarrow_{38} B)$$
$$= range(A \rightarrow_{53} B) = range(A)^2.range(B);$$

$$range(A \rightarrow_{71} B) = range(A \rightarrow_{94} B) = range(A \rightarrow_{119} B)$$
$$= range(A \rightarrow_{124} B) = range(A \rightarrow_{125} B) = range(A \rightarrow_{127} B)$$
$$= range(A \rightarrow_{129} B) = range(A \rightarrow_{130} B) = range(A \rightarrow_{132} B)$$
$$= range(A \rightarrow_{134} B) = range(A \rightarrow_{135} B) = range(A \rightarrow_{137} B)$$
$$= LCM(range(A), range(B)^2);$$

$$range(A \rightarrow_{114} B) = range(A \rightarrow_{115} B) = range(A \rightarrow_{117} B)$$
$$= range(A \rightarrow_{120} B) = range(A \rightarrow_{122} B) = LCM(range(A)^2, range(B));$$

$$range(A \rightarrow_{21} B) = range(A \rightarrow_{111} B) = range(A \rightarrow_{113} B)$$
$$= range(A \rightarrow_{116} B) = range(A \rightarrow_{118} B) = range(A \rightarrow_{121} B)$$
$$= range(A \rightarrow_{123} B) = LCM(range(A)^2, range(B)^3);$$

$$range(A \rightarrow_{75} B) = range(A \rightarrow_{126} B) = range(A \rightarrow_{128} B)$$
$$= range(A \rightarrow_{131} B) = range(A \rightarrow_{133} B) = range(A \rightarrow_{136} B)$$
$$= range(A \rightarrow_{138} B) = LCM(range(A)^3, range(B)^2).$$

In the same way we can check that all other operators, defined over CIFSs and having as parameters rational numbers, are also CIFSs.

Therefore, all operations and operators, applied over CIFSs, as a result keep the property of constructiveness.

When universe E is a finite set, its elements can be enumerated by some rational numbers in the interval $[0,1]$ and now, we can apply the results from [558, 555, 556].

3.   In [382], Pyotr Novikov introduced the concept of provable assertion (D-assertion), following an idea of Gödel, published in [240]. In logical level, one can assert that a given assertion is provable or unprovable. Now, from IF-point of view we can speak of degree of provability of a given assertion, degree of provability of the contrary assertion and degree of unprovability. These three estimations will be calculated constructively in all cases, whenever the assertion is proved constructively.

We give the following example that uses Dirichlet's principle. Let us have n balls placed in m boxes. Let our assertion be:

$T(m, n, s) = $ "there exists at least one box that contains at least s balls."

We can easily see that the IF-estimation function for this assertion gives the pair of values

$$V(T(m, n, s)) = \langle \max(0, \frac{n - m(s - 1)}{n + 1}), \min(1, \frac{s}{n + 1}) \rangle.$$

Obviously, number $\max(0, \frac{n-m(s-1)}{n+1})$ corresponds to the number of cases for which the assertion is valid, number $\min(1, \frac{s}{n+1})$ corresponds to the number of cases for which the assertion is not valid, and for $m \geq s$

$$n + 1 - (n - m(s - 1) + s) = ms - m - s + 1$$

is the number of cases, when the assertion is uncertain.

It is easily seen that

$$0 \leq \max(0, \frac{n - m(s - 1)}{n + 1}) + \min(1, \frac{s}{n + 1}) \leq 1.$$

Therefore, we can define the intuitionistic fuzzy degree of provability.

For example, if we have $m = 10$ boxes and $n = 0, 1$ or 2 balls, we can prove that there is no box containing $s = 3$ or more balls. If we have $n = 21, 22, ...$ balls, we can prove that we always have at least one box containing 3 balls. When we have $n = 3, 4, ..., 20$ balls, we cannot prove that there exists at least one box with 3 balls – this assertion may be true, but it may also be false. Hence, we are in a situation of uncertainly.

The text of the present Section is published in [84].

11.2   On Some Open and Solved Problems

In [39], it was mentioned that the IFS theory is a relatively new branch of the fuzzy set theory and so there are many unsolved or unformulated problems in it. In the author's opinion, when speaking of a new theory, it is harmful to discuss in advance whether it would be reasonable to define new concepts in it or not. If it is possible to define the new concepts correctly, it must be done. After some time, if they turn out to be useless, they will cease being objects of discussion. But after their definition, these new concepts may help the emergence of new ideas for development of the theory in general.

In the IFS theory, there are a bunch of open problems. Above, in Sections **2.4, 3.3, 3.8, 4.5, 5.2, 5.8**, we formulated thirteen of them.

The first list of problems was given in [39]. Here, we give them together with short discussion for the cases, when they are solved – fully of partially.

A lot of authors send me copies of their publications. Now, my collection contains more than 1000 papers, but, of course, I do not know all publications on IFSs and hence I will cite only those which I have available. On the other hand, and I know that there are a lot of other papers (on a part of them I was a referee), but I do not have their reprint-versions and so I cannot cite them.

Open problem 14. *To introduce an axiomatic system for IFS, and their extensions and modifications.* Some first attempts in this direction are published in [196, 274].

Open problem 15. *What other operations, relations, operators, norms and metrics (essential from the standpoint of the IFS applications) can be defined over the IFSs and over their extensions and what properties will they have?* The big part of the operations and relations introduced during the last years, are with the author's participation and they are mentioned in the book. Other operations are discussed in [111, 112, 113, 114, 115, 126, 127, 128, 129, 153, 158, 160, 161, 162, 167, 180, 181, 186, 198, 199, 200, 201, 202, 203, 204, 205, 215, 226, 239, 248, 249, 266, 267, 269, 273, 275, 280, 286, 310, 312, 317, 318, 326, 331, 338, 339, 343, 345, 347, 351, 366, 374, 376, 383, 384, 385, 389, 390, 391, 392, 438, 440, 455, 478, 479, 481, 482, 515, 518, 559, 567, 568, 571, 573, 597, 598, 600, 601, 602, 605, 609].

Open problem 16. *What other extensions and modifications of the IFSs can be introduced and what properties will they have? What are the connections between IFSs (and their modifications) and the other fuzzy set extensions?* During the last ten years a lot of IFS extensions and modifications were introduced. As we mentioned in the beginning of Chapter **10**, the author plans in the near future to write a book, devoted to them and a discussion on the research on this theme will be included there.

Open problem 17. *To develop a theory of IF-functions and integrals.* In this direction there are more than 10 publications after year 2000 – see, e.g., [125, 127, 130, 385, 456, 490, 491, 501, 519, 520, 521, 522, 523, 524].

Open problem 18. *To study IF algebraic objects.* Now, there are a lot of papers in this direction - see, e.g., [5, 6, 220, 252, 268, 282, 285, 287, 289, 291, 292, 300, 304, 306, 307, 308, 309, 310, 332, 336, 348, 363, 377, 405, 219, 191, 190, 152, 132, 7, 540, 541, 546, 574, 575, 587, 588, 589, 594, 595, 596, 603, 604].

Open problem 19. *To study the notion of IF linear programming.* Some first steps in this direction are already made in [123, 149].

Open problem 20. *To develop statistical and probabilistical tools for IFSs and IFLs.* I know the papers [150, 151, 185, 238, 246, 247, 277, 316, 316, 333, 334, 335, 369, 370, 402, 403, 404, 406, 407, 408, 409, 441, 463, 464].

Open problem 21. *To construct a theory of IF numbers (and IF complex numbers) and to study their properties.* The research on this theme started in the end of the 1990s (see, e.g. [64, 242, 244, 276, 350, 365, 380]).

Open problem 22. *To introduce elements of IFS in topology and geometry.* Already there are a lot of research - see [1, 2, 3, 135, 136, 137, 143, 175, 176, 177, 178, 179, 182, 206, 253, 254, 255, 257, 270, 283, 293, 298, 305, 322, 323, 324, 325, 353, 355, 356, 357, 358, 359, 360, 361, 362, 373, 375, 386, 387, 388, 400, 419, 420, 422, 423, 512, 544, 545, 547, 583, 599, 610], but all these publications are related only to the standard IF operations. It will be interesting, if in future the modal and extended modal operators be included in this research, too.

Open problem 23. *To develop algorithms for intuitionistic defuzzification and comparison.* A short remark of A. Ban, J. Kacprzyk and the author was published in [131].

Open problem 24. *To study the concepts of IF-information and IF-entropy.* Some first steps in this direction were made in [129, 156, 278, 279, 354, 466, 485, 487, 490, 560, 561, 562, 563, 564].

Open problem 25. *To develop a theory of IF neural networks, genetic algorithms, machine learning and decision making procedures.* Now there are a lot of results in each of these areas of IF-applications – see [37, 86, 95, 100, 101, 121, 134, 163, 172, 193, 213, 251, 259, 271, 272, 299, 330, 340, 341, 346, 349, 352, 371, 393, 394, 416, 427, 428, 429, 430, 431, 433, 435, 436, 437, 442, 443, 444, 445, 446, 447, 448, 449, 450, 451, 452, 457, 458, 459, 460, 461, 462, 467, 470, 473, 477, 483, 484, 456, 486, 488, 489, 490, 491, 492, 495, 501, 508, 509, 513, 514, 517, 525, 548, 549, 550, 569, 578, 579, 581, 611].

Open problem 26. *What real applications do IFSs have?* Some applications are discussed in [88, 134, 170, 195, 207, 208, 212, 214, 237, 275, 281, 313, 319, 327, 343, 374, 395, 396, 417, 418, 426, 434, 469, 494, 496, 497, 498, 499, 500, 503, 504, 505, 506, 507, 526, 562, 567, 577].

Open problem 27. *To develop an IF interpretation of the quantum logic.*

Open problem 28. *To develop an IF interpretation of the many-sorted logic.*

Open problem 29. *To develop IF interpretations of abductive and approximate reasoning.*

So far, I have not had information about publications on the last three problems.

Some time ago, I published the papers [43, 82, 85] with other problems in IFS theory. Below, the new problems, that are not included in the above list, are given and commented.

Open problem 30. *To develop efficient algorithms for construction of degree of membership and non-membership of a given IFS.*

Open problem 31. *Research on the possible geometrical interpretations of IFSs and their classification.*

Open problem 32. *What are the connections between the operators and operations?* For example, as we showed in Section **6.2**, if the IFSs A and B are given, then the equalities $P_B(A) = A \cup B$ and $Q_B(A) = A \cap B$ hold. Are there other similar representations?

Open problem 33. *To develop an IF Geometry.*

Open problem 34. *To develop entire General and Combinatorial IF Topologies.*

Open problem 35. *To develop algorithms for solving IF equations and inequalities and systems of IF equations and inequalities of algebraic, differential and other types.*

Open problem 36. *To define new logical operations in the frameworks of the IF logic and to study their properties.*

Open problem 37. *To develop IF Prolog and IF constraint logic programming.* Up to now, I know only the two research of mine [26, 27]).

Open problem 38. *To develop an IF approach to computational linquistics, including an approach to natural language semantics based on IF logic.*

Open problem 39. *To develop IF preference theory, IF utility theory and IF risk theory.*

Open problem 40. *To develop a general theory of IF systems.* The first step is made in [36].

Solving any of the above problems (as well as many other problems unformulated here) will promote the development of the IFS theory.

In conclusion, 12 years after publishing of [39], I repeat its final words. *The increasing interest in the theory and applications of the IFSs makes the author an optimist for the future of this extension of the fuzzy set theory.*

References

1. Abbas, S.: Intuitionistic supra fuzzy topological spaces. Chaos, Solutions and Fractals 21, 1205–1214 (2004)
2. Abbas, S., Krsteska, B.: Intuitionistic fuzzy strongly preirresolute continuous mappings. Notes on Intuitionistic Fuzzy Sets 13(1), 1–19 (2007), http://ifigenia.org/wiki/issue:nifs/13/1/1-19
3. Abbas, S., Krsteska, B.: Some properties of $(R, S) - T_0$ and $(R, S) - T_1$ spaces. In: IJMMS, pp. 1–19 (2008)
4. Abd-Allah, M.A., El-Saady, K., Ghareeb, A.: Rough intuitionistic fuzzy subgroup. Chaos, Solitons & Fractals 42(4), 2145–2153 (2009)
5. Akram, M., Shum, K.: Intuitionistic fuzzy topological BCC-algebras. Advances in Fuzzy Mathematics 1(1), 1–13 (2006)
6. Akram, M.: Intuitionistic fuzzy closed ideals in BCI-algebras. Int. Mathematical Forum 1(9), 445–453 (2006)
7. Akram, M., Dudek, W.: Interval-valued intuitionistic fuzzy Lie ideals of Lie algebras. World Applied Sciences Journal 7, 812–819 (2009)
8. Andonov, V.: On some properties of one Cartesian product over intuitionistic fuzzy sets. Notes on Intuitionistic Fuzzy Sets 14(1), 12–19 (2008), http://ifigenia.org/wiki/issue:nifs/14/1/12-19
9. Antonov, I.: On a new geometrical interpretation of the intuitionistic fuzzy sets. Notes on Intuitionistic Fuzzy Sets 1(1), 29–31 (1995), http://ifigenia.org/wiki/issue:nifs/1/1/29-31
10. Asparoukhov, O., Atanassov, K.: Intuitionistic fuzzy interpretation of confidential intervals of criteria for decision making. In: Lakov, D. (ed.) Proc. of the First Workshop on Fuzzy Based Expert Systems, Sofia, September 28-30, pp. 56–58 (1994)
11. Atanassov, K.: Intuitionistic fuzzy sets, VII ITKR's Session, Sofia (Deposed in Central Sci. - Techn. Library of Bulg. Acad. of Sci., 1697/84) (June 1983) (in Bulg.)
12. Atanassov, K.: Conditions in Generalized nets. In: Proc. of the XIII Spring Conf. of the Union of Bulg, Sunny Beach, pp. 219–226 (April 1984)
13. Atanassov, K.: Intuitionistic fuzzy relations. In: Proc. of the Third Int. Symp. on Automation and Sci. Instrumentation, Varna, vol. II, pp. 56–57 (October 1984)
14. Atanassov, K.: Intuitionistic fuzzy sets. Fuzzy Sets and Systems 20(1), 87–96 (1986)

15. Atanassov, K.: Generalized index matrices. Comptes Rendus de l'Academie Bulgare des Sciences 40(11), 15–18 (1987)
16. Atanassov, K.: Review and new results on intuitionistic fuzzy sets. Preprint IM-MFAIS-1-88, Sofia (1988)
17. Atanassov, K.: Two variants of intuitonistic fuzzy propositional calculus. Preprint IM-MFAIS-5-88, Sofia (1988)
18. Atanassov, K.: Two operators on intuitionistic fuzzy sets. Comptes Rendus de l'Academie bulgare des Sciences 41(5), 35–38 (1988)
19. Atanassov, K.: More on intuitionistic fuzzy sets. Fuzzy Sets and Systems 33(1), 37–45 (1989)
20. Atanassov, K.: Geometrical interpretation of the elements of the intuitionistic fuzzy objects. Preprint IM-MFAIS-1-89, Sofia (1989)
21. Atanassov, K.: Four new operators on intuitionistic fuzzy sets. Preprint IM-MFAIS-4-89, Sofia (1989)
22. Atanassov, K.: Temporal intuitionistic fuzzy sets. Comptes Rendus de l'Academie bulgare des Sciences 44(7), 5–7 (1991)
23. Atanassov, K.: Generalized Nets. World Scientific, Singapore (1991)
24. Atanassov, K.: Intuitionistic fuzzy sets and expert estimations. Busefal 55, 67–71 (1993)
25. Atanassov, K.: A universal operator over intuitionistic fuzzy sets. Comptes Rendus de l'Academie bulgare des Sciences 46(1), 13–15 (1993)
26. Atanassov, K.: Constraint logic programming and intuitionistic fuzzy logics. Busefal 56, 98–107 (1993)
27. Atanassov, K., Georgiev, C.: Intuitionistic fuzzy Prolog. Fuzzy Sets and Systems 53(1), 121–128 (1993)
28. Atanassov, K.: New operations defined over the intuitionistic fuzzy sets. Fuzzy Sets and Systems 61(2), 137–142 (1994)
29. Atanassov, K.: Remark on intuitionistic fuzzy expert systems. Busefal 59, 71–76 (1994)
30. Atanassov, K.: Intuitionistic fuzzy sets and expert estimations II. Busefal 59, 64–69 (1994)
31. Atanassov, K.: Index matrix representation of the intuitionistic fuzzy graphs. In: Fifth Sci. Session of the "Mathematical Foundation of Artificial Intelligence" Seminar, Preprint MRL-MFAIS-10-94, Sofia, October 5, pp. 36–41 (1994)
32. Atanassov, K.: Remark on the concept of intuitionistic fuzzy relation. In: Fifth Sci. Session of the "Mathematical Foundation of Artificial Intelligence" Seminar, Preprint MRL-MFAIS-10-94, Sofia, October 5, pp. 42–46 (1994)
33. Atanassov, K.: On intuitionistic fuzzy graphs and intuitionistic fuzzy relations. In: Proceedings of the VI IFSA World Congress, Sao Paulo, Brazil, vol. 1, pp. 551–554 (July 1995)
34. Atanassov, K.: An equality between intuitionistic fuzzy sets. Fuzzy Sets and Systems 79(3), 257–258 (1996)
35. Atanassov, K.: Some operators on intuitionistic fuzzy sets. In: Kacprzyk, J., Atanassov, K. (eds.) Proceedings of the First International Conference on Intuitionistic Fuzzy Sets, Sofia, October 18-19 (1997); Notes on Intuitionistic Fuzzy Sets 3(4), 28–33 (1997), http://ifigenia.org/wiki/issue:nifs/3/4/28-33
36. Atanassov, K.: Generalized Nets and Systems Theory. "Prof. M. Drinov" Academic Publishing House, Sofia (1997)

37. Atanassov, K.: Generalized Nets in Artificial Intelligence. Generalized nets and Expert Systems, vol. 1. Prof. M. Drinov" Academic Publishing House, Sofia (1998)
38. Atanassov, K.: Temporal intuitionistic fuzzy graphs. Notes on Intuitionistic Fuzzy Sets 4(4), 59–61 (1998), http://ifigenia.org/wiki/issue:nifs/4/4/59-61
39. Atanassov, K.: Intuitionistic Fuzzy Sets. Springer, Physica-Verlag, Heidelberg (1999)
40. Atanassov, K.: On four intuitionistic fuzzy topological operators. Mathware & Soft Computing 8, 65–70 (2001)
41. Atanassov, K.: Converse factor: definition, properties and problems. Notes on Number Theory and Discrete Mathematics 8(1), 37–38 (2002)
42. Atanassov, K.: Remark on an application of the intuitionistic fuzzy sets in number theory. Advanced Studies in Contemporary Mathematics 5(1), 49–55 (2002)
43. Atanassov, K.: Open problems in intuitionistic fuzzy sets theory. In: Proceedings of 6th Joint Conf. on Information Sciences, Research Triangle Park, North Carolina, USA, March 8-13, pp. 113–116 (2002)
44. Atanassov, K.: Remark on a property of the intuitionistic fuzzy interpretation triangle. Notes on Intuitionistic Fuzzy Sets 8(1), 34–36 (2002), http://ifigenia.org/wiki/issue:nifs/8/1/34-36
45. Atanassov, K.: On index matrix interpretations of intuitionistic fuzzy graphs. Notes on Intuitionistic Fuzzy Sets 8(4), 73–78 (2002), http://ifigenia.org/wiki/issue:nifs/8/4/73-78
46. Atanassov, K.: On the temporal intuitionistic fuzzy sets. In: Proc. of the Ninth International Conf., IPMU 2002, Annecy, France, July 1-5, vol. III, pp. 1833–(1837)
47. Atanassov, K.: Temporal intuitionistic fuzzy sets (review and new results). In: Proc. of the Thirty Second Spring of the Union of Bulg. Mathematicians, Sunny Beach, April, 5-8, pp. 79–88 (2003)
48. Atanassov, K.: New operators over intuitionistic fuzzy sets. In: Proc. of IPMU 2004, Perugie, Italy, vol. 2, pp. 1383–1387 (June 2004)
49. Atanassov, K.: On the modal operators defined over the intuitionistic fuzzy sets. Notes on Intuitionistic Fuzzy Sets 10(1), 7–12 (2004), http://ifigenia.org/wiki/issue:nifs/10/1/07-12
50. Atanassov, K., Answer to Dubois, D., Gottwald, S., Hajek, P., Kacprzyk, J., Prade's, H., paper : Terminological difficulties in fuzzy set theory – the case of "intuitionistic fuzzy sets". Fuzzy Sets and Systems 156(3), 496–499 (2005)
51. Atanassov, K.: Intuitionistic fuzzy implications and Modus Ponens. Notes on Intuitionistic Fuzzy Sets 11(1), 1–5 (2005), http://ifigenia.org/wiki/issue:nifs/11/1/01-05
52. Atanassov, K.: On some types of intuitionistic fuzzy negations. Notes on Intuitionistic Fuzzy Sets 11(4), 170–172 (2005), http://ifigenia.org/wiki/issue:nifs/11/4/170-172
53. Atanassov, K.: On one type of intuitionistic fuzzy modal operators. Notes on Intuitionistic Fuzzy Sets 11(5), 24–28 (2005), http://ifigenia.org/wiki/issue:nifs/11/5/24-28
54. Atanassov, K.: On some intuitionistic fuzzy negations. In: Proc. of the First Int. Workshop on IFSs, Banska Bystrica, September 22 (2005); Notes on Intuitionistic Fuzzy Sets 11(6), 13–20 (2005), http://ifigenia.org/wiki/issue:nifs/11/6/13-20

55. Atanassov, K.: On some intuitionistic fuzzy implications. Comptes Rendus de l'Academie Bulgare Des Sciences 59(1), 19–24 (2006)

56. Atanassov, K.: A new intuitionistic fuzzy implication from a modal type. Advanced Studies in Contemporary Mathematics 12(1), 117–122 (2006)

57. Atanassov, K.: On intuitionistic fuzzy negations and De Morgan Laws. In: Proc. of Eleventh International Conf., IPMU 2006, Paris, July 2-7, pp. 2399–2404 (2006)

58. Atanassov, K.: On eight new intuitionistic fuzzy implications. In: Proc. of 3rd Int. IEEE Conf. "Intelligent Systems", IS 2006, London, September 4-6, pp. 741–746 (2006)

59. Atanassov, K.: On intuitionistic fuzzy negations. In: Computational Intelligence, Theory and Applications, pp. 159–167. Springer, Berlin (2006)

60. Atanassov, K.: On the implications and negations over intuitionistic fuzzy sets. In: Proc. of the Scientific Conference of Burgas Free University 2006, Burgas, June 9-11, vol. 3, pp. 374–384 (2006)

61. Atanassov, K.: The most general form of one type of intuitionistic fuzzy modal operators. Notes on Intuitionistic Fuzzy Sets 12(2), 36–38 (2006), http://ifigenia.org/wiki/issue:nifs/12/2/36-38

62. Atanassov, K.: On two sets of intuitionistic fuzzy negations. Issues in Intuitionistic Fuzzy Sets and Generalized Nets 3, 35–39 (2006)

63. Atanassov, K.: Remark and open problems on intuitionistic fuzzy sets and complex analysis. Notes on Intuitionistic Fuzzy Sets 13(4), 45–48 (2007), http://ifigenia.org/wiki/issue:nifs/13/4/45-48

64. Atanassov, K.: Remark on intuitionistic fuzzy numbers. Notes on Intuitionistic Fuzzy Sets 13(3), 29–32 (2007), http://ifigenia.org/wiki/issue:nifs/13/3/29-32

65. Atanassov, K.: On Generalized Nets Theory. "Prof. M. Drinov" Academic Publ. House, Sofia (2007)

66. Atanassov, K.: Some properties of the operators from one type of intuitionistic fuzzy modal operators. Advanced Studies in Contemporary Mathematics 15(1), 13–20 (2007)

67. Atanassov, K.: On intuitionistic fuzzy implication $\rightarrow^{\varepsilon}$ and intuitionistic fuzzy negation $\neg^{\varepsilon}$. Issues in Intuitionistic Fuzzy Sets and Generalized Nets 6, 6–19 (2008)

68. Atanassov, K.: Intuitionistic fuzzy implication $\rightarrow^{\varepsilon,\eta}$ and intuitionistic fuzzy negation $\neg^{\varepsilon,\eta}$. Developments in Fuzzy Sets, Intuitionistic Fuzzy Sets, Generalized Nets and Related Topics 1, 1–10 (2008)

69. Atanassov, K.: 25 years of intuitionistic fuzzy sets or the most interesting results and the most important mistakes of mine. In: Advances in Fuzzy Sets, Intuitionistic Fuzzy Sets, Generalized Nets and Related Topics. vol. I: Foundations, pp. 1–35. Academic Publishing House EXIT, Warszawa (2008)

70. Atanassov, K.: The most general form of one type of intuitionistic fuzzy modal operators. Part 2. Notes on Intuitionistic Fuzzy Sets 14(1), 27–32 (2008), http://ifigenia.org/wiki/issue:nifs/14/1/27-32

71. Atanassov, K.: Remark on operation "subtraction" over intuitionistic fuzzy sets. Notes on Intuitionistic Fuzzy Sets 15(3), 20–24 (2009), http://ifigenia.org/wiki/issue:nifs/15/3/20-24

72. Atanassov, K.: Intuitionistic fuzzy subtractions $-'^{\varepsilon,\eta}$ and $-''^{\varepsilon,\eta}$. In: Developments in Fuzzy Sets, Intuitionistic Fuzzy Sets, Generalized Nets and Related Topics. vol. I: Foundations, pp. 1–10. SRI Polish Academy of Sciences, Warsaw (2010)

73. Atanassov, K.: On the new intuitionistic fuzzy operator $x_{a,b,c,d,e,f}$. Notes on Intuitionistic Fuzzy Sets 16(2), 35–38 (2010), http://ifigenia.org/wiki/issue:nifs/16/2/35-38

74. Atanassov, K.: New ways for altering the intuitionistic fuzzy experts' estimations. Issues in Intuitionistic Fuzzy Sets and Generalized Nets 8, 24–29 (2010)

75. Atanassov, K.: On index matrices, Part 1: Standard cases. Advanced Studies in Contemporary Mathematics 20(2), 291–302 (2010)

76. Atanassov, K.: On index matrices, Part 2: Intuitionistic fuzzy case. Proceedings of the Jangjeon Mathematical Society 13(2), 121–126 (2010)

77. Atanassov, K.: A New Approach to the Distances between Intuitionistic Fuzzy Sets. In: Hüllermeier, E., Kruse, R., Hoffmann, F. (eds.) IPMU 2010, Part I. CCIS, vol. 80, pp. 581–590. Springer, Heidelberg (2010)

78. Atanassov, K.: Remark on equalities between intuitionistic fuzzy sets. Notes on Intuitionistic Fuzzy Sets 16(3), 40–41 (2010), http://ifigenia.org/wiki/issue:nifs/16/3/40-41

79. Atanassov, K.: On two topological operators over intuitionistic fuzzy sets. Issues in Intuitionistic Fuzzy Sets and Generalized Nets 8, 1–7 (2010)

80. Atanassov, K.: Cantor's norms for intuitionistic fuzzy sets. Issues in Intuitionistic Fuzzy Sets and Generalized Nets 8, 36–39 (2010)

81. Atanassov, K.: On the intuitionistic fuzzy implications and negations. Part 1. In: Cornelis, C., et al. (eds.) 35 Years of Fuzzy Set Theory - Celebratory Volume Dedicated to the Retirement of Etienne E. Kerre, pp. 19–38. Springer, Berlin (2010)

82. Atanassov, K.: On some open and solved problems in the theory and applications of intuitionistic fuzzy sets. Issues in Intuitionistic Fuzzy Sets and Generalized Nets 9, 1–34 (2011)

83. Atanassov, K.: Definitions and properties of some intuitionistic fuzzy subtraction. Issues in Intuitionistic Fuzzy Sets and Generalized Nets 9, 35–56 (2011)

84. Atanassov, K.: The intuitionistic fuzzy sets as constructive objects. In: Recent Advances in Fuzzy Sets, Intuitionistic Fuzzy Sets, Generalized Nets and Related Topics. vol. I: Foundations, pp. 1–23. SRI Polish Academy of Sciences, Warsaw (2011)

85. Atanassov, K.: On some open and solved problems in the theory and applications of intuitionistic fuzzy sets. Issues in Intuitionistic Fuzzy Sets and Generalized Nets 9, 1–34

86. Atanassov, K., Aladjov, H.: Generalized Nets in Artificial Intelligence. In: Atanassov, K., Aladjov, H. (eds.) Generalized Nets and Machine Learning, vol. 2, "Prof. M. Drinov" Academic Publishing House, Sofia (2001)

87. Atanassov, K., Ban, A.: On an operator over intuitionistic fuzzy sets. Comptes Rendus de l'Academie bulgare des Sciences 53(5), 39–42 (2000)

88. Atanassov, K., Chakarov, V., Shannon, A., Sorsich, J.: Generalized Net Models of the Human Body. "Prof. M. Drinov" Academic Publishing House, Sofia (2008)

89. Atanassov, K., Dimitrov, D.: On the negations over intuitionistic fuzzy sets. Part 1. Annual of "Informatics" Section Union of Scientists in Bulgaria 1, 49–58 (2008)

90. Atanassov, K., Dimitrov, D.: On one of Baczyncki-Jayaram's problems. Cybernetics and Information Technologies 9(2), 14–20 (2009)

91. Atanassov, K., Dimitrov, D.: Extension of one of Baczyncki-Jayaram's problems. Comptes Rendus de l'Academie bulgare des Sciences 62(11), 1377–1386 (2009)

92. Atanassov, K., Dimitrov, D.: Intuitionistic fuzzy implications and axioms for implications. Notes on Intuitionistic Fuzzy Sets 16(1), 10–20 (2010), http://ifigenia.org/wiki/issue:nifs/16/1/10-20

93. Atanassov, K., Gargov, G.: Interval valued intuitionistic fuzzy sets. Fuzzy Sets and Systems 31(3), 343–349 (1989)

94. Atanassov, K., Gargov, G.: Intuitionistic fuzzy logic. Comptes Rendus de l'Academie bulgare des Sciences 43(3), 9–12 (1990)

95. Atanassov, K., Gluhchev, G., Hadjitodorov, S., Kacprzyk, J., Shannon, A., Szmidt, E., Vassilev, V.: Generalized Nets Decision Making and Pattern Recognition. Warsaw School of Information Technology, Warszawa (2006)

96. Atanassov, K., Hinde, C.: On intuitionistic fuzzy negations and intuitionistic fuzzy extended modal operators. In: Part 3. Advances in Fuzzy Sets, Intuitionistic Fuzzy Sets, Generalized Nets and Related Topics. vol. I: Foundations, pp. 37–42. Academic Publishing House EXIT, Warszawa (2008)

97. Atanassov, K., Kolev, B.: On an intuitionistic fuzzy implication from a probabilistic type. Advanced Studies in Contemporary Mathematics 12(1), 111–116 (2006)

98. Atanassov, K., Pasi, G., Yager, R.: Intuitionistic fuzzy interpretations of multi-criteria multi-person and multi-measurement tool decision making. International Journal of Systems Science 36(14), 859–868 (2005)

99. Atanassov, K., Riečan, B.: On two operations over intuitionistic fuzzy sets. Journal of Applied Mathematics, Statistics and Informatics 2(2), 145–148 (2006)

100. Atanassov, K., Sotirov, S., Kodogiannis, V.: Intuitionistic fuzzy estimations of the Wi-Fi connections. In: First Int. Workshop on Generalized Nets, Intuitionistic Fuzzy Sets and Knowledge Engineering, London, September 6-7, pp. 75–80 (2006)

101. Atanassov, K., Sotirova, E., Orozova, D.: Generalized net model of an intuitionistic fuzzy expert system with frame-type data bases and different forms of hypotheses estimations. Annuals of Burgas Free University XXII, 257–262

102. Atanassov, K., Stoeva, S.: Intuitionistic fuzzy sets. In: Proc. of Polish Symp. on Interval & Fuzzy Mathematics, Poznan, pp. 23–26 (August 1983)

103. Atanassov, K., Stoeva, S.: Intuitionistic L-fuzzy sets. In: Trappl, R. (ed.) Cybernetics and Systems Research, vol. 2, pp. 539–540. Elsevier Sci. Publ., Amsterdam (1984)

104. Atanassov, K., Szmidt, E., Kacprzyk, J.: On intuitionistic fuzzy multi-dimensional sets. Issues in Intuitionistic Fuzzy Sets and Generalized Nets 7, 1–6 (2008)

105. Atanassov, K., Szmidt, E., Kacprzyk, J., Rangasamy, P.: On intuitionistic fuzzy multi-dimensional sets. In: Part 2. Advances in Fuzzy Sets, Intuitionistic Fuzzy Sets, Generalized Nets and Related Topics. vol. I: Foundations, pp. 43–51. Academic Publishing House EXIT, Warszawa (2008)

106. Atanassov, K., Szmidt, E., Kacprzyk, J.: On intuitionistic fuzzy multi-dimensional sets. In: Part 3. Developments in Fuzzy Sets, Intuitionistic Fuzzy Sets, Generalized Nets and Related Topics. vol. I: Foundations, pp. 19–26. SRI Polish Academy of Sciences, Warsaw (2010)

107. Atanassov, K., Szmidt, E., Kacprzyk, J.: On intuitionistic fuzzy multi-dimensional sets. Part 4. Notes on Intuitionistic Fuzzy Sets 17(2), 1–7 (2011)

108. Atanassov, K., Trifonov, T.: Towards combining two kinds of intuitionistic fuzzy sets. Notes on Intuitionistic Fuzzy Sets 11(2), 1–11 (2005),
http://ifigenia.org/wiki/issue:nifs/11/2/01-11

109. Atanassov, K., Trifonov, T.: On a new intuitionistic fuzzy implication of Gödel's type. Proceedings of the Jangjeon Mathematical Society 8(2), 147–152 (2005)

110. Atanassov, K., Trifonov, T.: Two new intuitionistic fuzzy implications. Advanced Studies in Contemporary Mathematics 13(1), 69–74 (2006)

111. Atanassova, L.: On some properties of intuitionistic fuzzy negation $\neg_@$. Notes on Intuitionistic Fuzzy Sets 15(1), 32–35 (2009),
http://ifigenia.org/wiki/issue:nifs/15/1/32-35

112. Atanassova, L.: On an intuitionistic fuzzy subtraction, generated by an implication from Kleene-Dienes type. Notes on Intuitionistic Fuzzy Sets 15(4), 30–32 (2009), http://ifigenia.org/wiki/issue:nifs/15/4/30-32

113. Atanassova, L.: A new intuitionistic fuzzy implication. Cybernetics and Information Technologies 9(2), 21–25 (2009)

114. Atanassova, L.: On intuitionistic fuzzy subtractions $-'_@$ and $-''_@$. Issues in Intuitionistic Fuzzy Sets and Generalized Nets 8, 19–23 (2010)

115. Atanassova, L.: On intuitionistic fuzzy subtractions $-'_{19}$ and $-''_{19}$. In: Developments in Fuzzy Sets, Intuitionistic Fuzzy Sets, Generalized Nets and Related Topics. vol. I: Foundations, pp. 27–32. SRI PAS, Warsaw (2010)

116. Atanassova, L.: Equalities of extended intuitionistic fuzzy modal operators. Issues in Intuitionistic Fuzzy Sets and Generalized Nets 9, 57–60 (2011)

117. Atanassova, L., Atanassov, K.: Intuitionistic Fuzzy Interpretations of Conway's Game of Life. In: Dimov, I., Dimova, S., Kolkovska, N. (eds.) NMA 2010. LNCS, vol. 6046, pp. 232–239. Springer, Heidelberg (2011)

118. Atanassova, L., Atanassov, K.: Intuitionistic fuzzy interpretations of Conway's game of life. Part 2: Topological transformations of the game field. In: Proc. of the 2nd Workshop on Generalized Nets, Intuitionistic Fuzzy Sets and Knowledge Engineering, London, July 9, pp. 83–88 (2010)

119. Atanassova, L., Atanassov, K.: Intuitionistic fuzzy interpretations of Conway's game of life. Part 3: Modal and extended modal transformations of the game field. In: Recent Advances in Fuzzy Sets, Intuitionistic Fuzzy Sets, Generalized Nets and Related Topics. vol. I: Foundations, pp. 25–31. SRI Polish Academy of Sciences, Warsaw (2011)

120. Atanassova, L., Gluhchev, G., Atanassov, K.: On intuitionistic fuzzy histograms. Notes in Intuitionistic Fuzzy Sets 16(4), 32–36 (2010)

121. Atanassova, V.: Strategies for decision making in the conditions of intuitionistic fuzziness. In: Reusch, B. (ed.) Computational Intelligence, Theory and Applications, pp. 263–269. Springer, Berlin (2005)

122. Atanassova, V.: Representation of fuzzy and intuitionistic fuzzy data by Radar charts. Notes on Intuitionistic Fuzzy Sets 16(1), 21–26 (2010),
http://ifigenia.org/wiki/issue:nifs/16/1/21-26

123. Bablu, J.: Fuzzy and Intuitionistic Fuzzy Linear Programming with Application in Some Transportation Models. PhD Thesis, Bengal Engineering and Science University, Howrah, India (2009)

124. Baczynski, M., Jayaram, B.: Fuzzy Implications. Springer, Berlin (2008)

125. Ban, A.: Applications of conormed-seminormed fuzzy integrals. Analele Universitatii din Oradea. Fascicola Matematica Tom VI, 109–118 (1997)

126. Ban, A.: A family of monotone measures on intuitionistic fuzzy sets with respect to Frank intuitionistic fuzzy t-norms. In: Atanassov, K., Kacprzyk, J., Krawczak, M., Szmidt, E. (eds.) Issues in the Representation and Processing of Uncertain Imprecise Information: Fuzzy Sets, Intuitionistic Fuzzy Sets, Generalized Nets, and Related Topics, pp. 58–71. Akademicka Oficyna Wydawnictwo EXIT, Warsaw (2005)

127. Ban, A.: Intuitionistic fuzzy-valued fuzzy measures and integrals. In: Atanassov, K., Kacprzyk, J., Krawczak, M., Szmidt, E. (eds.) Issues in the Representation and Processing of Uncertain Imprecise Information: Fuzzy Sets, Intuitionistic Fuzzy Sets, Generalized Nets, and Related Topics, pp. 25–57. Akademicka Oficyna Wydawnictwo EXIT, Warsaw (2005)

128. Ban, A.: Intuitionistic Fuzzy Measures, Theory and Applications. Nova Sci. Publishers, New York (2006)

129. Ban, A., Gal, S.: Decomposable measures and information measures for intuitionistic fuzzy sets. Fuzzy Sets and Systems 123(1), 103–117 (2001)

130. Ban, A., Fechete, I.: Componentwise decomposition of some lattice-valued fuzzy integrals. Information Sciences 177, 1430–1440 (2007)

131. Ban, A., Kacprzyk, J., Atanassov, K.: On de-I-fuzzyfication of intuitionistic fuzzy sets. Comptes Rendus de l'Academie bulgare des Sciences 61(12), 1535–1540 (2008)

132. Banerjee, B., Basnet, D.: Intuitionistic fuzzy subrings and ideals. The Journal of Fuzzy Mathematics 11(1), 139–155 (2003)

133. Barwise, J. (ed.): Handbook of Mathematical Logic. North-Holland, Amsterdam (1977)

134. Batchkova, I., Toneva, G.: Neural intuitionistic fuzzy generalized nets for on-line system identification of coupled distillation columns. In: Matko, D., Music, G. (eds.) Proc. of DYCOMANS Phase 2 Workshop 1 "Techniques for Supervisory Management Systems", Bled (Slovenia), May 12-14, pp. 213–220 (1999)

135. Bayhan, S., Çoker, D.: On fuzzy separation axioms in intuitionistic fuzzy topological spaces. Busefal 67, 77–87 (1996)

136. Bayhan, S., Çoker, D.: On separation axioms in intuitionistic topological spaces. Int. J. of Mathematics and Mathematical Sciences 27(10), 621–630 (2001)

137. Bayhan, S., Çoker, D.: Pairwise separation axioms in intuitionistic fuzzy topological spaces. Hacettepe Journal of Mathematics and Statistics 34, 101–114 (2005)

138. de Berg, M., Cheong, O., van Kreveld, M., Overmars, M.: Computational Geometry: Algorithms and Applications. Springer, Berlin (2008)

139. Bolk, L., Borowik, P.: Many-Valued Logics. Springer, Berlin (1992)

140. Bordogna, G., Fedrizzi, M., Pasi, G.: A linguistic modeling of consensus in Group Decision Making based on OWA operators. IEEE Trans. on System Man and Cybernetics 27(1), 126–132 (1997)

141. Brouwer, L.E.J.: Collected Works, vol. 1. North Holland, Amsterdam (1975)

142. Van Dalen, D. (ed.): Brouwer's Cambridge Lectures on Intuitionism. Cambridge Univ. Press, Cambridge (1981)
143. Brown, L., Diker, M.: Ditopological texture spaces and intuitionistic sets. Fuzzy Sets and Systems 98(2), 217–224 (1998)
144. Buckley, J.: Fuzzy Probabilities. Springer, Berlin (2003)
145. Buckley, J.: Fuzzy Statistics. Springer, Berlin (2004)
146. Buckley, J.: Fuzzy Probability and Statistics. Springer, Berlin (2006)
147. Buckley, J.: Fundamentals of Statistics with Fuzzy Data. Springer, Berlin (2006)
148. Buhaescu, T.: Some observations on intuitionistic fuzzy relations. In: Itinerant Seminar on Functional Equations, Approximation and Convexity, Cluj-Napoca, pp. 111–118 (1989)
149. Buhaescu, T.: Linear programming with intuitionistic fuzzy objective. In: International Colloquy the Risk in Contemporary Economie, Galati, Romania, November 10-11, pp. 29–30 (1995)
150. Burduk, R.: Probability Error in Global Optimal Hierarchical Classifier with Intuitionistic Fuzzy Observations. In: Corchado, E., Wu, X., Oja, E., Herrero, Á., Baruque, B. (eds.) HAIS 2009. LNCS, vol. 5572, pp. 533–540. Springer, Heidelberg (2009)
151. Burduk, R.: Probability Error in Bayes Optimal Classifier with Intuitionistic Fuzzy Observations. In: Kamel, M., Campilho, A. (eds.) ICIAR 2009. LNCS, vol. 5627, pp. 359–368. Springer, Heidelberg (2009)
152. Burillo, P., Bustince, H.: Estructuras algebraicas para conjuntos intuicionistas fuzzy. In: II Congreso Espanol Sobre Tecnologias y Logica Fuzzy, Boadilla del Monte, Noviembre 2-4, pp. 135–146 (1992)
153. Burillo, P., Bustince, H.: Informational energy on intuitionistic fuzzy sets and on interval-valued intuitionistic fuzzy sets. Relationship between the measures of information. In: Lakov, D. (ed.) Proc. of the First Workshop on Fuzzy Based Expert Systems, Sofia, September 28-30, pp. 46–49 (1994)
154. Burillo, P., Bustince, H.: Intuitionistic fuzzy relations. Part I, Mathware and Soft Computing 2(1), 5–38 (1995)
155. Burillo, P., Bustince, H.: Intuitionistic fuzzy relations. Part II, Mathware and Soft Computing 2(2), 117–148 (1995)
156. Burillo, P., Bustince, H.: Entropy on intuitionistic fuzzy sets and on interval-valued fuzzy sets. Fuzzy Sets and Systems 78(3), 305–316 (1996)
157. Bustince, H.: Conjuntos Intuicionistas e Intervalo-valorados Difusos: Propiedades y Construccion. Relaciones Intuicionistas y Estructuras. Ph.D., Univ. Publica de Navarra, Pamplona (1994)
158. Bustince, H.: Numerical information measurements in interval-valued intuitionistic fuzzy sets (IVIFS). In: Lakov, D. (ed.) Proc. of the First Workshop on Fuzzy Based Expert Systems, Sofia, September 28-30, pp. 50–52 (1994)
159. Bustince, H., Burillo, P.: Vague sets are intuitionistic fuzzy sets. Fuzzy Sets and Systems 79(3), 403–405 (1996)
160. Bustince, H., Burillo, P.: Perturbation of intuitionistic fuzzy relations. International Journal of Uncertainty, Fuzziness and Knowledge-Based Systems 9(1), 81–103 (2001)
161. Bustince, H., Mohedano, V., Barrenechea, E., Pagola, M.: A study of the intuitionistic fuzzy S-implications. First properties. In: Procesos de Toma De Decisiones, Modelado y Agregacion de Preferencias, TIC 2002, 11492-E, pp. 141–150 (2002)

162. Chen, C., Zehua, L.V., Pei-yu, Q., Hui, X.: New method to measure similarity between intuitionistic fuzzy sets based on normal distribution functions. Mini-Micro Systems 28(3), 500–503 (2007)

163. Chen, D., Lei, Y., Ye, T.: Aggregation and application of the information about attribute weights with different forms in group making. Journal of Air Force Engineering University 7(6), 51–53 (2006)

164. Chen, G.H., Huang, Z.G.: Linear programming method for multiattribute group decision making using IF sets. Information Sciences 180(9), 1591–1609 (2010)

165. Chen, S.J., Hwang, C.L.: Fuzzy Multiple Attribute Decision Making - Methods and Applications. Springer (1992)

166. Chen, T.-Y.: A note on distances between intuitionistic fuzzy sets and/or interval valued fuzzy sets based on the Hausdorff metric. Fuzzy Sets and Systems 158(22), 2523–2525 (2007)

167. Chen, T.-Y.: Remarks on the subtraction and division operations over intuitionistic fuzzy sets and interval-valued fuzzy sets. International Journal of Fuzzy Systems 9(3), 169–172 (2007)

168. Chen, W.J.: Intuitionistic fuzzy Lie sub-superalgebras and ideals of Lie superalgebras. In: Proc. of the 2nd International Joint Conference on Computational Sciences and Optimization, Sanya, April 24-26, vol. 2, pp. 841–845 (2009)

169. Chountas, P., Alzebdi, M., Shannon, A., Atanassov, K.: On intuitionistic fuzzy trees. Notes on Intuitionistic Fuzzy Sets 15(2), 30–32 (2009), http://ifigenia.org/wiki/issue:nifs/15/2/30-32

170. Chountas, P., Kolev, B., Rogova, E., Tasseva, V., Atanassov, K.: Generalized Nets in Artificial Intelligence. vol. 4: Generalized nets, Uncertain Data and Knowledge Engineering. "Prof. M. Drinov" Academic Publishing House, Sofia (2007)

171. Chountas, P., Shannon, A., Rangasamy, P., Atanassov, K.: On intuitioniostic fuzzy trees and their index matrix interpretation. Notes on Intuitionistic Fuzzy Sets 15(4), 52–56 (2009), http://ifigenia.org/wiki/issue:nifs/15/4/52-56

172. Chountas, P., Sotirova, E., Kolev, B., Atanassov, K.: On intuitionistic fuzzy expert system with temporal parameters, using intuitionistic fuzzy estimations. In: Computational Intelligence, Theory and Applications, pp. 241–249. Springer (2006)

173. Christofides, N.: Graph Theory. Academic Press, New York (1975)

174. Cui, L.C., Li, Y.M., Zhang, X.H.: Intuitionistic fuzzy linguistic quantifiers based on intuitionistic fuzzy-valued fuzzy measures and integrals. International Journal of Uncertainty, Fuzziness and Knowledge-Based Systems 17(3), 427–448 (2009)

175. Çoker, D.: An introduction to fuzzy subspaces in intuitionistic fuzzy topological spaces. The Journal of Fuzzy Mathematics 4(4), 749–764 (1996)

176. Çoker, D.: An introduction to intuitionistic topological spaces. BUSEFAL 81, 51–56 (2000)

177. Çoker, D., Demirci, M.: An introduction to intuitionistic fuzzy topological spaces in Sostak's sense. BUSEFAL 67, 67–76 (1996)

178. Çoker, D., Haydar, A.: On fuzzy compactness in intuitionistic fuzzy topological spaces. The Journal of Fuzzy Mathematics 3(4), 899–909 (1995)

179. Çoker, D., Haydar, A., Turanli, N.: A Tychonoff theorem in intuitionistic fuzzy topological spaces. IJMMS 70, 3829–3837 (2004)
180. Cornelis, C., Deschrijver, G., Kerre, E.: Implication on intuitionistic fuzzy and interval-valued fuzzy set theory: construction, classification. Application. Int. Journal of Approximate Reasoning 35, 55–95 (2004)
181. Coşkun, E.: Systems on intuitionistic fuzzy special sets and intuitionistic fuzzy special measures. Information Sciences 128, 105–118 (2000)
182. Coşkun, E., Çoker, D.: On neighborhood structures in intuitionistic topological spaces. Mathematica Balkanica 12(3-4), 283–293 (1998)
183. Craig, W.: Three uses of the Herbrand-Gentzen theorem in relating model theory and proof theory. J. Symbolic Logic 22, 269–285 (1957)
184. Cristea, I., Davvaz, B.: Atanassov's intuitionistic fuzzy grade of hypergroups. Information Sciences 180(8), 1506–1517 (2010)
185. Čunderlikova-Lendelová, K., Riečan, B.: The probability theory on B-structures. Developments in Fuzzy Sets, Intuitionistic Fuzzy Sets, Generalized Nets and Related Topics 1, 33–60 (2008)
186. Çuvalcioğlu, G.: Some properties of $E_{\alpha,\beta}$ operator. Advanced Studies in Contemporary Mathematics 14(2), 305–310 (2007)
187. Çuvalcioğlu, G.: Some properties of $E_{\alpha,\beta}$ operator. Advanced Studies in Contemporary Mathematics 14(2), 305–310 (2007)
188. Danchev, S.: A new geometrical interpretation of some concepts in the intuitionistic fuzzy logics. Notes on Intuitionistic Fuzzy Sets 1(2), 116–118 (1995)
189. Danchev, S.: A generalization of some operations defined over the intuitionistic fuzzy sets. Notes on Intuitionistic Fuzzy Sets 2(1), 1–3 (1996)
190. Davvaz, S., Corsini, P., Leoreanu-Fotea, V.: Atanassov's intuitionistic (S,T)-fuzzy n-ary sub-hypergroups. Information Sciences 179, 654–666 (2009)
191. Davvaz, B., Dudek, W.: Intuitionistic fuzzy H_v-ideals. Int. J. of Mathematics and Mathematical Sciences 206, 1–11 (2006)
192. Davvaz, B., Leoreanu-Fotea, V.: Intuitionistic Fuzzy n-ary Hypergroups. Journal of Multiple-Valued Logic and Soft Computing 16(1-2), 87–103 (2010)
193. De, S.K., Biswas, R., Roy, A.: Multicriteria decision making using intuitionistic fuzzy set theory. The Journal of Fuzzy Mathematics 6(4), 837–842 (1998)
194. De, S.K., Biswas, R., Roy, A.R.: Some operations on intuitionistic fuzzy sets. Fuzzy sets and Systems 114(4), 477–484 (2000)
195. De, S.K., Biswas, R., Roy, A.R.: An Application of intuitionistic fuzzy sets in medical diagnosis. Fuzzy Sets and Systems 117(2), 209–213 (2001)
196. Demirci, M.: Axiomatic theory of intuitionistic fuzzy sets. Fuzzy Sets and Systems 110(2), 253–266 (2000)
197. Dencheva, K.: Extension of intuitionistic fuzzy modal operators ⊞ and ⊠ . In: Proceedings of the Second Int. IEEE Symposium: Intelligent Systems, Varna, June 22-24, vol. 3, pp. 21–22 (2004)
198. Deschrijver, G.: Generators of t-norms in interval-valued fuzzy set theory. In: Proc. of Fourth Conf. of the European Society for Fuzzy Logic and Technology EUSFLAT, Barcelona, September 7-9, pp. 253–258 (2005)
199. Deschrijver, G.: The Archimedian property for t-norms in interval-valued fuzzy set theory. Fuzzy Sets and Systems 157, 2311–2327 (2006)
200. Deschrijver, G.: Representations of triangular norms in intuitionistic L-fuzzy set theory. In: Proc. of Eleventh International Conf., IPMU 2006, Paris, July 2-7, pp. 2348–2353 (2006)

201. Deschrijver, G.: Representations of triangular norms in intuitionistic L-fuzzy set theory. In: Proc. of Eleventh International Conf., IPMU 2006, Paris, July 2-7, pp. 2348–2353 (2006)

202. Deschrijver, G., Cornelis, C., Kerre, E.: On the representation of intuitionistic fuzzy t-norms and t-conorms. IEEE Transactions on Fuzzy Systems 12(1), 45–61 (2004)

203. Deschrijver, G., Kerre, E.: A method for constructing non t-representable intuitionistic fuzzy t-norms satisfying the residuation principle. In: Proc. of the Third Conf. of the European Society for Fuzzy Logic and Technology, EUSFLAT 2003, Zittau, September 10-12, pp. 164–167 (2003)

204. Deschrijver, G., Kerre, E.: Classes of intuitionistic fuzzy T-norms satisfying the residuation principle. Int. Journal of Uncertainty, Fuzziness and Knowledge-Based Systems 11(6), 691–709 (2003)

205. Deschrijver, G., Kerre, E.: Triangular norms and related operators in L*-fuzzy set theory. In: Klement, E., Mesiar, R. (eds.) Logical, Algebraic, Analytic, and Probabilistic Aspects of Triangular Norms, pp. 231–259. Elsevier, Amsterdam (2005)

206. Diker, M.: Connectedness in ditopological texture spaces. Fuzzy Sets and Systems 108(2), 223–230 (1999)

207. Dimitrov, D.: Interpretation patterns and markets. Fuzzy Economic Review 6(1), 89–107 (2001)

208. Dimitrov, D.: The Paretian liberal in an intuitionistic fuzzy context. In: Proc. of the Int. Conf. "Logic, Game Theory and Social Choice", Saint-Peterburg, pp. 70–73 (2001)

209. Dimitrov, D.: New intuitionistic fuzzy implications and their corresponding negation. Issues in Intuitionistic Fuzzy Sets and Generalized Nets 6, 36–42 (2008)

210. Dimitrov, D.: IFSTool - software for intuitionistic fuzzy sets. Issues in Intuitionistic Fuzzy Sets and Generalized Nets 9, 61 (2011)

211. Dimitrov, D., Atanassov, K.: On intuitionistic logic axioms and intuitionistic fuzzy implications and negations. Developments in Fuzzy Sets, Intuitionistic Fuzzy Sets, Generalized Nets and Related Topics 1, 61–68 (2008)

212. Dimitrov, D., Atanassov, K., Shannon, A., Bustince, H., Kim, S.-K.: Intuitionistic fuzzy sets and economic theory. In: Lakov, D. (ed.) Proceedings of The Second Workshop on Fuzzy Based Expert Systems, FUBEST 1996, Sofia, October 9-11, pp. 98–102 (1996)

213. Dimitrov, I., Jonova, I., Bineva, V., Sotirova, E., Sotirov, S.: Generalized net model of the students' knowledge assessments using multilayer perceptron with intuitionistic fuzzy estimations. In: Eleventh Int. Workshop on Generalized Nets and Second Int. Workshop on Generalized Nets, Intuitionistic Fuzzy Sets and Knowledge Engineering, London, July 9-10, pp. 27–34 (2010)

214. Dimitrova, S., Dimitrova, L., Kolarova, T., Petkov, P., Atanassov, K., Christov, R.: Generalized net models of the activity of NEFTOCHIM Petrochemical Combine in Bourgas. In: Atanassov, K. (ed.) Applications of Generalized Nets, pp. 208–213. World Scientific, Singapore (1993)

215. Drigas, P., Pekala, B.: Properties of decomposable operations on some extension of the fuzzy set theory. In: Advances in Fuzzy Sets, Intuitionistic Fuzzy Sets, Generalized Nets and Related Topics. vol. I: Foundations, pp. 105–118 (2008)

216. Dubois, D., Gottwald, S., Hajek, P., Kacprzyk, J., Prade, H.: Terminological difficulties in fuzzy set theory - the case of "intuitionistic fuzzy sets". Fuzzy Sets and Systems 156(3), 485–491 (2005)

217. Dubois, D., Prade, H.: Fuzzy Sets and Systems: Theory and Applications. Acad. Press, New York (1980)

218. Dubois, D., Prade, H.: Theorie des possibilites, Paris, Masson (1988)

219. Dudek, W., Davvaz, B., Bae Jun, Y.: On intuitionistic fuzzy sub-hyperquasigroups of hyperquasigroups. Information Sciences 170, 251–262 (2005)

220. Dudek, W., Bae Jun, Y.: Intuitionistic fuzzy BCC-ideals of BCC-algebras. Italian Journal of Pure and Applied Mathematics (16), 9–22 (2004)

221. Dudek, W.A., Zhan, J.M., Davvaz, B.: Intuitionistic (S,T)-fuzzy hyperquasigroups. Soft Computing 12(12), 1229–1238 (2008)

222. Dworniczak, P.: Some remarks about the L. Atanassova's paper "A new intuitionistic fuzzy implication". Cybernetics and Information Technologies 10(3), 3–9 (2010)

223. Dworniczak, P.: On one class of intuitionistic fuzzy implications. Cybernetics and Information Technologies 10(4), 13–21 (2010)

224. Dworniczak, P.: On the basic properties of the negations generated by some parametric intuitionistic fuzzy implications. Notes on Intuitionistic Fuzzy Sets 17(1), 23–29 (2011)

225. Dworniczak, P.: On some two-parametric intuitionistic fuzzy implication. Notes on Intuitionistic Fuzzy Sets 17(2), 8–16 (2011)

226. Fan, J.: Similarity measures on vague values and vague sets. Systems Engineering - Theory & Practice 26(8), 95–100 (2006)

227. Feys, R.: Modal logics, Gauthier, Paris (1965)

228. Fishburn, P.C.: A comparative analysis of Group Decision Methods. Systems Research & Behavioral Science 16(6), 538–544 (1971)

229. Fishburn, P.C.: Utility and Decision Making under Uncertainty. Mathematical Social Sciences 8, 253–285 (1984)

230. Fodor, J.C., Rubens, M.: Fuzzy Preference Modelling and Multicriteria Decision Support. Kluwer Academic Publisher, Dordrecht (1994)

231. Fol, A.: Words and Patterns. Kliment Ochridski Publ. House, Sofia (1994) (in Bulgarian)

232. Fraenkel, A., Bar-Hillel, Y.: Foundations of Set Theory. North-Holland Publ. Co., Amsterdam (1958)

233. Gardner, M.: Mathematical Games: The fantastic combinations of John Conway's new solitaire game "Life". Scientific American 223, 120–123 (1970)

234. Gargov, G.: Knowledge, uncertainty and ignorance in logic: bilattices and beyond. Journal of Applied Non-Classical Logics 9(2-3), 195–283 (1999)

235. Georgiev, P., Todorova, L., Vassilev, P.: Generalized net model of algorithm for intuitionistic fuzzy estimations. Notes on Intuitionistic Fuzzy Sets 13(2), 61–70 (2007), http://ifigenia.org/wiki/issue:nifs/13/2/61-70

236. Georgiev, K.: A simplification of the neutrosophic sets, neutrosophic logic and intuitionistic fuzzy sets. Notes on Intuitionistic Fuzzy Sets 11(2), 28–31 (2005), http://ifigenia.org/wiki/issue:nifs/11/2/28-31

237. Georgieva, O., Pencheva, T., Krawczak, M.: An application of generalized nets with intuitionistic fuzzy sets for modelling of biotechnological processes with distributed parameters. Issues in Intuitionistic Fuzzy Sets and Generalized Nets 3, 5–10 (2006)

238. Gerstenkorn, T., Manko, J.: Probability of fuzzy intuitionistic sets. BUSEFAL 45, 128–136 (1990)
239. Gerstenkorn, T., Tepavcevic, A.: Lattice valued intuitionistic fuzzy relations and Application. In: Atanassov, K., Hryniewicz, O., Kacprzyk, J. (eds.) Soft Computing Foundations and Theoretical Aspects, pp. 221–234. Academicka Oficyna Wydawnicza EXIT, Warsaw (2004)
240. Gödel, K.: Eine Interpretation des intuitionistischen Aussagenkalkülus. Ergebn. Math. Kolloquium 4, 39–40 (1933)
241. Gorzalczany, M.: A method of inference in approximate reasoning based on interval-valued fuzzy sets. Fuzzy Sets and Systems 21(1), 1–17 (1987)
242. Grzegorzewski, P.: Distances and orderings in a family of intuitionistic fuzzy numbers. In: Proc. of the Third Conf. of the European Society for Fuzzy Logic and Technology, EUSFLAT 2003, Zittau, September 10-12, pp. 223–227 (2003)
243. Grzegorzewski, P.: The Hamming Distance Between Intuitionistic Fuzzy Sets. In: De Baets, B., Kaynak, O., Bilgiç, T. (eds.) IFSA 2003. LNCS, vol. 2715, pp. 35–38. Springer, Heidelberg (2003)
244. Grzegorzewski, P.: Intuitionistic fuzzy numbers - principles, metrics and ranking. In: Atanassov, K., Hryniewicz, O., Kacprzyk, J. (eds.) Soft Computing Foundations and Theoretical Aspects, pp. 235–249. Academicka Oficyna Wydawnicza EXIT, Warsaw (2004)
245. Grzegorzewski, P.: On measuring distances between intuitionistic fuzzy sets. In: Atanassov, K., Kacprzyk, J., Krawczak, M. (eds.) Issues in Intuitionistic Fuzzy Sets and Generalized Nets, vol. 2, pp. 107–115. Wydawnictwo WSISiZ, Warsaw (2004)
246. Grzegorzewski, P., Mrowka, E.: Independence, conditional probability and the bayes formula for intuitionistic fuzzy events. In: Proc. of the Ninth International Conf., IPMU 2002, Annecy, France, July 1-5, vol. III, pp. 1851–1858 (2002)
247. Grzegorzewski, P., Mrowka, E.: Probability of intuitionistic fuzzy events. In: Grzegorzewski, P., Hryniewicz, O., Gil, M. (eds.) Soft Methods in Probability, Statistics and Data Analysis, pp. 105–115. Physica-Verlag, Heidelberg (2002)
248. Grzegorzewski, P., Mrowka, E.: Subsethood measure for intuitionistic fuzzy sets. In: Proceedings of 2004 IEEE Int. Conf. on Fuzzy Systems, Budapest, July 25-29, vol. 1, pp. 139–142 (2004)
249. Guo, X., Li, J.: Entropy, distance measure and similarity measure of intuitionistic fuzzy sets and their relations. Mathematics in Practice and Theory 37(4), 109–113 (2007)
250. Ha, M.H., Yang, L.Z.: Intuitionistic fuzzy Riemann integral. In: Proc. of the 21st Chinese Control and Decision Conference, Guilin, June 17-19, pp. 3783–3787 (2009)
251. Hadjitodorov, S.: An intuitionistic fuzzy version of the nearest prototype classification method, based on a moving-pattern procedure. Int. J. General Systems 30(2), 155–165 (2001)
252. Han, Y., Chen, S., Chen, S.: The TD-disturbing-valued fuzzy normal subgroup. Fuzzy Systems and Mathematics 5, 25–29 (2006)
253. Hanafy, I., Abd El Aziz, A.M., Salman, T.: Semi θ-compactness in intuitionistic fuzzy topological spaces. Proyecciones Journal of Mathematics 25(1), 31–45 (2006)
254. Hanafy, I.: Completely continuous functions in intuitionistic fuzzy topological spaces. Czechoslovak Mathematical Journal 53, 793–803 (2003)

255. Hanafy, I.: On fuzzy γ-open sets and fuzzy γ-continuity in intuitionistic fuzzy topological spaces. The Journal of Fuzzy Mathematics 10(1), 9–19 (2002)

256. Barwise, J. (ed.): Handbook of Mathematical Logic. North-Holland Publ. Co., Amsterdam (1977)

257. Hazra, H., Mondal, T., Samanta, S.: Connectedness in topology of intuitionistic fuzzy sets. Advanced Studies in Contemporary Mathematics 7(2), 119–134 (2003)

258. Herrera, F., Herrera-Viedma, E., Verdegay, J.L.: A linguistic decision process in group decision making. Group Decision and Negotiation 5, 165–176 (1996)

259. Herrera, F., Martinez, L., Sanchez, P.: Managigng non-homogeneous information in group decision making. European Journal of Operational Research 166, 115–132 (2005)

260. Heyting, A.: Intuitionism. An Introduction. North-Holland, Amsterdam (1956)

261. Hinde, C., Atanassov, K.: On intuitionistic fuzzy negations and intuitionistic fuzzy modal operators. Notes on Intuitionistic Fuzzy Sets 13(4), 41–44 (2007), http://ifigenia.org/wiki/issue:nifs/13/4/41-44

262. Hinde, C., Atanassov, K.: On intuitionistic fuzzy negations and intuitionistic fuzzy modal operators with contradictory evidence. In: Proceedings of the 9th WSEAS Int. Conf. on Fuzzy Systems, Sofia, May 2-4, pp. 166–170 (2008)

263. Hinde, C., Atanassov, K.: On intuitionistic fuzzy negations and intuitionistic fuzzy extended modal operators. In: Kacprzyk, J., Atanassov, K. (eds.) Part 1. Proceedings of the 12th International Conference on Intuitionistic Fuzzy Sets, Sofia, May 17-18, vol. 1, pp. 7–11 (2008); Notes on Intuitionistic Fuzzy Sets 14(1), 7–11(2008), http://ifigenia.org/wiki/issue:nifs/14/1/7-11

264. Hinde, C., Atanassov, K.: On intuitionistic fuzzy negations and intuitionistic fuzzy extended modal operators. In: Part 2. Proceedings of the 4th International IEEE Conf. Intelligent Systems, Varna, September 6-8, vol. 2, pp. 13-19–13-20 (2008)

265. Hinde, C., Patching, R.: Inconsistent intuitionistic fuzzy sets. Developments in Fuzzy Sets, Intuitionistic Fuzzy Sets, Generalized Nets and Related Topics 1, 133–153 (2008)

266. Hinde, C., Patching, R., McCoy, S.: Inconsistent intuitionistic fuzzy sets and mass assignment. Developments in Fuzzy Sets, Intuitionistic Fuzzy Sets, Generalized Nets and Related Topics 1, 155–174 (2008)

267. Homenda, W.: Balanced norms: from triangular norms towards iterative operators. In: Atanassov, K., Hryniewicz, O., Kacprzyk, J. (eds.) Soft Computing Foundations and Theoretical Aspects, pp. 251–262. Academicka Oficyna Wydawnicza EXIT, Warsaw (2004)

268. Hong, Y., Fang, X.: Characterizing intraregular semigroups by intuitionistic fuzzy sets. Mathware & Soft Computing 12(2), 121–128 (2005)

269. Hong, D.H., Kim, C.: A note on similarity measures between vague sets and between elements. Information Sciences 115, 83–96 (1999)

270. Hongliang, L., Dexue, Z.: On the embedding of top in the category of stratified L-topologiacl spaces. Chinese Annals of Mathematics, Series A 26(2), 219–228 (2005)

271. Hristova, M., Sotirova, E.: Generalized net model of a two-contour model for quality control of e-learning. Issues in Intuitionistic Fuzzy Sets and Generalized Nets 7, 35–44 (2008)

272. Hristova, M., Sotirova, E.: Multifactor method of teaching quality estimation at universities with intuitionistic fuzzy evaluation. Notes on Intuitionistic Fuzzy Sets 14(2), 80–83 (2008),
http://ifigenia.org/wiki/issue:nifs/14/2/80-83

273. Huang, G.: Similarity measure between vague sets and elements based on weighted factor. Computer Engineering and Applications 43(14), 177–179, 229 (2007)

274. Huawen, L.: Axiomatic construction for intuitionistic fuzzy sets. The Journal of Fuzzy Mathematics 8(3), 645–650 (2000)

275. Huawen, L.: Similarity measures between intuitionistic fuzzy sets and between elements and Application to pattern recognition. The Journal of Fuzzy Mathematics 13(1), 85–94 (2005)

276. Huawen, L., Kaiquan, S.: Intuitionistic fuzzy numbers and intuitionistic distribution numbers. The Journal of Fuzzy Mathematics 8(4), 909–918 (2000)

277. Hung, W.-L.: Using statistical viewpoint in developing correlation of intuitionistic fuzzy sets. International Journal of Uncertainty, Fuzziness and Knowledge-Based Systems 9, 509–516 (2001)

278. Hung, W.-L.: A note on entropy of intuitionistic fuzzy sets. Int. Journal of Uncertainty, Fuzziness and Knowledge-Based Systems 11(5), 627–633 (2003)

279. Hung, W.-L.: A note on entropy of intuitionistic fuzzy sets. International Journal of Uncertainty, Fuzziness and Knowledge-Based Systems 11, 627–633 (2003)

280. Hung, W.-L., Yang, M.-S.: On similarity measures between intuitionistic fuzzy sets. Int. Journal of Intelligent Systems 23, 364–383 (2008)

281. Hung, W.-L., Yang, M.-S.: On the J-divergence of intuitionistic fuzzy sets with its application to pattern recognition. Information Sciences 178(6), 1641–1650 (2008)

282. Hur, K., Jang, S.Y., Kang, H.W.: Intuitionistic fuzzy ideals of a ring. Journal of the Korea Society of Mathematical Education, Series B: Pure and Applied Mathematics 12(3), 193–209 (2005)

283. Hur, K., Kim, J.H., Ryou, J.H.: Intuitionistic fuzzy topological spaces. Journal of the Korea Society of Mathematical Education, Series B: Pure and Applied Mathematics 11(3), 243–265 (2004)

284. Hwang, C., Lin, M.J.: Group Decision Making under Multiple Criteria. Methods and Applications. Lecture Notes in Economics and Mathematical Systems. Springer (1987)

285. Ibrahimoglu, I., Çoker, D.: On intuitionistic fuzzy subgroups and their products. BUSEFAL 70, 16–21 (1997)

286. Janis, V.: t-norm based cuts of intuitionistic fuzzy sets. Information Sciences 180, 1134–1137 (2010)

287. Jeong, E.C., Jun, Y.B., Kim, H.S.: Intuitionistic fuzzy (+)-closed subsets of IF-algebras. Far East J. Math. Sci. (FJMS) 13(2), 227–251 (2004)

288. Ju, H.M., Yuan, X.H.: Fuzzy connectedness in interval-valued fuzzy topological spaces. In: Proc. of the 2nd International Conference on Intelligent Information Technology and Security Informatics, Moscow, January 23-25, pp. 31–33 (2009)

289. Jun, Y.B.: Intuitionistic fuzzy subsemigroups and subgroups associated by intuitionistic fuzzy graphs. Commun. Korean Math. Soc. 21(3), 587–593 (2006)

290. Jun, Y.B.: Intuitionistic fuzzy transformation semigroups. Information Sciences 179(24), 4284–4291 (2009)

291. Jun, Y.B., Kim, K.H.: Intuitionistic fuzzy ideals of BCK-algebras. Internat. J. Math. & Math. Sci. 24(12), 839–849 (2000)
292. Jun, Y., Öztürk, M., Park, C.: Intuitionistic nil radicals of intuitionistic fuzzy ideals and Euclidean intuitionistic fuzzy ideal in rings. Information Sciences 177(21), 4662–4677 (2007)
293. Jun, Y.B., Song, S.Z.: On intuitionistic fuzzy topological lattice implication algebras. The Journal of Fuzzy Mathematics 10(4), 913–920 (2002)
294. Kacprzyk, J., Fedrizzi, M., Nurmi, H.: Group decision making and consensus under fuzzy preferences and fuzzy majority. Fuzzy Sets and Systems 49, 21–31 (1992)
295. Kacprzyk, J., Nurmi, H., Fedrizzi, M. (eds.): Consensus Under Fuzziness. Kluwer Academic Publishers, Dordrecht (1997)
296. Kacprzyk, J., Szmidt, E.: Intuitionistic fuzzy relations and measures of consensus. In: Bouchon-Meunier, B., Guitierrez-Rios, J., Magdalena, L., Yager, R. (eds.) Technologies for Constructing Intelligent Systems, vol. 2, pp. 261–274. Physica-Verlag, Heidelberg (2002)
297. Kandel, A., Byatt, W.: Fuzzy sets, fuzzy algebra and fuzzy statistics. Proc. of the IEEE 66(12), 1619–1639 (1978)
298. Kandil, A., Tantawy, O., Wafaie, M.: On flou (intuitionisic) topological spaces. The Journal of Fuzzy Mathematics 15(2), 471–492 (2007)
299. Kangas, A., Kangas, J.: Probability, possibility and evidence: approaches to consider risk and uncertainty in forestry decision making. Forest Policy and Economics 6, 169–188 (2004)
300. Kang, H.W., Hur, K., Ryou, J.H.: t-intuitionistic fuzzy subgroupoids. J. Korea Soc. Math. Educ. Ser. B: Pure and Appl. Math. 10(4), 233–244 (2003)
301. Kaufmann, A.: Introduction a la Theorie des Sour-ensembles Flous, Paris, Masson (1977)
302. Kelley, J.: General Topology. D. van Nostrand Co., Toronto (1957)
303. Khan, A., Jun, Y.B., Shabir, M.: Ordered semigroups characterized by their intuitionistic fuzzy bi-ideals. Iranian Journal of Fuzzy Systems 7(2), 55–69 (2010)
304. Kim, K.H.: On intuitionistic Q-fuzzy semiprime ideals in semigroups. In: Advances in Fuzzy Mathematics, vol. 1(1), pp. 15–21 (2006)
305. Kim, Y.C., Abbas, S.E.: Connectedness in intuitionistic fuzzy topological spaces. Commun. Korean Math. Soc. 20(1), 117–134 (2005)
306. Kim, K., Jun, Y., Kim, B.: On intuitionistic fuzzy po-ideals of ordered semigroups. East-West J. of Mathematics 3(1), 23–32 (2001)
307. Kim, K.H., Jun, Y.B.: Intuitionistic fuzzy interior ideals of semigroups. Int. J. of Mathematics and Mathematical Sciences 27(5), 261–267 (2001)
308. Kim, K.H., Jun, Y.B.: Intuitionistic fuzzy ideals of semigroups. Indian J. Pure Appl. Math. 33, 443–449 (2002)
309. Kim, K.H., Lee, J.G.: On intuitionistic fuzzy Bi-ideals of semigroups. Turk. J. Math. 29, 201–210 (2005)
310. Kim, K.H., Lee, J.G.: Intuitionistic (T,S)-normed fuzzy ideals of Γ-rings. Int. Mathematical Forum 3(3), 115–123 (2008)
311. Klir, G., Yuan, B.: Fuzzy Sets and Fuzzy Logic. Prentice Hall, New Jersey (1995)
312. Kolev, B.: Intuitionistic fuzzy relational databases and translation of the intuitionistic fuzzy SQL. In: Proceedings of the 6th International FLINS Conference Applied Computational Intelligence, Blankenberge, September 1-3, pp. 189–194 (2004)

313. Kolev, B., El-Darzi, E., Sotirova, E., Petronias, I., Atanassov, K., Chountas, P., Kodogianis, V.: Generalized Nets in Artificial Intelligence. Generalized nets, Relational Data Bases and Expert Systems, vol. 3. "Prof. M. Drinov" Academic Publishing House, Sofia (2006)

314. Kondakov, N.: Logical Dictionary, Moskow, Nauka (1971) (in Russian)

315. Koshelev, M., Kreinovich, V., Rachamreddy, B., Yasemis, H., Atanassov, K.: Fundamental justification of intuitionistic fuzzy logic and interval-valued fuzzy methods. Notes on Intuitionistic Fuzzy Sets 4(2), 42–46 (1998), http://ifigenia.org/wiki/issue:nifs/4/2/42-46

316. Krachounov, M.: Intuitionistic probability and intuitionistic fuzzy sets. In: First Int. Workshop on Intuitionistic Fuzzy Sets, Generalized Nets and Knowledge Engineering, London, September 6-7, pp. 18–24 (2006)

317. Kral, P.: T-operations and cardinalities of IF-sets. In: Hryniewicz, O., Kacprzyk, J., Kuchta, D. (eds.) Issues in Soft Computing, Decisions and Operational Research, pp. 219–228. Akademicka Oficyna Wydawnictwo EXIT, Warsaw (2005)

318. Kral, P.: Cardinalities and new measure of entropy for IF sets. In: Atanassov, K., Kacprzyk, J., Krawczak, M., Szmidt, E. (eds.) Issues in the Representation and Processing of Uncertain Imprecise Information: Fuzzy Sets, Intuitionistic Fuzzy Sets, Generalized Nets, and Related Topics, pp. 209–216. Akademicka Oficyna Wydawnictwo EXIT, Warsaw (2005)

319. Krawczak, M., Sotirov, S., Atanassov, K.: Multilayer Neural Networks and Generalized Nets. Warsaw School of Information Technology, Warsaw (2010)

320. Kreinovich, V., Mukaidono, M., Atanassov, K.: From fuzzy values to intuitionistic fuzzy values, to intuitionistic fuzzy intervals etc.: can we get an arbitrary ordering? Notes on Intuitionistic Fuzzy Sets 5(3), 11–18 (1999), http://ifigenia.org/wiki/issue:nifs/5/3/11-18

321. Kreinovich, V., Nguyen, H., Wu, B., Atanassov, K.: Fuzzy justification of heuristic methods in inverse problems and in numerical computations, with applications to detection of business cycles from fuzzy and intuitionistic fuzzy data. Notes on Intuitionistic Fuzzy Sets 4(2), 47–56 (1998), http://ifigenia.org/wiki/issue:nifs/4/2/47-56

322. Krsteska, B., Abbas, S.: Intuitionistic fuzzy strongly propen (preclosed) mappings. Mathematica Moravica 10, 47–53 (2006)

323. Krsteska, B., Abbas, S.: Intuitionistic fuzzy strongly irresolute precontinuous mappings in Coker's spaces. Kragujevac Journal of Mathematics 30, 243–252 (2007)

324. Krsteska, B., Ekici, E.: Intuitionistic fuzzy almost strong precontinuity in Coker's sense. IJPAM 43, 69–84 (2008)

325. Krsteska, B., Ekici, E.: Intuitionistic fuzzy contra strongly precontinuous mappings. Filomat 21, 273–284 (2007)

326. de Kumar, S., Biswas, R., Roy, A.R.: Some operations on intuitionistic fuzzy sets. Fuzzy Sets and Systems 114(4), 477–484 (2000)

327. Kuppannan, J., Parvathi, R., Thirupathi, D., Palaniappan, N.: Intuitionistic fuzzy approach to enhance text documents. In: Proc. of 3rd Int. IEEE Conf. Intelligent Systems, IS 2006, September 4-6, pp. 733–737 (2006)

328. Kuratowski, K.: Topology, vol. 1. Acad. Press, New York (1966)

329. Kuratowski, K., Mostowski, A.: Set Theory. North-Holland Publ. Co., Amsterdam (1967)

330. Lei, Y., Wang, B., Wang, Y.: Techniques for battlefield situation assessment based on intuitionistic fuzzy decision. Acta Electronica Sinica 34(12), 2175–2179 (2006)
331. Lei, Y., Zhao, Y., Wang, T., Wang, J., Shen, X.: On the measurement of similarity on semantic match for intuitionistic fuzzy. Journal of Air Force Engineering University 6(2), 83–87 (2005)
332. Lemnaouar, Z., Abdelaziz, A.: On the representation of L-M algebra by intuitionistic fuzzy subsets. Arima 4, 72–85 (2006)
333. Lendelova, K.: Conditional IF-probability. In: Lawry, J., et al. (eds.) Soft Methods for Integrated Uncertainty Modelling, pp. 275–283. Springer, Berlin (2006)
334. Lendelova, K.: Probability of L-posets. In: Proc. of Fourth Conf. of the European Society for Fuzzy Logic and Technology EUSFLAT, Barcelona, September 7-9, pp. 320–324 (2005)
335. Lendelova, K., Michalikova, A., Riečan, B.: Representation of probability on triangle. In: Hryniewicz, O., Kacprzyk, J., Kuchta, D. (eds.) Issues in Soft Computing, Decisions and Operational Research, pp. 235–242. Akademicka Oficyna Wydawnictwo EXIT, Warsaw (2005)
336. Lendelova, K., Petrovicova, J.: Representation of IF-probability on MV-algebras. Soft Computing 10(7), 564–566 (2006)
337. Leyendekkers, J., Shannon, A., Rybak, J.: Pattern Recognition: Modular Rings & Integer Structure. Raffles KvB Monograph No. 9, North Sydney (2007)
338. Li, D.-F.: Some measures of dissimilarity in intuitionistic fuzzy structures. Journal of Computer and Systems Sciences 68, 115–122 (2004)
339. Li, D.-F.: Representation of level sets and extension principles for Atanassov's intuitionistic fuzzy sets and algebraic operations. Uncertainty- Critical Review IV, 63–74 (2010)
340. Li, D.-F.: Multiattribute decision making models and methods using intuitionistic fuzzy sets. Journal of Computer and System Sciences 70, 73–85 (2005)
341. Li, D.-F.: Extension of the LIMMAP for multiattribute decision making under Atanassov's intuitionistic fuzzy environment. Fuzzy Optimization and Decision Making 7(1), 17–34 (2008)
342. Li, D.-F.: Linear programming method for MADM with interval-valued intuitionistic fuzzy sets. Expert Systems with Applications 37(8), 5939–5945 (2010)
343. Li, D.-F., Cheng, C.T.: New similarity measures of intuitionistic fuzzy sets and Application to pattern recognition. Pattern Recognition Letters 23, 221–225 (2002)
344. Li, D.-F., Nan, J.X.: A nonlinear programming approach to matrix games with payoffs of Atanassov's intuitionistic fuzzy sets. International Journal of Uncertainty, Fuzziness and Knowledge-Based Systems 17(4), 585–607 (2009)
345. Li, Y., Olson, D., Qin, Z.: Similarity measures between intuitionistic fuzzy (vague) sets: a comparative analysis. Pattern Recognition Letters 28, 278–285 (2007)
346. Li, D.-F., Yang, J.-B.: A multiattribute decision making approach uning intuitionistic fuzzy sets. In: Proc. of the Third Conf. of the European Society for Fuzzy Logic and Technology, EUSFLAT 2003, Zittau, September 10-12, pp. 183–186 (2003)

347. Liang, Z., Shi, P.: Similarity measures on intuitionistic fuzzy sets. Pattern Recognition Letters 24, 2687–2693 (2003)

348. Lin, M.: Intuitionistic fuzzy group and its induced quotient group. Journal of Xiamen University 45(2), 157–161 (2006)

349. Lin, L., Yuan, X.-H., Xia, Z.-Q.: Multicriteria fuzzy decision-making methods based on intuitionistic fuzzy sets. Journal of Computer and System Sciences 73, 84–88 (2007)

350. Liu, F., Yuan, X.: Fuzzy number intuitionistic fuzzy set. Fuzzy Systems and Mathematics 21(1), 88–91 (2007)

351. Liu, H.-W.: New similarity measures between intuitionistic fuzzy sets and between elements. Mathematical and Computer Modelling 42, 61–70 (2005)

352. Liu, H.-W., Wang, G.: Multi-criteria decision-making methods based on intuitionistic fuzzy sets. European Journal of Operational Research 179(1), 220–233 (2007)

353. Liu, H., Wenqing, S.: On the definition of intuitionistic fuzzy topology. BUSEFAL 78, 25–28 (1999)

354. Liu, X.-D., Zheng, S.-H., Xiong, F.-L.: Entropy and Subsethood for General Interval-Valued Intuitionistic Fuzzy Sets. In: Wang, L., Jin, Y. (eds.) FSKD 2005. LNCS (LNAI), vol. 3613, pp. 42–52. Springer, Heidelberg (2005)

355. Lupiáñez, F.G.: Hausdorffness in intuitionistic fuzzy topological spaces. Mathware & Soft Computing 10(1), 17–22 (2003)

356. Lupiáñez, F.G.: Hausdorffness in intuitionistic fuzzy topological spaces. The Journal of Fuzzy Mathematics 12(3), 521–525 (2004)

357. Lupiáñez, F.G.: Separation in intuitionistic fuzzy topological spaces. Int. Journal of Pure and Applied Mathematics 17(1), 29–34 (2004)

358. Lupiáñez, F.G.: Intuitionistic fuzzy topological operators and topology. Int. Journal of Pure and Applied Mathematics 17(1), 35–40 (2004)

359. Lupiáñez, F.G.: On intuitionistic fuzzy topological spaces. Kybernetes: The International Journal of Systems & Cybernetics 35(5), 743–747 (2006)

360. Lupiáñez, F.G.: Nets and filters in intuitionistic fuzzy topological spaces. Information Sciences 176, 2396–2404 (2006)

361. Lupiáñez, F.G.: Covering properties in intuitionistic fuzzy topological spaces. Kybernetes 36(5/6), 749–753 (2007)

362. Lupiáñez, F.G.: Some recent results on Atanassov's intuitionistic fuzzy topological spaces. In: Proc. of 8th Intern. FLINS Conf. Computational Intelligence in Decision and Control, pp. 229–234. World Scientific, New Jersey (2008)

363. Ma, X.: Intuitionistic fuzzy filters of BCI-algebras. Journal of Hubei Institute for Nationalities 25(1), 39–41 (2007)

364. Magrez, P., Smets, P.: Fuzzy Modus Ponens: a new model suitable for applications in knowledge-based systems. International Journal of Intelligent Systems 4, 181–200 (1989)

365. Mahapatra, G., Mahapatra, B.: Intuitionistic fuzzy fault tree analysis using intuitionistic fuzzy numbers. Int. Mathematical Forum 5(21), 1015–1024 (2010)

366. Manko, J.: Effectiveness and relevancy measures under modal cardinality for intuitionistic fuzzy sets. In: Szczepaniak, P., Kacprzyk, J., Niewiadomski, A. (eds.) Advances in Web Intelligence, pp. 286–292. Springer, Berlin (2005)

367. Marazov, I.: Visible Mith. Christo Botev Publ. House, Sofia (1992) (in Bulgarian)

368. Martine-Löf, P.: Notes on Constructive Mathematics. Almqvist & Wiksell, Stockholm (1970)

369. Mazukerova, P.: M-state and M-probability sets and mass assignment. Developments in Fuzzy Sets, Intuitionistic Fuzzy Sets, Generalized Nets and Related Topics 1, 237–242 (2008)

370. Mazurekova, P., Valencakova, V.: Conditional M-probability. In: Proc. of 8th Intern. FLINS Conf. Computational Intelligence in Decision and Control, pp. 331–336. World Scientific, New Jersey (2008)

371. Melo-Pinto, P., Kim, T., Atanassov, K., Sotirova, E., Shannon, A., Krawczak, M.: Generalized net model of e-learning evaluation with intuitionistic fuzzy estimations. In: Issues in the Representation and Processing of Uncertain and Imprecise Information, Warsaw, pp. 241–249 (2005)

372. Mendelson, E.: Introduction to Mathematical Logic. D. Van Nostrand, Princeton (1964)

373. Min, W.K., Park, C.-K.: Some results on intuitionistic fuzzy topological spaces defined by intuitionistic gradation of openness. Commun. Korean Math. Soc. 20(4), 791–801 (2005)

374. Mitchell, H.B.: On the Dengfeng-Chuntian similarity measure and its application to pattern recognition. Pattern Recognit. Lett. 24, 3101–3104 (2003)

375. Mondal, T.K., Samanta, S.: Topology of interval-valued intuitionistic fuzzy sets. Fuzzy Sets and Systems 119(3), 483–494 (2001)

376. Mrówka, E., Grzegorzewski, P.: Friedman's test with missing observations. In: Proc. of Fourth Conf. of the European Society for Fuzzy Logic and Technology EUSFLAT/LFA, Barcelona, September 7-9, pp. 627–632 (2005)

377. Muthuraj, R.: Some Characterizations of intuitionistic fuzzy sub-bigroups. PhD Thesis, Alagappa University, Karaikudi, India (2008)

378. Nagell, T.: Introduction to Number Theory. John Wiley & Sons, New York (1950)

379. Nehi, H.M.: A new ranking method for intuitionistic fuzzy numbers. International Journal of Fuzzy Systems 12(1), 80–86 (2010)

380. Nikolova, M.: Sequences of intuitionistic fuzzy numbers. Notes on Intuitionistic Fuzzy Sets 8(3), 26–30 (2002),
http://ifigenia.org/wiki/issue:nifs/8/3/26-30

381. Nikolova, M., Nikolov, N., Cornelis, C., Deschrijver, G.: Survey of the research on intuitionistic fuzzy sets. Advanced Studies in Contemporary Mathematics 4(2), 127–157 (2002)

382. Novikov, P.: Constructive Mathematical Logic from Point of View of Classical One. Moskow, Nauka (1977) (in Russian)

383. Nowak, P.: Monotone measures of intuitionistic fuzzy sets. In: Proc. of the Third Conf. of the European Society for Fuzzy Logic and Technology, EUSFLAT 2003, Zittau, September 10-12, pp. 172–176 (2003)

384. Nowak, P.: Construction of monotone measures of intuitionistic fuzzy sets. In: Atanassov, K., Hryniewicz, O., Kacprzyk, J. (eds.) Soft Computing Foundations and Theoretical Aspects, pp. 303–317. Academicka Oficyna Wydawnicza EXIT, Warsaw (2004)

385. Nowak, P.: Integral with respect to a monotone measure on intuitionistic fuzzy sets. In: de Baets, B., de Caluwe, R., de Tre, G., Fodor, J., Kacprzyk, J., Zadrożny, S. (eds.) Current Issues in Data and Knowledge Engineering, pp. 378–387. Academicka Oficyna Wydawnicza EXIT, Warsaw (2004)

386. Ozcelik, A., Narli, S.: On submaximality in intuitionistic topological spaces. Int. J. of Computational and Mathematical Sciences 1(2), 139–141 (2007)

387. Özçağ, S., Çoker, D.: On connectedness in intuitionistic fuzzy special topological spaces. Int. Journal of Math. & Math. Sci. 21(1), 33–40 (1998)

388. Özçağ, S., Çoker, D.: A note on connectedness in intuitionistic fuzzy special topological spaces. Int. Journal of Mathematics and Math. Sciences 23(1), 45–54 (2000)

389. Pankowska, A.: Examples of Applications of IF-sets with triangular norms to group decision making problems. In: Atanassov, K., Kacprzyk, J., Krawczak, M., Szmidt, E. (eds.) Issues in the Representation and Processing of Uncertain Imprecise Information: Fuzzy Sets, Intuitionistic Fuzzy Sets, Generalized Nets, and Related Topics, pp. 291–306. Akademicka Oficyna Wydawnictwo EXIT, Warsaw (2005)

390. Pankowska, A., Wygralak, M.: A general concept of IF-Sets with triangular norms. In: Atanassov, K., Hryniewicz, O., Kacprzyk, J. (eds.) Soft Computing Foundations and Theoretical Aspects, pp. 319–335. Academicka Oficyna Wydawnicza EXIT, Warsaw (2004)

391. Pankowska, A., Wygralak, M.: General IF-sets with triangular norms and their Applications to group decision making. Information Sciences 176, 2713–2754 (2006)

392. Radzikowska, A.M.: Rough Approximation Operations Based on IF Sets. In: Rutkowski, L., Tadeusiewicz, R., Zadeh, L.A., Żurada, J.M. (eds.) ICAISC 2006. LNCS (LNAI), vol. 4029, pp. 528–537. Springer, Heidelberg (2006)

393. Pankowska, A., Wygralak, M.: Algorithms of group decision making based on generalized IF-sets. In: Hryniewicz, O., Kacprzyk, J., Koronacki, J., Wierzchon, S. (eds.) Issues in Intelligent Systems - Paradigms, pp. 185–197. Wydawnictwo WSISiZ, Warsaw (2005)

394. Pankowska, A., Wygralak, M.: General IF-sets with triangular norms and their Applications to group decision making. Information Sciences 176, 2713–2754 (2006)

395. Parvathi, R.: Theory of operators on intuitionistic fuzzy sets of second type and their Applications to image processing. PhD Thesis, Dept. of Mathematics, Alagappa Univ., Karaikudi, India (2005)

396. Parvathi, R., Karunambigai, M.G.: Intuitionistic fuzzy graphs. In: Computational Intelligence, Theory and Applications, pp. 139–150. Springer, Berlin (2006)

397. Parvathi, R., Vassilev, P., Atanassov, K.: New topological operators over intuitionistic fuzzy sets. Advanced Studies in Contemporary Mathematics, vol. 18(1), pp. 49–57 (2009)

398. Parvathi, R., Vassilev, P., Atanassov, K.: New topological operators over intuitionistic fuzzy sets. Submitted to Advanced Studies in Contemporary Mathematics

399. Pencheva, T.: Intuitionistic fuzzy logic in generalized net model of an advisory system for yeast cultivation on-line control. Notes on Intuitionistic Fuzzy Sets 15(4), 45–51 (2009),
http://ifigenia.org/wiki/issue:nifs/15/4/45-51

400. Ramadan, A., Abbas, S., Abd El-Latif, A.: Compactness in intuitionistic fuzzy topological spaces. Int. Journal of Mathematics and Math. 1, 19–32 (2005)

401. Rasiowa, H., Sikorski, R.: The Mathematics of Metamathematics. Pol. Acad. of Sci., Warsaw (1963)

402. Rencova, M., Riečan, B.: Probability on IF-sets: an elementary approach. In: First Int. Workshop on Intuitionistic Fuzzy Sets, Generalized Nets and Knowledge Engineering, London, September 6-7, pp. 8–17 (2006)

403. Riečan, B.: A descriptive definition of the probability on intuitionistic fuzzy sets. In: Proc. of the Third Conf. of the European Society for Fuzzy Logic and Technology, EUSFLAT 2003, Zittau, September 10-12, pp. 210–213 (2003)

404. Riečan, B.: On the probability theory on the Atanassov sets. In: Proc. of 3rd Int. IEEE Conf. Intelligent Systems, IS 2006, London, September 4-6, pp. 730–732 (2006)

405. Riečan, B.: On IF-sets and MV-algebras. In: Proc. of Eleventh International Conf., IPMU 2006, Paris, July 2-7, pp. 2405–2407 (2006)

406. Riečan, B.: On two ways for the probability theory on IF-sets. In: Lawry, J., et al. (eds.) Soft Methods for Integrated Uncertainty Modelling, pp. 285–290. Springer, Berlin (2006)

407. Riečan, B.: Probability theory on intuitionistic fuzzy events. Dept. of Math., Matej Bel Univ., Banska Bystrica, Preprint 4/2007 (February 1, 2007)

408. Riečan, B.: Probability Theory on IF Events. In: Aguzzoli, S., Ciabattoni, A., Gerla, B., Manara, C., Marra, V. (eds.) ManyVal 2006. LNCS (LNAI), vol. 4460, pp. 290–308. Springer, Heidelberg (2007)

409. Riečan, B.: M-probability theory on IF-events. Dept. of Math., Matej Bel Univ., Banska Bystrica, Preprint 6/2007 (April 20, 2007)

410. Riečan, B., Atanassov, K.: A set-theoretical operation over intuitionistic fuzzy sets. Notes on Intuitionistic Fuzzy Sets 12(2), 24–25 (2006), http://ifigenia.org/wiki/issue:nifs/12/2/24-25

411. Riečan, B., Atanassov, K.: n-extraction operation over intuitionistic fuzzy sets. Notes on Intuitionistic Fuzzy Sets 12(4), 9–11 (2006), http://ifigenia.org/wiki/issue:nifs/12/4/9-11

412. Riečan, B., Atanassov, K.: On intuitionistic fuzzy level operators. Notes on Intuitionistic Fuzzy Sets 16(3), 42–44 (2010), http://ifigenia.org/wiki/issue:nifs/16/3/42-44

413. Riečan, B., Atanassov, K.: Operation division by n over intuitionistic fuzzy sets. Notes on Intuitionistic Fuzzy Sets 16(4), 1–4 (2010), http://ifigenia.org/wiki/issue:nifs/16/4/1-4

414. Riečan, B., Boyadzhieva, D., Atanassov, K.: On intuitionistic fuzzy subtraction, related to intuitionistic fuzzy negation $\neg_{11}$. Notes on Intuitionistic Fuzzy Sets 15(4), 9–14 (2009), http://ifigenia.org/wiki/issue:nifs/15/4/9-14

415. Riečan, B., Rencova, M., Atanassov, K.: On intuitionistic fuzzy subtraction, related to intuitionistic fuzzy negation $\neg_4$. Notes on Intuitionistic Fuzzy Sets 15(4), 15–18 (2009), http://ifigenia.org/wiki/issue:nifs/15/4/15-18

416. Rodriguez, J., Vitoriano, B., Montero, J.: Decision making and Atanassov's approach to fuzzy sets. In: Advances in Fuzzy Sets, Intuitionistic Fuzzy Sets, Generalized Nets and Related Topics, Foundations, vol. I, pp. 203–213 (2008)

417. Roeva, O.: Generalized net model of oxygen control system using intuitionistic fuzzy logic. In: Proceedings of the 1st International Workshop on Generalized Nets, Intuitionistic Fuzzy Sets and Knowledge Engineering, London, September 6-7, pp. 49–55 (2006)

418. Roeva, O., Pencheva, T., Bentes, I., Nascimento, M.M.: Modelling of temperature control system in fermentation processes using generalized nets and intuitionistic fuzzy logics. Notes on Intuitionistic Fuzzy Sets 11(4), 151–157 (2005), http://ifigenia.org/wiki/issue:nifs/11/4/151-157

419. Samanta, S., Mondal, T.K.: Intuitionistic gradation of openness: intuitionistic fuzzy topology. BUSEFAL 73, 8–17 (1997)

420. Samanta, S., Chattopadhyay, K., Mukherjee, U., Mondal, T.K.: Role of clans in the proximities of intuitionistic fuzzy sets. BUSEFAL 74, 31–37 (1998)

421. Schwartz, L.: Analyse Mathematique. Hermann, Paris (1967)

422. Seenivasan, V., Balasubramanian, G.: Invertible fuzzy topological spaces. Italian Journal of Pure and Applied Mathematics (22), 223–230 (2007)

423. Seenivasan, V., Balasubramanian, G.: On upper and lower α-irresolute fuzzy multifunctions. Italian Journal of Pure and Applied Mathematics (23), 25–36 (2008)

424. Shannon, A., Atanassov, K.: A first step to a theory of the intuitionistic fuzzy graphs. In: Lakov, D. (ed.) Proc. of the First Workshop on Fuzzy Based Expert Systems, Sofia, September 28-30, pp. 59–61 (1994)

425. Shannon, A., Atanassov, K.: Intuitionistic fuzzy graphs from $\alpha-$, β - and (α, β)-levels. Notes on Intuitionistic Fuzzy Sets 1(1), 32–35 (1995), http://ifigenia.org/wiki/issue:nifs/1/1/32-35

426. Shannon, A., Atanassov, K., Orozova, D., Krawczak, M., Sotirova, E., Melo-Pinto, P., Petrounias, I., Kim, T.: Generalized nets and information flow within a university. Warsaw School of Information Technology, Warsaw (2007)

427. Shannon, A., Dimitrakiev, D., Sotirova, E., Krawczak, M., Kim, T.: Towards a model of the digital university: generalized net model of a lecturer's evaluation with intuitionistic fuzzy estimations. Cybernetics and Information 9(2), 69–78 (2009)

428. Shannon, A., Kerre, E., Szmidt, E., Sotirova, E., Petrounias, I., Kacprzyk, J., Atanassov, K., Krawczak, M., Melo-Pinto, P., Mellani, S., Kim, T.: Intuitionistic fuzzy estimation and generalized net model of e-learning within a university local network. Advanced Studies in Contemporary Mathematics 9(1), 41–46 (2004)

429. Shannon, A., Langova-Orozova, D., Sotirova, E., Atanassov, K., Melo-Pinto, P., Kim, T.: Generalized net model with intuitionistic fuzzy estimations of the process of obtaining of scientific titles and degrees. Notes on Intuitionistic Fuzzy Sets 11(3), 95–114 (2005), http://ifigenia.org/wiki/issue:nifs/11/3/95-114

430. Shannon, A., Langova-Orozova, D., Sotirova, E., Atanassov, K., Melo-Pinto, P., Kim, T.: Generalized net model for adaptive electronic assessment, using intuitionistic fuzzy estimations. In: Computational Intelligence, Theory and Applications, pp. 291–297. Springer (2005)

431. Shannon, A., Orozova, D., Sotirova, E., Atanassov, K., Krawszak, M., Melo-Pinto, P., Kim, T.: System for electronic student-teacher interactions with intuitionistic fuzzy estimations. Notes on Intuitionistic Fuzzy Sets 13(2), 81–87 (2007), http://ifigenia.org/wiki/issue:nifs/13/2/81-87

432. Shannon, A., Rangasamy, P., Atanassov, K., Chountas, P.: On intuitionistic fuzzy trees. In: Developments in Fuzzy Sets, Intuitionistic Fuzzy Sets, Generalized Nets and Related Topics, Foundations, vol. I, pp. 177–184 (2010)

433. Shannon, A., Riečan, B., Orozova, D., Sotirova, E., Atanassov, K., Krawczak, M., Georgiev, P., Nikolov, R., Sotirov, S., Kim, T.: A method for ordering of university subjects using intuitionistic fuzzy evaluations. Notes on Intuitionistic Fuzzy Sets 14(2), 84–87 (2008),
http://ifigenia.org/wiki/issue:nifs/14/2/84-87

434. Shannon, A., Roeva, O., Pencheva, T., Atanassov, K.: Generalized Nets Modelling of Biotechnological Processes. "Prof. M. Drinov" Academic Publishing House, Sofia (2004)

435. Shannon, A., Sotirova, E., Petrounias, I., Atanassov, K., Krawczak, M., Melo-Pinto, P., Kim, T.: Generalized net model of lecturers' evaluation of student work with intuitionistic fuzzy estimations. In: Second International Workshop on Intuitionistic Fuzzy Sets, Banska Bystrica, Slovakia, December 3 (2006); Notes on Intuitionistic Fuzzy Sets 12(4), 22–28 (2006),
http://ifigenia.org/wiki/issue:nifs/12/4/22-28

436. Shannon, A., Sotirova, E., Kacprzyk, J., Krawczak, M., Atanassov, K., Melo-Pinto, P., Kim, T.: A generalized net model of the university electronic archive with intuitionistic fuzzy estimations of the searched information. Issues in Intuitionistic Fuzzy Sets and Generalized Nets 4, 75–88 (2007)

437. Shannon, A., Sotirova, E., Atanassov, K., Krawczak, M., Melo-Pinto, P., Kim, T.: Generalized net model for the reliability and standardization of assessments of student problem solving with intuitionistic fuzzy estimations. In: Developments in Fuzzy Sets, Generalized Nets and Related Topics. Applications. System Research Institute, Polish Academy of Sciences, vol. 2, pp. 249–256 (2008)

438. Shyi-Ming, C.: Similarity measures between vague sets and between elements. IEEE Transactions on Systems, Man, and Cybernetics - Part B: Cybernetics 27(1), 153–158 (1997)

439. Smarandache, F.: Neutrosophic set - a generalization of the intuitionistic fuzzy set. International Journal of Pure and Applied Mathematics 24(3), 287–297 (2005)

440. Song, Y., Zhang, Q., Zhou, X.: Supplier selection model based on distance measure between intuitionistic fuzzy sets. In: Proc. of IEEE International Conference on Systems, Man and Cybernetics, vol. 5, pp. 3795–3799

441. Sotirov, S.: Determining of intuitionistic fuzzy sets in estimating probability of spam in the e-mail by the help of the neural networks. Issues in Intuitionistic Fuzzy Sets and Generalized Nets 4, 43–48 (2007)

442. Sotirov, S.: Intuitionistic fuzzy estimations for connections of the transmit routines of the bluetooth interface. Advanced Studies in Contemporary Mathematics 15(1), 99–108 (2007)

443. Sotirov, S.: Method for determining of intuitionistic fuzzy sets in discovering water floods by neural networks. Issues in Intuitionistic Fuzzy Sets and Generalized Nets 4, 9–14 (2007)

444. Sotirov, S., Atanassov, K.: Intuitionistic fuzzyfFeed forward neural network. Part I, Cybernetics and Information Technologies 9(2), 62–68 (2009)

445. Sotirov, S., Bobev, V.: Generalized net model of the Ro-Ro traffic with intuitionistic fuzzy estimation. Notes on Intuitionistic Fuzzy Sets 14(4), 27–33 (2008), http://ifigenia.org/wiki/issue:nifs/14/4/27-33

446. Sotirov, S., Dimitrov, A.: Neural network for defining intuitionistic fuzzy estimation in petroleum recognition. Issues in Intuitionistic Fuzzy Sets and Generalized Nets 8, 74–78 (2010)

447. Sotirov, S., Kodogiannis, V., Blessing, R.E.: Intuitionistic fuzzy estimations for connections with low rate wireless personal area networks. In: First Int. Workshop on Generalized Nets, Intuitionistic Fuzzy Sets and Knowledge Engineering, London, September 6-7, pp. 81–87 (2006)

448. Sotirov, S., Sotirova, E., Krawczak, M.: Application of data mining in digital university: multilayer perceptron for lecturer's evaluation with intuitionistic fuzzy estimations. Issues in Intuitionistic Fuzzy Sets and Generalized Nets 8, 102–108 (2010)

449. Sotirov, S., Sotirova, E., Orozova, D.: Neural network for defining intuitionistic fuzzy sets in e-learning. Notes on Intuitionistic Fuzzy Sets 15(2), 33–36 (2009), http://ifigenia.org/wiki/issue:nifs/15/2/33-36

450. Sotirov, S., Sotirova, E., Zhelev, Y., Zheleva, M.: Generalized net model of process of the European awareness scenario workshop method with intuitionistic fuzzy estimation. Notes on Intuitionistic Fuzzy Sets 14(4), 20–26 (2008), http://ifigenia.org/wiki/issue:nifs/14/4/20-26

451. Sotirova, E., Atanassov, K., Sotirov, S., Borisova, L., Vanev, P., Vitanov, I.: Generalized net modeling of the process of introducing new product to the market and intuitionistic fuzzy estimating its effectiveness. Notes on Intuitionistic Fuzzy Sets 12(3), 41–45 (2006), http://ifigenia.org/wiki/issue:nifs/12/3/41-45

452. Sotirova, E., Orozova, D.: Generalized net model of the phases of the data mining process. In: Developments in Fuzzy Sets, Intuitionistic Fuzzy Sets, Generalized Nets and Related Topics. Foundation and Applications, Warsaw, pp. 247–260 (2010)

453. Srivastava, A.K., Tiwari, S.P.: IF-topologies and IF-automata. Soft Computing 14(6), 571–578 (2010)

454. Stamenov, A.: A property of the extended intuitionistic fuzzy modal operator $F_{\alpha,\beta}$. In: Proceedings of the 2nd Int. IEEE Symposium: Intelligent Systems, Varna, June 22-24, vol. 3, pp. 16–17 (2004)

455. Szmidt, E.: Applications of Intuitionistic Fuzzy Sets in Decision Making. D.Sc. dissertation, Technical University, Sofia (2000)

456. Szmidt, E., Baldwin, J.: Intuitionistic fuzzy set functions, mass assignment theory, possibility theory and histograms. In: 2006 IEEE World Congress on Computational Intelligence, pp. 237–243 (2006)

457. Szmidt, E., Kacprzyk, J.: Group decision making via intuitionistic fuzzy sets. In: Lakov, D. (ed.) Proceedings of The Second Workshop on Fuzzy Based Expert Systems FUBEST 1996, Sofia, October 9-11, pp. 107–112 (1996)

458. Szmidt, E., Kacprzyk, J.: Intuitionistic fuzzy sets for more realistic group decision making. In: Proceedings of International Conference Transitions to Advanced Market Institutions and Economies, Warsaw, June 18-21, pp. 430–433 (1997)

459. Szmidt, E., Kacprzyk, J.: Applications of intuitionistic fuzzy sets in decision making. In: Actas del VIII Congreso Espanol Sobre Technologias y Logica Fuzzy, Pamplona, September 8-10, pp. 143–148 (1998)

460. Szmidt, E., Kacprzyk, J.: Group decision making under intuitionistic fuzzy preference relations. In: Proceedings of 7th International Conference Information Processing and Management of Uncertainty in Knowledge-Based Systems, Paris, July 6-10, pp. 172–178 (1998)

461. Szmidt, E., Kacprzyk, J.: Decision making in an intuitionistic fuzzy environment. In: Proceedings of EUROFUSE-SIC 1999, Budapest, May 25-28, pp. 292–297 (1999)

462. Szmidt, E., Kacprzyk, J.: Probability of intuitionistic fuzzy events and their Applications in decision making. In: Proc. of 1999 EUSFLAT-ESTYLF Joint Conference, Palma, September 22-25, pp. 457–460 (1999)

463. Szmidt, E., Kacprzyk, J.: A concept of a probability of an intuitionistic fuzzy event. In: Proc. of 1999 IEEE Int. Fuzzy Systems Conference FUZZ-IEEE 1999, Seoul, Korea, August 22-25, pp. III-1346–III-1349 (1999)

464. Szmidt, E., Kacprzyk, J.: Probability of intuitionistic fuzzy events and their Applications in decision making. In: Proc. of 1999 Eusflat-Estylf Joint Conference, Palma, September 22-25, pp. 457–460 (1999)

465. Szmidt, E., Kacprzyk, J.: Distances between intuitionistic fuzzy sets. Fuzzy Sets and Systems 114(3), 505–518 (2000)

466. Szmidt, E., Kacprzyk, J.: Entropy for intuitionistic fuzzy sets. Fuzzy Sets and Systems 118(3), 467–477 (2001)

467. Szmidt, E., Kacprzyk, J.: Intuitionistic fuzzy sets for the softening of group decision making models. In: Proceedings of the 14th Int. Conf. on Systems Theory, Wroclaw, September 11-14, pp. 237–249 (2001)

468. Szmidt, E., Kacprzyk, J.: Distance from consensus under intuitionistic fuzzy preferences. In: Proceedings of EUROFUSE Workshop on Preference Modelling and Applications EUROFUSE-PM 2001, Granada, April 25-27 (2001)

469. Szmidt, E., Kacprzyk, J.: An intuitionistic fuzzy set based approach to intelligent data analysis: an Application to medical diagnosis. In: Abraham, A., Jain, L., Kacprzyk, J. (eds.) Recent Advances in Intelligent Paradigms and Applications, pp. 57–70. Springer, Berlin (2002)

470. Szmidt, E., Kacprzyk, J.: Using intuitionistic fuzzy sets in group decision making. Control and Cybernetics 31(4), 1037–1053 (2002)

471. Szmidt, E., Kacprzyk, J.: Analysis of agreements in a group of experts via distances between intuitionistic fuzzy preferences. In: Proc. of the Ninth International Conf. IPMU 2002, Annecy, France, July 1-5, vol. III, pp. 1859–1865 (2002)

472. Szmidt, E., Kacprzyk, J.: Group agreement analysis via distances and entropy of intuitionistic fuzzy sets. In: Proceedings of Int. Conf. on Fuzzy Information Processing: Theories and Applications, Beijing, China, March 1-4, vol. I, pp. 631–635 (2003)

473. Szmidt, E., Kacprzyk, J.: A concept of similarity for intuitionistic fuzzy sets and its use in group decision making. In: Proceedings of 2004 IEEE Int. Conf. on Fuzzy Systems, Budapest, July 25-29, vol. 2, pp. 1129–1134 (2004)

474. Szmidt, E., Kacprzyk, J.: Similarity of intuitionistic fuzzy sets and the Jaccard coefficient. In: Proceedings of Tenth Int. Conf. IPMU 2004, Perugia, July 4-9, vol. 2, pp. 1405–1412 (2004)

475. Szmidt, E., Kacprzyk, J.: New measures of entropy for intuitionistic fuzzy sets. Notes on Intuitionistic Fuzzy Sets 11(2), 12–20 (2005), http://ifigenia.org/wiki/issue:nifs/11/2/12-20

476. Szmidt, E., Kacprzyk, J.: Distances between intuitionistic fuzzy sets and their applications in reasoning. In: Halgamude, S., Wang, L. (eds.) Computational Intelligence for Modelling and Prediction, pp. 101–116. Springer, Berlin (2005)

477. Szmidt, E., Kacprzyk, J.: A New Concept of a Similarity Measure for Intuitionistic Fuzzy Sets and Its Use in Group Decision Making. In: Torra, V., Narukawa, Y., Miyamoto, S. (eds.) MDAI 2005. LNCS (LNAI), vol. 3558, pp. 272–282. Springer, Heidelberg (2005)

478. Szmidt, E., Kacprzyk, J.: Similarity measures for intuitionistic fuzzy sets. In: Atanassov, K., Kacprzyk, J., Krawczak, M., Szmidt, E. (eds.) Issues in the Representation and Processing of Uncertain Imprecise Information: Fuzzy Sets, Ntuitionistic Fuzzy Sets, Generalized Nets, and Related Topics, pp. 355–372. Akademicka Oficyna Wydawnictwo EXIT, Warsaw (2005)

479. Szmidt, E., Kacprzyk, J.: An Application of Intuitionistic Fuzzy Set Similarity Measures to a Multi-criteria Decision Making Problem. In: Rutkowski, L., Tadeusiewicz, R., Zadeh, L.A., Żurada, J.M. (eds.) ICAISC 2006. LNCS (LNAI), vol. 4029, pp. 314–323. Springer, Heidelberg (2006)

480. Szmidt, E., Kacprzyk, J.: Distances between intuitionistic fuzzy sets: straightforward approaches not work. In: Proc. of 3rd Int. IEEE Conf. "Intelligent Systems" IS 2006, London, September 4-6, pp. 716–721 (2006)

481. Szmidt, E., Kacprzyk, J.: A Similarity Measure for Intuitionistic Fuzzy Sets and Its Application in Supporting Medical Diagnostic Reasoning. In: Rutkowski, L., Siekmann, J.H., Tadeusiewicz, R., Zadeh, L.A. (eds.) ICAISC 2004. LNCS (LNAI), vol. 3070, pp. 388–393. Springer, Heidelberg (2004)

482. Szmidt, E., Kacprzyk, J.: A new measure of entropy and its connection with a similarity measure for intuitionistic fuzzy sets. In: Proc. of Fourth Conf. of the European Society for Fuzzy Logic and Technology EUSFLAT, Barcelona, September 7-9, pp. 461–466 (2005)

483. Szmidt, E., Kacprzyk, J.: An Application of Intuitionistic Fuzzy Set Similarity Measures to a Multi-criteria Decision Making Problem. In: Rutkowski, L., Tadeusiewicz, R., Zadeh, L.A., Żurada, J.M. (eds.) ICAISC 2006. LNCS (LNAI), vol. 4029, pp. 314–323. Springer, Heidelberg (2006)

484. Szmidt, E., Kacprzyk, J.: Atanassov's intuitionistic fuzzy sets as a promising tool for extended fuzzy decision making models. In: Bustince, H., Herrera, F., Montero, J. (eds.) Fuzzy Sets and Their Extensions: Representation, Aggregation and Models, pp. 335–355. Springer, Berlin (2008)

485. Szmidt, E., Kacprzyk, J.: Amount of information and its realibility in the ranking of Atanassov's intuitionistic fuzzy alternativies. In: Rakus-Andersson, E., Yager, R., Ichalkaranje, N., Jain, L. (eds.) Recent Advances in Decision Making, pp. 7–19. Springer, Berlin (2009)

486. Szmidt, E., Kacprzyk, J.: A model of case based reasoning using intuitionistic fuzzy sets. In: Proc. of the 2006 IEEE World Congress on Computational Intelligence, pp. 8428–8435 (2006)

487. Szmidt, E., Kacprzyk, J.: Entropy and similarity for intuitionistic fuzzy sets. In: Proc. of the 11th Int. Conf. IPMU, Paris, pp. 2375–2382 (2006)

488. Szmidt, E., Kacprzyk, J.: Classification of imbalanced and overlapping classes using intuitionistic fuzzy sets. In: Proc. of the 3rd International IEEE Conference Intelligent Systems IS 2006, London, pp. 722–727 (2006)

489. Szmidt, E., Kacprzyk, J.: Classification with Nominal Data Using Intuitionistic Fuzzy Sets. In: Melin, P., Castillo, O., Aguilar, L.T., Kacprzyk, J., Pedrycz, W. (eds.) IFSA 2007. LNCS (LNAI), vol. 4529, pp. 76–85. Springer, Heidelberg (2007)

490. Szmidt, E., Kacprzyk, J.: Some Problems with Entropy Measures for the Atanassov Intuitionistic Fuzzy Sets. In: Masulli, F., Mitra, S., Pasi, G. (eds.) WILF 2007. LNCS (LNAI), vol. 4578, pp. 291–297. Springer, Heidelberg (2007)

491. Szmidt, E., Kacprzyk, J.: A new similarity measure for intuitionistic fuzzy sets: straightforward approaches not work. In: Proc. of the 2007 IEEE Conference on Fuzzy Systems, Imperial Colledge, London, UK, July 23-26, pp. 481–486 (2007)

492. Szmidt, E., Kacprzyk, J.: A new approach to ranking alternatives using intuitionistic fuzzy sets. In: Ruan, D., Montero, J., Lu, J., Martinez, L., D'hondt, P., Kerre, E.E. (eds.) Computational Intelligence in Decision and Control, pp. 265–270. World Scientific (2008)

493. Szmidt, E., Kacprzyk, J.: Dilemmas with distances between intuitionistic fuzzy sets. In: Chountas, P., Petrounias, I., Kacprzyk, J. (eds.) Intelligent Techniques and Tools for Novel System Architectures, pp. 415–430. Springer, Heidelberg (2008)

494. Szmidt, E., Kacprzyk, J.: Intuitionistic fuzzy sets - a prospective tool for text categorization. In: Atanassov, K., Chountas, P., Kacprzyk, J., et al. (eds.) Developments in Fuzzy Sets, Intuitionistic Fuzzy Sets, Generalized Nets and Related Topics, Applications, vol. II, pp. 281–300. Academic Publisching House EXIT; Systems Research Institute PAS, Warsaw (2008)

495. Szmidt, E., Kacprzyk, J.: Intuitionistic fuzzy sets as a promising tool for extended fuzzy decision making models. In: Bustince, H., Herrera, F., Montero, J. (eds.) Fuzzy Sets and Their Extensions: Representation, Aggregation and Models, pp. 330–355. Springer, Berlin (2008)

496. Szmidt, E., Kacprzyk, J.: Using Intuitionistic Fuzzy Sets in Text Categorization. In: Rutkowski, L., Tadeusiewicz, R., Zadeh, L.A., Zurada, J.M. (eds.) ICAISC 2008. LNCS (LNAI), vol. 5097, pp. 351–362. Springer, Heidelberg (2008)

497. Szmidt, E., Kacprzyk, J.: Dealing with typical values by using Atanassov's intuitionistic fuzzy sets. In: Proceedings of 2008 IEEE World Congress on Computational Intelligence, Hong Kong, June 1-6 (2008)

498. Szmidt, E., Kacprzyk, J.: On Some typical values for Atanassov's intuitionistic fuzzy sets. In: Proc. of the 4th International IEEE Conference Intelligent Systems, Varna, Bulgaria, September 6-8, vol. I, II and III, pp. 2–7 (2008)

499. Szmidt, E., Kacprzyk, J.: Ranking alternatives expressed via Atanassov's intuitionistic fuzzy sets. In: Information Processing and Management of Uncertainty in Knowledge-Based Systems, IPMU 2008, Malaga, June 22-27, pp. 1604–1611. University of Malaga, Spain (2008)

500. Szmidt, E., Kacprzyk, J.: Amount of information and its reliability in the ranking of Atanassov's intuitionistic fuzzy alternatives. In: Rakus-Anderson, E., Yager, R., Ichalkaranje, N., Jain, L. (eds.) Recent Advances in Decision Making, pp. 7–19. Springer, Berlin (2009)

501. Szmidt, E., Kacprzyk, J.: Analysis of similarity measures for Atanassov's intuitionistic fuzzy sets. In: Proceedings IFSA/EUSFLAT 2009, Lizbona, Portugalia, Lipca 20-24, pp. 1416–1421 (2009)

502. Szmidt, E., Kacprzyk, J.: Some remarks on the Hausdorff distance between Atanassov's intuitionistic fuzzy sets. In: Preference Modelling and Decision Analysis, EUROFUSE WORKSHOP 2009, Pamplona (Spain), September 16-18, pp. 311–316. Public University of Navarra (2009)

503. Szmidt, E., Kacprzyk, J.: Dealing with typical values via Atanassov's intuitionistic fuzzy sets. International Journal of General Systems 39(5) (2010)

504. Szmidt, E., Kacprzyk, J.: Correlation of Intuitionistic Fuzzy Sets. In: Hüllermeier, E., Kruse, R., Hoffmann, F. (eds.) IPMU 2010. LNCS(LNAI), vol. 6178, pp. 169–177. Springer, Heidelberg (2010)

505. Szmidt, E., Kacprzyk, J.: On an Enhanced Method for a More Meaningful Ranking of Intuitionistic Fuzzy Alternatives. In: Rutkowski, L., Scherer, R., Tadeusiewicz, R., Zadeh, L.A., Zurada, J.M. (eds.) ICAISC 2010. LNCS(LNAI), vol. 6113, pp. 232–239. Springer, Heidelberg (2010)

506. Szmidt, E., Kacprzyk, J.: The Spearman rank correlation coefficient between intuitionistic fuzzy sets. In: Proceedings of the Conference: 2010 IEEE International Conference on Intelligent Systems, London, UK, July 7-9, pp. 276–280 (2010)

507. Szmidt, E., Kreinovich, V.: Symmetry between true, false, and uncertain: An explanation. Notes on Intuitionistic Fuzzy Sets 15(4), 1–8 (2009), http://ifigenia.org/wiki/issue:nifs/15/4/1-8

508. Szmidt, E., Kukier, M.: A new approach to classification of imbalanced classes via Atanassov's intuitionistic fuzzy sets. In: Wang, H.-F. (ed.) Intelligent Data Analysis: Developing New Methodologies Through Pattern Discovery and Recovery, pp. 85–101. Idea Group (2008)

509. Szmidt, E., Kukier, M.: Atanassov's intuitionistic fuzzy sets in classification of imbalanced and overlapping classes. In: Chountas, P., Petrounias, I., Kacprzyk, J. (eds.) Intelligent Techniques and Tools for Novel System Architectures, pp. 455–471. Springer, Berlin (2008)

510. Takeuti, G., Titani, S.: Intuitionistic fuzzy logic and intuitionistic fuzzy set theory. The Journal of Symbolic Logic 49(3), 851–866 (1984)

511. Tan, C.Q., Chen, X.H.: Intuitionistic fuzzy Choquet integral operator for multi-criteria decision making. Expert Systems with Applications 37(1), 149–157 (2010)

512. Tan, E., Aslim, G.: On fuzzy continuous functions in intuitionistic fuzzy topological spaces. BUSEFAL 81, 32–41 (2000)

513. Tan, C., Zhang, Q.: Aggregation of opinion in group decision making based on intuitionistic fuzzy distances. Mathematics in Practice and Theory 36(2), 119–124 (2006)

514. Tan, C., Zhang, Q.: Intuitionistic fuzzy sets method for fuzzy multiple attribute decision making. Fuzzy Systems and Mathematics 20(4), 71–76 (2006)

515. Tanaka, H., Sugihara, K., Maeda, Y.: Non-addaptive measures by interval probability functions. Information Sciences 164, 209–227 (2004)

516. Tanev, D.: On an intuitionistic fuzzy norm. Notes on Intuitionistic Fuzzy Sets 1(1), 25–26 (1995), http://ifigenia.org/wiki/issue:nifs/1/1/25-26

517. Tang, Z., Liang, J., Li, S.: New fuzzy multiple objectives decision making based on vague sets. Computing Technology and Automation 25(4), 31–34 (2006)

518. Tarassov, V.: Lattice products, bilattices and some extensions of negations, triangular norms and triangular conorms. In: Proceedings of the Int. Conf. on Fuzzy Sets and Soft Computing in Economics and Finance FSSCEF 2004, Saint-Peterburg, June 17-20, vol. I, pp. 272–282 (2004)

519. Tcvetkov, R.: Some ways and means to define addition and multiplication operations between intuitionistic fuzzy sets. Notes on Intuitionistic Fuzzy Sets 9(3), 22–25 (2003), http://ifigenia.org/wiki/issue:nifs/9/3/22-25

520. Tcvetkov, R.: Derivatives related to intuitionistic fuzzy sets. Notes on Intuitionistic Fuzzy Sets 10(3), 44–46 (2004),
http://ifigenia.org/wiki/issue:nifs/10/3/44-46

521. Tcvetkov, R.: Some new relations between intuitionistic fuzzy sets. In: Proceedings of the Second International IEEE Conference Intelligent systems, Varna, vol. 3, pp. 23–26 (2004)

522. Tcvetkov, R.: Extended Sugeno integrals and integral topological operators over intuitionistic fuzzy sets. In: First Int. Workshop on Intuitionistic Fuzzy Sets, Generalized Nets and Knowledge Engineering, London, September 6-7, pp. 133–144 (2006)

523. Tcvetkov, R.: Extended Choquet integrals and intuitionistic fuzzy sets. Notes on Intuitionistic Fuzzy Sets 13(2), 18–22 (2007),
http://ifigenia.org/wiki/issue:nifs/13/2/18-22

524. Tcvetkov, R., Szmidt, E., Kacprzyk, J.: On some issues related to the distances between the Atanasov intuitionistic fuzzy sets. Cybernetics and Information Technologies 9(2), 54–61 (2009)

525. Tenekedjiev, K.: Hurwitz alpha-expected utility criterion for decisions with partially quantified uncertainty. In: First Int. Workshop on Intuitionistic Fuzzy Sets, Generalized Nets and Knowledge Engineering, London, September 6-7, pp. 56–75 (2006)

526. Thilagavathi, S., Parvathi, R., Karunambigai, M.G.: Operations on intuitionistic fuzzy graphs II. Developments in Fuzzy Sets, Intuitionistic Fuzzy Sets, Generalized Nets and Related Topics 1, 319–336 (2008)

527. Todorova, L.: Determining the specificity, sensitivity, positive and negative predictive values in intuitionistic fuzzy logic. Notes on Intuitionistic Fuzzy Sets 14(2), 73–79 (2008),
http://ifigenia.org/wiki/issue:nifs/14/2/73-79

528. Todorova, L., Antonov, A., Hadjitodorov, S.: Intuitionistic fuzzy Voronoi diagrams – definition and properties. Notes on Intuitionistic Fuzzy Sets 10(4), 56–60 (2004), http://ifigenia.org/wiki/issue:nifs/10/4/56-60

529. Todorova, L., Antonov, A.: Intuitionistic fuzzy Voronoi diagrams – definition and properties II. Notes on Intuitionistic Fuzzy Sets 11(3), 88–90 (2005), http://ifigenia.org/wiki/issue:nifs/11/3/88-90

530. Todorova, L., Atanassov, K.: An example for a difference between ordinary (crisp), fuzzy and intuitionistic fuzzy sets. In: Kacprzyk, J., Atanassov, K. (eds.) Proceedings of the Sixth International Conference on Intuitionistic Fuzzy Sets, Varna, September 13-14. Notes on Intuitionistic Fuzzy Sets, vol. 8(3), pp. 31–33 (2002),
http://ifigenia.org/wiki/issue:nifs/8/3/31-33

531. Todorova, L., Atanassov, K., Hadjitodorov, S., Vassilev, P.: On an intuitionistic fuzzy approach for decision making in medicine: Part 1. Bioautomation International Journal 6, 92–101 (2007)

532. Todorova, L., Atanassov, K., Hadjitodorov, S., Vassilev, P.: On an intuitionistic fuzzy approach for decision making in medicine: Part 2. Bioautomation International Journal 7, 64–69 (2007)

533. Todorova, L., Atanassov, K., Szmidt, E., Kacprzyk, J.: Intuitionistic fuzzy generalized net for decision making with Voronoi's diagrams. Issues in Intuitionistic Fuzzy Sets and Generalized Nets 4, 15–26 (2007)

534. Todorova, L., Dantchev, S., Atanassov, K., Tasseva, V., Georgiev, P.: On aggregating multiple fuzzy values into a single intuitionistic fuzzy estimate. In: Proceedings of the 3rd International IEEE Conference Intelligent Systems, pp. 738–740 (September 2006)

535. Todorova, L., Vassilev, P.: Algorithm for clustering data set represented by intuitionistic fuzzy estimates. Bioautomation International Journal 14(1), 61–68 (2010)

536. Todorova, L., Vassilev, P.: Application of K-nearest neighbor rule in the case of intuitionistic fuzzy sets for pattern recognition. Bioautomation International Journal 13(4), 265–270 (2009)

537. Todorova, L., Vassilev, P.: A note on a geometric interpretation of the intuitionistic fuzzy set operators $P_{\alpha,\beta}$, and $Q_{\alpha,\beta}$. Notes on Intuitionistic Fuzzy Sets 15(4), 25–29 (2009),
    http://ifigenia.org/wiki/issue:nifs/15/4/25-29

538. Todorova, L., Vassilev, P., Georgiev, P.: Generalized net model of aggregation algorithm for intuitionistic fuzzy estimates of classification. Issues in Intuitionistic Fuzzy Sets and Generalized Nets 5, 54–63 (2007)

539. Todorova, L., Vassilev, P., Szmidt, E., Hadjitodorov, S.: Spatial interpretation of Operators over Intuitionistic Fuzzy Sets. Issues in Intuitionistic Fuzzy Sets and Generalized Nets 8, 8–18 (2010)

540. Torkzadeh, L., Zahedi, M.M.: Intuitionistic fuzzy commutative hyper k-ideals. In: Proceedings of the 11th Int. Fuzzy Systems Association World Congress "Fuzzy Logic, Soft Computing and Computational Intelligence", Beijing, July 28-31, vol. 1, pp. 301–306 (2005)

541. Torkzadeh, L., Zahedi, M.M.: Intuitionistic fuzzy commutative hyper K-ideals. J. Appl. Math. & Computing 21(1), 451–467 (2006)

542. Trifonov, T.: Verifying universal properties of intuitionistic fuzzy connectives with Mathematica. Notes on Intuitionistic Fuzzy Sets 17(4) (2011) (in press)

543. Trifonov, T., Atanassov, K.: On some intuitionistic properties of intuitionistic fuzzy implications and negations. In: Computational Intelligence, Theory and Applications, pp. 151–158. Springer, Berlin (2006)

544. Turanh, N.: An overwiew of intuitionistic fuzzy supratopological spaces. Hacettepe Journal of Mathematics and Statistics 32, 17–26 (2003)

545. Turanli, N., Çoker, D.: Fuzzy connectedness in intuitionistic fuzzy topological spaces. Fuzzy Sets and Systems 116(3), 369–375 (2000)

546. Uçkun, M., Öztürk, M.A., Jun, Y.B.: Intuitionistic fuzzy sets in gamma-semigroups. Bull. Korean Math. Soc. 44(2), 359–367 (2007)

547. Uma, M., Roja, E., Balasubramanian, G.: On pairwise fuzzy pre-basically and pre-extremally disconnected spaces. Italian Journal of Pure and Applied Mathematics (23), 75–84 (2008)

548. Valchev, D., Sotirov, S.: Intuitionistic fuzzy detection Of signal availabilty in multipath wireles chanels. Notes on Intuitionistic Fuzzy Sets 15(2), 24–29 (2009), http://ifigenia.org/wiki/issue:nifs/15/2/24-29

549. Vardeva, I., Sotirov, S.: Sotirov, Generalized net model of SSL with intuitionistic fuzzy estimations. Notes on Intuitionistic Fuzzy Sets 13(2), 48–53 (2007), http://ifigenia.org/wiki/issue:nifs/13/2/48-53

550. Vardeva, I., Sotirov, S.: Intuitionistic fuzzy estimations of damaged packets with multilayer perceptron. In: Proc. of the 10th International Workshop on Generalized Nets, Sofia, pp. 63–69 (2009)

551. Vasilev, T.: Four equalities connected with intuitionistic fuzzy sets. Notes on Intuitionistic Fuzzy Sets 14(3), 1–4 (2008),
     http://ifigenia.org/wiki/issue:nifs/14/3/1-4
552. Vassilev, P.: A note on the intuitionistic fuzzy set operator $F_{\alpha,\beta}$. In: Proceedings of the 2nd Int. IEEE Symposium: Intelligent Systems, Varna, June 22-24, vol. 3, pp. 18–20 (2004)
553. Vassilev, P.: On reassessment of expert evaluations in the case of intuitionistic fuzziness. Advanced Studies in Contemporary Mathematics 20(4), 569–574 (2010)
554. Vassilev, P.: A Note on the Extended Modal Operator $G_{\alpha,\beta}$. Notes on Intuitionistic Fuzzy Sets 16(2), 12–15 (2010),
     http://ifigenia.org/wiki/issue:nifs/16/2/12-15
555. Vassilev, P.: Intuitionsitic fuzzy sets and fuzzy sets depending on Q2 - norms (Part 1). In: Advances in Fuzzy Sets, Intuitionistic Fuzzy Sets, Generalized Nets and Related Topics. Foundations, vol. I, pp. 237–250. EXIT, Warsaw (2008)
556. Vassilev, P.: Subnorms, Non-Archimedean Field Norms and Their Corresponding d-FS and d-IFS. Cybernetics and Information Technologies 9(2), 30–37 (2009)
557. Vassilev, P., Parvathi, R., Atanassov, K.: Note on intuitionistic fuzzy sets of p-th type. Issues in Intuitionistic Fuzzy Sets and Generalized Nets 6, 43–50 (2008)
558. Vassilev, P., Vassilev-Missana, M.: Intuitionistic fuzzy sets and fuzzy sets depending on Q-norms. Notes on Intuitionistic Fuzzy Sets 14(1), 1–6 (2008),
     http://ifigenia.org/wiki/issue:nifs/14/1/1-6
559. Vijayabalaji, S., Thillagovindan, N., Jun, Y.B.: Intuitionistic fuzzy n-normed linear space. Bull. Korean Math. Soc. 44(2), 291–308 (2007)
560. Vlachos, I., Sergiadis, G.: On the entropy of intuitionistic fuzzy events. In: Proc. of Int. Conf. on Computational Intelligence for Modelling, Control and Automation, CIMCA 2005, art. no. 1631460, pp. 153–138 (2005)
561. Vlachos, I., Sergiadis, G.: Inner product based entropy in the intuitionistic fuzzy setting. International Journal of Uncertainty, Fuzziness and Knowledge-Based Systems 14(3), 351–366 (2006)
562. Vlachos, I., Sergiadis, G.: Intuitionistic fuzzy information - Applications to pattern recognition. Pattern Recognition Letters 28, 197–206 (2007)
563. Vlachos, I., Sergiadis, G.: Subsethood, entropy, and cardinality for interval-valued fuzzy sets - An algebraic derivation. Fuzzy Sets and Systems 158, 1384–1396 (2007)
564. Wang, Y., Lei, Y.: A technique for constructing intuitionistic fuzzy entropy. Kongzhi Juece (Control and decision) 22(12), 1390–1394
565. Wagn, W., Xin, X.: Distance measure between intuitionistic fuzzy sets. Pattern Recognition Letters 26(13), 2063–2069 (2005)
566. Weisstein, E., Hull, C.: From MathWorld – A Wolfram Web Resource,
     http://mathworld.wolfram.com/ConvexHull.htm
567. Wen-Liang, H., Yang, M.-S.: Similarity measures of intuitionistic fuzzy sets based on Hausdorff dispance. Pattern Recognition Letters 25, 1603–1611 (2004)
568. Wen-Liang, H., Yang, M.-S.: Similarity measures of intuitionistic fuzzy sets based on Lp metric. Int. Journal of Approximate Reasoning 46(1), 120–136 (2007)

569. Whalen, T.: Soft decision analysis, Issues in Information Technology. In: Kacprzyk, J., Krawczak, M., Zadrożny, S. (eds.) Akademicka Oficyna Wydawnicza, pp. 225–246. EXIT, Warsaw (2002)

570. Wikipedia contributors: Conway's Game of Life. Wikipedia, The Free Encyclopedia (March 18, 2011), http://en.wikipedia.org/w/index.php?title=Conway's_Game_of_Life&oldid=419530565 (accessed March 20, 2011)

571. Wu, D., Mendel, J.: Uncertainty measures for interval type-2 fuzzy sets. Information Sciences 177(23), 5378–5393 (2007)

572. Wu, L.B., Huo, L.Y., Li, J.T.: ERP selection model based on intuitionistic fuzzy valued Sugeno integral. In: Proc. of International Conference on Engineering and Business Management, Chengdu, March 25-27, vol. 1-8, pp. 1273–1277 (2010)

573. Wygralak, M.: I-fuzzy sets with triangular norms, their hesitation areas and cardinalities. In: Atanassov, K., Kacprzyk, J., Krawczak, M., Szmidt, E. (eds.) pp. 397–408. Akademicka Oficyna Wydawnictwo EXIT, Warsaw (2005)

574. Li, X.: The intuitionistic fuzzy normal subgroup and its some equivalent propositions. BUSEFAL 82, 40–44 (2000)

575. Xu, C.-Y.: Homomorphism of intuitionistic fuzzy groups. In: Proceedings of the Sixth Int. Conf. on Machine Learning and Cybernetics, ICMLC 2007, pp. 1178–1183 (2007)

576. Xu, C.Y.: New structures of intuitionistic fuzzy groups. In: Proc. of the 4th International Conference on Intelligent Computing, Shanghai, September 15-18, pp. 145–152 (2008)

577. Xu, Z.: On similarity measuring of interval-valued intuitionistic fuzzy sets and their Applications to pattern recognition. Journal of Southeast University (English Edition) 23(1), 139–143 (2007)

578. Xu, Z.S.: Models for multiple attribute decision making with intuitionistic fuzzy sets based on Lp metric. Int. Journal of Uncertainty, Fuzziness and Knowledge-Based Systems 15(3), 285–297 (2007)

579. Xu, Z.: Multi-person multi-attribute decision making models under intuitionistic fuzzy environment. Fuzzy Optimization and Decision Making 6(3), 221–236 (2007)

580. Xu, Z.S.: Choquet integrals of weighted intuitionistic fuzzy information. Information Sciences 180(5), 726–736 (2010)

581. Xu, Z., Yager, R.: Dynamic intuitionistic fuzzy multi-attribute decision making. Int. Journal of Approximate Reasoning 48(1), 246–262 (2008)

582. Yager, R.: Non-numeric multi-criteria multi-person decision making. International Journal of Group Decision Making and Negotiation 2, 81–93 (1993)

583. Yan, C.H., Wang, X.: Intuitionistic I-fuzzy topological spaces. Czechoslovak Math. Journal 60(135), 233–252 (2010)

584. Yang, L.Z., Ha, M.H.: A New Similarity measure between intuitionistic fuzzy sets based on a Choquet integral model. In: Proc. of the 5th International Conference on Fuzzy Systems and Knowledge Discovery, Jinan, October 18-20, pp. 116–121 (2008)

585. Yang, Y., Chiclana, F.: Intuitionistic fuzzy sets: Spherical representation and distances. Int. J. of Intelligent Systems 24, 399–420 (2009)

586. Ye, F.: An extended TOPSIS method with interval-valued intuitionistic fuzzy numbers for virtual enterprise partner selection. Expert Systems with Applications 37(10), 7050–7055 (2010)

587. Yon, Y.H., Jun, Y.B., Kim, K.H.: Intuitionistic fuzzy R-subgroups of near-rings. Soochow Journal of Mathematics 27(3), 243–253 (2001)
588. Yon, Y.H., Kim, K.H.: On intuitionistic fuzzy filters and ideals of lattices. Far East J. Math. Sci. 1(3), 429–442 (1999)
589. Yong, U.C.: A note on fuzzy ideals in gamma-near-rings. Far East Journal of Mathematical Sciences 26(3), 513–520 (2007)
590. Yosida, K.: Functional analysis. Springer, Berlin (1965)
591. Yuan, X.H., Li, H.X., Lee, E.S.: On the definition of the intuitionistic fuzzy subgroups. Computers & Mathematics with Applications 59(9), 3117–3129 (2010)
592. Zadeh, L.: The Concept of a Linguistic Variable and its Application to Approximate Reasoning. American Elsevier Publ. Co., New York (1973)
593. Zadeh, L.: Fuzzy sets. Information and Control 8, 338–353 (1965)
594. Zahedi, M., Torkzadeh, I.: Intuitionistic fuzzy dual positive implicative hyper K-ideals. Proc. of World Academy of Science, Engineering and Technology, 57–60 (2005)
595. Zarandi, A., Saeid, A.B.: Intuitionistic fuzzy ideals of Bg-Algebras. Proc. of World Academy of Science, Engineering and Technology 5, 187–189 (2005)
596. Zedam, L., Amroune, A.: On the representation of L-M-algebra by intuitionistic fuzzy subsets. International Journal of Mathematics and Analysis 1(2), 189–200 (2007)
597. Zeng, W., Guo, P.: Normalized distance, similarity measure, inclusion measure and entropy of interval-valued fuzzy sets and their relationship. Information Sciences 178(5), 1334–1342 (2008)
598. Zeng, W., Li, H.: Note on "Some operations on intuitionistic fuzzy sets". Fuzzy Sets and Systems 157(3), 990–991 (2006)
599. Zeng, D., Wu, Z.: Topological space of interval valued intuitionistic fuzzy sets. Journal of Chengdu University of Information Technology 21(4), 583–585 (2006)
600. Xu, Z.: Intuitionistic preference relations and their Application in group decision making. Information Sciences 177(11), 2363–2379
601. Xu, Z.: Some similarity measures of intuitionistic fuzzy stes and their applications to multiple attribure decision making. Fuzzy Optimization and Decision Making 6, 109–121 (2007)
602. Xu, Z.: On Correlation Measures of Intuitionistic Fuzzy Sets. In: Corchado, E., Yin, H., Botti, V., Fyfe, C. (eds.) IDEAL 2006. LNCS, vol. 4224, pp. 16–24. Springer, Heidelberg (2006)
603. Zhan, J., Tan, Z.: Intuitionistic M-fuzzy groups. Soochow Journal of Mathematics 30(1), 85–90 (2004)
604. Zhan, J., Tan, Z.: Intuitionistic fuzzy deductive systems in Hilbert algebras. Southeast Asian Bulletin of Mathematics 29, 813–826 (2005)
605. Zhang, C., Fu, H.: Similarity measures on three kinds of fuzzy sets. Pattern Recognition Letters 27, 1307–1317 (2006)
606. Zhang, K.M., Bai, Y.: Intuitionistic fuzzy sublattice group. In: Proc. of International Conference on Intelligent Human-Machine Systems and Cybernetics, Hangzhou, August 26-27, vol. 1, pp. 145–148 (2009)
607. Zhang, K.M., Bai, Y.: Intuitionistic fuzzy subgroups and its characterizations. In: Proc. of the 7th International Conference on Machine Learning and Cybernetics, Kunming, July 12-15, vol. 1-7, pp. 533–535 (2008)

608. Zhang, X.X., Yue, G.X., Teng, Z.: Possibility degree of interval-valued intuitionistic fuzzy numbers and its application. In: Proc. of the International Symposium on Information Proceeding (ISIP 2009), Huangshan, August 21-23, pp. 33–36 (2009)

609. Zhou, Y.: New approach for measuring similarity between intuitionistic fuzzy sets. In: Atanassov, K., Kacprzyk, J., Krawczak, M., Szmidt, E. (eds.) Issues in the Representation and Processing of Uncertain Imprecise Information: Fuzzy Sets, Intuitionistic Fuzzy Sets, Generalized Nets, and Related Topics, pp. 420–430. Akademicka Oficyna Wydawnictwo EXIT, Warsaw (2005)

610. Zhou, L., Wu, W.-Z., Zhang, W.-X.: On intuitionistic fuzzy rough sets and their topological structures. Int. Journal of General Systems 38(6), 589–616 (2009)

611. Zhou, Z., Wu, Q.-Z., Liu, F.-X., Guan, G.-X.: Multicriteria decision-making based on interval-valued vague sets. Transaction of Beijing Institute of Technology 26(8), 693–696 (2006)

612. Zimmermann, H.-J.: Fuzzy Set Theory and its Applications. Kluwer Academic Publishers, Boston (1993)

Index